**BUSINESS REPLY MAIL**

FIRST- CLASS MAIL     PERMIT NO. 47     EDWARDSVILLE, KS

NO POSTAGE
NECESSARY
IF MAILED
IN THE
UNITED STATES

POSTAGE WILL BE PAID BY ADDRESSEE

MEDI-SIM,  INC.
P.O. BOX 13267
EDWARDSVILLE KS 66113-9989

# BIOLOGY

Fourth Edition

## Sylvia S. Mader

***Part One***
*The Cell*

***Part Two***
*Genetic Basis of Life*

***Part Three***
*Evolution and Diversity*

***Part Four***
*Plant Structure and Function*

***Part Five***
*Animal Structure and Function*

***Part Six***
*Behavior and Ecology*

**WCB** **Wm. C. Brown Publishers**
Dubuque, Iowa • Melbourne, Australia • Oxford, England

**Book Team**

Editor *Kevin Kane*
Developmental Editor *Carol Mills*
Production Editors *Anne E. Scroggin/Ann Fuerste*
Designer *Mark Elliot Christianson*
Art Editor *Miriam J. Hoffman*
Photo Editor *Lori Gockel*
Permissions Editor *Karen Storlie*
Art Processor *Andréa Lopez-Meyer*

## Wm. C. Brown Publishers
A Division of Wm. C. Brown Communications, Inc.

Vice President and General Manager *Beverly Kolz*
National Sales Manager *Vincent R. Di Blasi*
Assistant Vice President, Editor-in-Chief *Edward G. Jaffe*
Marketing Manager *Paul Ducham*
Advertising Manager *Amy Schmitz*
Managing Editor, Production *Colleen A. Yonda*
Manager of Visuals and Design *Faye M. Schilling*
Design Manager *Jac Tilton*
Art Manager *Janice Roerig*
Publishing Services Manager *Karen J. Slaght*
Permissions/Records Manager *Connie Allendorf*

## Wm. C. Brown Communications, Inc.

Chairman Emeritus *Wm. C. Brown*
Chairman and Chief Executive Officer *Mark C. Falb*
President and Chief Operating Officer *G. Franklin Lewis*
Corporate Vice President, President of WCB Manufacturing *Roger Meyer*

Wild horses gallop across the Patagonian plains of South America's Argentine coast at the foot of the Fitzroy Mountains. In 1832, Charles Darwin came here aboard the HMS Beagle, captained by Robert Fitzroy. During his 5-year, around-the-world voyage on the Beagle, Darwin gathered data that would later support his theory of evolution.

Much has been learned since Darwin's day about the 65-million-year evolution of the horse, which began with a much smaller ancestor that had many toes and teeth with low crowns. As grasslands replaced a forestlike environment, the horse evolved into a much larger animal with fewer toes and teeth with high crowns. Today, the horse has only one toe and is highly adapted for running fast and far, allowing it to seek new sources of food and water when needed.

**Cover Credits**
***Biology*** Front cover photo on complete text and background photo for all parts © Galen Rowell/Mountain Light Photography
***Part 1: The Cell*** Front cover inset photo © Carolina Biological Supply Co.
***Part 2: Genetic Basis of Life*** Front cover inset photo © Gunter Ziesler/Peter Arnold, Inc.
***Part 3: Evolution and Diversity*** Front cover inset photo © Douglas Faulkner/Photo Researchers, Inc.
***Part 4: Plant Structure and Function*** Front cover inset photo © Astrid and Hahns-Frieder Michler/ Science Photo Library/Photo Researchers, Inc.
***Part 5: Animal Structure and Function*** Front cover inset photo © Runk/Schoenberger/Grant Heilman
***Part 6: Behavior and Ecology*** Front cover inset photo © Carl R. Sams II/Peter Arnold, Inc.

Copy Edited by Laura Beaudoin
Photo Research by Michelle Oberhoffer

The Credits section for this book begins on page C-1 and is considered an extension of the copyright page.

Library of Congress Catalog Card Number: 91-76060

ISBN ***Biology*** Casebound 0-697-12383-9
ISBN ***Biology*** Paper binding 0-697-15096-8
ISBN ***Part 1: The Cell*** 0-697-15098-4
ISBN ***Part 2: Genetic Basis of Life*** 0-697-15099-2
ISBN ***Part 3: Evolution and Diversity*** 0-697-15100-X
ISBN ***Part 4: Plant Structure and Function*** 0-697-15101-8
ISBN ***Part 5: Animal Structure and Function*** 0-697-15102-6
ISBN ***Part 6: Behavior and Ecology*** 0-697-15103-4
ISBN ***Biology*** Boxed set: 0-697-15097-6

Printed in the United States of America by Wm. C. Brown Communications, Inc., 2460 Kerper Boulevard, Dubuque, IA 52001

10   9   8   7   6   5   4   3   2   1

# PUBLISHER'S NOTE TO THE INSTRUCTOR

| Binding Option | Description | ISBN |
|---|---|---|
| *Biology*, casebound | The full-length text (chapters 1–50), with hard-cover binding. | 0-697-12383-9 |
| *Biology*, paperbound | The full-length text, paperback covered and available at a significantly reduced price, when compared with the casebound version. | 0-697-15096-8 |
| Part 1 *The Cell*, paperbound | Part 1 features the first unit or chapters 1–9 of the text, covering the scientific method, evolution and unity of life, basic chemistry, cell biology, photosynthesis, and respiration. This paperback option is available at a significantly reduced price when compared with both the full-length casebound and paperbound versions. | 0-697-15098-4 |
| Part 2 *Genetic Basis of Life* | Part 2 features chapters 10–18 and covers cell reproduction, Mendelian and molecular genetics, gene activity, and recombinant DNA and biotechnology. This paperback is also available at a significantly reduced price when compared with the full-length versions of the text. | 0-697-15099-2 |
| Part 3 *Evolution and Diversity* | Part 3 features chapters 19–28 on Darwin and evolution, genetics in evolution and diversity. Paperbound, it is available for a fraction of the full-length casebound or paperbound prices. | 0-697-15100-X |
| Part 4 *Plant Structure and Function* | Part 4 features chapters 29–32 on plant structure and function. Paperbound, it is also available for a fraction of the full-length casebound or paperbound prices. | 0-697-15101-8 |
| Part 5 *Animal Structure and Function* | Part 5 features chapters 33–44 on animal systems and reproduction and development. Paperbound, it also sells for a fraction of the full-length book price. | 0-697-15102-6 |
| Part 6 *Behavior and Ecology* | Part 6 features chapters 45–50 on behavior and ecology. Paperbound, it sells for a fraction of the full-length book price. | 0-697-15103-4 |
| *Biology*, the Boxed Set | The entire text, offered in an attractive, boxed set of all six paperbound "splits." It is available at the same price as the full-length casebound text. | 0-697-15097-6 |

## Binding Options and Recycled Paper

*Biology*—in all its numerous binding options (listed here)—is printed on **recycled paper stock**. All of its ancillaries, as well as all advertising pieces will be printed on recycled paper, subject to market availability.

Our goal in offering the text and its ancillary package on **recycled paper** is to take an important first step toward minimizing the environmental impact of our products. If you have any questions about recycled paper use, *Biology*, its package, any of its binding options, or any of our other biology texts, feel free to call us at 1-800-331-2111. Thank you.

Kevin Kane
*Executive Editor*
*Life Sciences*

# BRIEF CONTENTS

# CONTENTS

# 6

## Membrane Structure and Function 84

# 7

## Cellular Energy 104

# 8

## Photosynthesis 121

# 9

## Glycolysis and Cellular Respiration 137

# READINGS

# THE BIOLOGY LEARNING SYSTEM

## 3

### Basic Chemistry

Your study of this chapter will be complete when you can

1. name and describe the subatomic particles of an atom, indicating which one accounts for the occurrence of isotopes;
2. describe and discuss the energy levels (electron shells) of an atom, including the orbitals of the first 2 levels;
3. draw a simplified atomic structure of any atom with an atomic number less than 20;
4. distinguish between ionic and covalent reactions, and draw representative atomic structures for ionic and covalent molecules;
5. tell which atom has been reduced and which has been oxidized in a particular reaction;
6. describe the chemical properties of water, and explain their importance for living things;
7. define an acid and a base; describe the pH scale, and state the significance of buffers.

A hippopotamus keeps cool by remaining in water. Water has many biological functions, both without and within organisms. Cells are largely composed of this inorganic molecule, which facilitates chemical reactions and helps maintain a normal body temperature because it is slow to heat.

24

## Concept Summaries

At the ends of major sections within each chapter, concept summaries briefly highlight key concepts in the section, helping students focus their study efforts on the basics.

## Bold-Faced Key Terms

Important terms are boldfaced and defined on first mention.

## Behavioral Objectives

Each chapter begins with a list of behavioral objectives designed to help students identify the major concepts of the chapter. Their study of the chapter is complete when they can satisfy these objectives.

## Text Line Art

Graphic diagrams placed within textual passages help clarify difficult concepts and enhance learning.

HAVE YOU THANKED A GREEN PLANT TODAY? Plants dominate our environment, but most of us spend little time thinking about the various services they perform for us and for all living things. Chief among these is their ability to carry on photosynthesis, during which they use sunlight (*photo*) as a source of solar energy to produce food (*synthesis*):

Other organisms, called algae, are also photosynthetic. Algae (chap. 24) are a diverse group, but many are water dwelling, microscopic organisms related to plants.

The organic food produced by photosynthesis is not only used by plants themselves, it is also the ultimate source of food for all other living things (fig. 8.1). Also, when organisms convert carbohydrate energy into ATP energy they make use of the oxygen ($O_2$) given off by photosynthesis. Nearly all living things are dependent on atmospheric oxygen derived from photosynthesis.

At one time in the distant past, plant and animal matter accumulated without decomposing. This matter became the fossil fuels (e.g., coal, oil, and gas) that we burn today for energy. This source of energy, too, is the product of photosynthesis.

Photosynthesis is absolutely essential for the continuance of life because it is the source of food and oxygen for nearly all living things.

### Sunlight

Photosynthesis is an energy transformation in which solar energy in the form of light is converted to chemical energy within carbohydrate molecules. Therefore, we will begin our discussion of photosynthesis with the energy source—sunlight.

Solar radiation can be described in terms of its energy content and its wavelength. The energy comes in discrete packets called **photons**. So, in other words, you can think of radiation as photons that travel in waves:

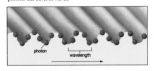

Figure 8.2*a* illustrates that solar radiation, or the **electromagnetic spectrum**, can be divided on the basis of wavelength—gamma rays have the shortest wavelength and radio waves have the longest

**Figure 8.1**
This squirrel is a herbivore. It feeds directly on plant material produced by a photosynthesizer. Carnivores, such as a hawk that may feed on this squirrel, are also dependent, although indirectly, on food produced by photosynthesizers.

wavelength. The energy content of photons is inversely proportional to the wavelength of the particular type of radiation; that is, short-wavelength radiation has photons of a higher energy content than long-wavelength radiation. High-energy photons, such as those of short-wavelength ultraviolet radiation, are dangerous to cells because they can break down organic molecules. Low-energy photons, such as those of infrared radiation, do not damage cells because they only increase the vibrational or rotational energy of molecules; they do not break bonds. But photosynthesis utilizes only the portion of the electromagnetic spectrum known as **visible light.** (It is called visible light because it is the part of the spectrum that allows us to see.) Photons of visible light have just the right amount of energy to promote electrons to a higher electron shell in atoms without harming cells.

We have mentioned previously that only about 42% of the solar radiation that hits the earth's atmosphere ever reaches the surface, and most of this radiation is within the visible-light range.

122

The Cell

## Readings

Throughout *Biology*, selected readings reinforce major concepts in the book. Most readings are written by the author, but a few are excerpted from popular magazines. A reading may provide insight into the process of science or show how a particular kind of scientific knowledge is applicable to the students' everyday lives.

## Study Questions

These questions appear at the end of each chapter. They call for specific, short essay answers that challenge students' mastery of the chapter's basic concepts.

## Objective Questions

Multiple-choice questions at the end of each chapter test basic recall of the chapter's key points. The answers to the objective questions are listed in appendix D.

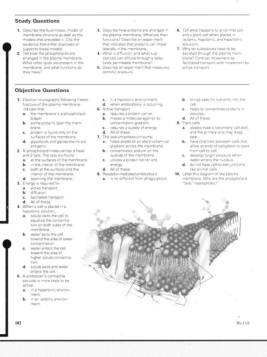

## Practice Problems

Practice problems at the end of passages in the genetics chapters help students master basic genetic quantifications. Answers to these problems are in appendix D.

## Selected Key Terms

A selected list of bold-faced, key terms from the chapter appears at the end of each chapter. Each term is accompanied by its phonetic spelling and a page number indicating where the term is introduced and defined in the chapter.

## Critical Thinking Case Studies

Each part ends with a case study designed to help students think critically by participating in the process of science. At many institutions, instructors are encouraged to develop the writing skills of their students. In such cases, instructors could require students to write out their answers to the questions in each case study. Suggested answers for each of these questions appear in the Instructor's Manual for the text.

## Concepts and Critical Thinking

These end-of-chapter questions require students to apply the chapter's basic concepts to biological concerns. Writing the answers to these or the Study Questions fulfills any Writing-Across-the-Curriculum requirement.

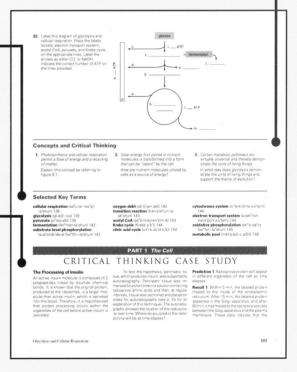

## Chapter Summaries

At the end of each chapter is a summary. This listing of important points should help students more readily identify important concepts and better facilitate their learning of chapter content.

## Further Readings

A list of readings at the end of each part suggests references that can be used for further study of topics covered in the chapters of that part. The items listed in this section were carefully chosen for readability and accessibility.

# PREFACE

*Biology* is an introductory college text that covers the basic concepts and principles of general biology. While the text is clear and straightforward enough to be read by liberal arts students, it is also comprehensive and authoritative, so it can just as easily be read by science majors. The text strives to use other forms of life, in addition to humans, as examples, making frequent reference to plants and invertebrates. The text takes a hierarchical approach, proceeding from the chemistry of the cell up to the organization of the biosphere.

## Science Process and Critical Thinking

*Biology* stresses the process of science in many ways. Chapter 2 is entirely devoted to describing the scientific method with examples. And throughout the text, experiments that have led us to our present level of knowledge are described. *New to this edition are the "How Do You Do That?" boxes, which help to emphasize technological literacy.* The boxes are meant to take the mystery out of laboratory procedures and show students that everyone is capable of utilizing the methods by which data are collected.

Understanding the concepts of biology and engaging in the process of science encourages one to think critically. *New to this edition, each chapter ends with a list of the major concepts within the chapter and asks critical thinking questions based on those concepts.* Those questions allow students to see that thinking conceptually leads to predictions and deductions concerning particular aspects of biology. Suggested answers to the end-of-chapter critical thinking questions appear in appendix D. Each major part in the text ends with a critical thinking case study. As before, the case studies were written by Dr. Robert Allen and myself. They provide students with an opportunity to use scientific methodology as they think critically. Suggested answers to the case study questions appear in the Instructor's Manual for the text.

## Writing Across the Curriculum

Students need a chance to practice their writing skills and at the same time test their ability to fulfill the behavioral objectives that begin each chapter. *This is provided by the "Writing Across the Curriculum Study Questions" at the close of the chapter,* which is based on an educational approach to improving the skills of students. Instructors who have their students write out the answers to the study questions and critical thinking questions will have answered the challenge of this new approach.

The objective questions prepare students to take examinations that contain recall-based questions. These, too, have been improved, and instructors will appreciate the addition of questions that require students to complete and/or label diagrams.

## Organization of the Text

Each chapter in *Biology* introduces particular concepts and describes biological experiments that have contributed to our present level of knowledge. As in the third edition, the chapters are of a suitable length to be read in one sitting. *A new chapter on animal organization and homeostasis has been added to part 5.* This chapter includes a discussion of animal tissues. The text has the following parts, which have been revised as discussed.

### Introduction

*Chapter 1, which concerns the characteristics of life and introduces important biological concepts, was completely rewritten to increase student interest.* The chapter shows that the unity of life is due to descent from a common ancestor and that diversity is due to adaptations to particular ways of life in an ecosystem.

Chapter 2 thoroughly explains scientific methods and gives examples of both experimental and observational biological research. The biological understanding of the word *theory* is also fully explained.

### Part 1 The Cell

Part 1 concerns cell structure and function including energy metabolism. *Membrane protein diversity is more thoroughly discussed and illustrated in this edition.* The first of the 3 chapters devoted to energetics gives an overview. Instructors can use just this one chapter or proceed on to cellular respiration and/or photosynthesis.

## Part 2 Genetic Basis of Life

There is a strong historical emphasis in this part, but practical aspects are not neglected. Students are given an opportunity to test their ability to do problems as they proceed. New advances in the field of genetics appear daily, and this part has been thoroughly updated to reflect our changing knowledge. *The very latest that is known about human genetic disorders is included. The biotechnology chapter includes a discussion of new techniques in gene therapy in humans and the human genome project.* The presentation is at an appropriate level for the beginning student.

## Part 3 Evolution and Diversity

*The evolution chapters were completely reorganized and rewritten with the valuable help of many instructors and scholars who specialize in this area.* The geological time scale was completely updated. Chapter 22, concerning the origin of life, was also rewritten and now includes several hypotheses on this topic. The diversity chapters were revised and the most up-to-date classification system is used. Instructors will appreciate a new expanded table that compares the major features of the plant divisions.

## Part 4 Plant Structure and Function

Four chapters are devoted to flowering plant anatomy and physiology. The first chapter provides a foundation for the others on nutrition and transport, reproduction, and growth and development.

## Part 5 Animal Structure and Function

*This part begins with a new chapter on animal organization and homeostasis.* There is a discussion of animal tissues and organ systems before homeostasis is discussed in some detail. *The comparative section that begins each of the animal physiology chapters has been rewritten for clarity and improvement. A new extended section on AIDS is included in the reproduction chapter.*

## Part 6 Behavior and Ecology

*A new animal behavior chapter takes a more modern approach to this topic.* The ecology chapters have a logical sequence from populations to communities to the biosphere. Environmental concerns are emphasized in the last 2 chapters. Some instructors may wish to begin the year's work with this part, which is certainly a workable alternative.

# Aids to the Reader

*Biology* was written so that students can enjoy, appreciate, and come to understand the concepts of biology and the scientific process. The following text features are especially designed to assist student learning.

## Text Introduction

The introduction section (chapters 1 and 2) lays a foundation upon which the rest of the text depends. The first chapter reviews the characteristics of life and presents the fundamental concepts of biology. The second chapter explains the scientific process in detail and gives examples of studies done by biologists. Various other experiments are described throughout the text.

## Study Objectives

Each chapter begins with a list of objectives designed to guide students as they study the chapter. Their study of the chapter is complete when they can satisfy these objectives, therefore, they help students prepare for an examination.

## Boxed Readings

Throughout *Biology,* selected boxed readings reinforce major concepts in the book. Most readings are of general interest and heighten student involvement by expanding on a topic discussed in the chapter.

## "How Do You Do That?" Boxes

New to this edition are the "How Do You Do That?" boxes, which describe laboratory methods. They are designed to take the mystery out of the manner in which biological data are collected.

## Drawings, Photographs, and Tables

The drawings, photographs, and tables in *Biology* have been designed to help students learn basic biological concepts as well as the specific content of the chapters, and are consistent with multicultural educational goals. Often it is easier to understand a given process by studying a drawing, especially when it is carefully coordinated with the text, as is the case here. The photographs were selected not only to please the eye but also to emphasize specific points in the text. The tables summarize and list important information, making it readily available for efficient study.

## Chapter Summaries

The summary is a numbered series of statements that follow the organization of the chapter. The summary helps students identify the concepts of the chapter and facilitates their learning of chapter content.

## Chapter Questions

Three kinds of questions—study questions, objective questions, and critical thinking questions—appear at the close of each chapter. They allow students to test their ability to fulfill the study objectives. *Writing across the curriculum* recognizes that students need an opportunity to practice writing in all courses. The study questions review the chapter, and their sequence follows that of the chapter. When students write out the answers to the study questions, they are writing while studying biology. The critical thinking questions are based on biological concepts that pertain to the chapter. They show that knowledge of a biological concept allows one to reason about some particular aspect of biology. Writing out the answers to the critical thinking questions also fulfills any writing across the curriculum requirements. The objective questions allow students to test their ability to answer recall-based, multiple choice questions. The objective questions have been expanded in this edition to include questions that require the completing or labeling of diagrams. Answers to the objective questions and critical thinking questions appear in appendix D.

## Selected Key Terms

Each chapter ends with a selected key term list. Key terms are boldfaced in the chapter, defined in context, and also appear in the glossary. Especially significant key terms appear in the selected key term list. Each term is accompanied by its phonetic spelling and is referenced to the page on which it is introduced and defined.

## Further Readings

The list of readings at the end of each part suggests references that can be used for further study of the topics covered in the chapters of that part. The items listed in this section were carefully chosen for readability and accessibility.

## Critical Thinking Case Studies

Each part ends with a case study designed to help students think critically by participating in the process of science. At many institutions, instructors are encouraged to develop the writing skills of their students. In such cases, instructors could require students to write out their answers to the questions in each case study. Suggested answers for each of these questions appear in the Instructor's Manual.

## Glossary

The glossary contains the terms that are boldfaced in the text. Many of the glossary entries are accompanied by a phonetic spelling and a definition, and each is referenced to the page on which it was first introduced and defined in the text.

# ADDITIONAL AIDS

## Instructor's Manual/Test Item File

Revised by Jean Helgeson, the Instructor's Manual/Test Item File is designed to assist instructors as they plan and prepare for classes using *Biology*. For each chapter in the text, the Instructor's Manual provides a chapter outline, key terms, an extended lecture outline, enrichment ideas, and a listing of selected films. The Test Item File contains approximately 40 multiple choice, true/false, and critical thinking essay questions per text chapter.

Suggested answers for the critical thinking case studies that appear at the end of each part in the text are placed at the end of the corresponding parts in the Instructor's Manual.

In addition, the Instructor's Manual includes an answer key for the Critical Thinking Case Study Workbook by Robert Allen, along with listings of the transparencies and micrograph slides available to instructors using *Biology,* fourth edition.

## Student Study Guide

The Student Study Guide that accompanies the text was revised by Jay Templin. For each text chapter, there is a corresponding study guide chapter that includes a list of study objectives, a chapter review, vocabulary terms that are page-referenced to the text, learning activities with text page references that correlate to the study objectives and require students to apply textual information, critical thinking questions, and chapter test questions with an answer key providing the student immediate feedback.

## Micrograph Slides

A boxed set of 99 slides includes all photomicrographs and electron micrographs printed in the text.

## Laboratory Manual

The Laboratory Manual that accompanies *Biology* was thoroughly revised by Kenneth Kilborn and myself. Its 34 exercises provide enough variety to meet the needs of a broad spectrum of class designs. Student aids include a list of learning objectives at the beginning of each exercise and numerous full-color illustrations throughout. Each exercise includes an introduction, and ample space is provided for students to record their observations as the lab proceeds. A laboratory review, consisting of a series of questions, ends each exercise. Answers to the review questions are provided in an appendix.

## Customized Laboratory Manual

The Laboratory Manual's 34 exercises are available as individual "lab separates," so instructors can custom-tailor the manual to their particular course needs. The separates, which are published in one color, can be individually selected at a greatly reduced price and will be collated and bound by ▧ on request.

## Laboratory Resource Guide

Helpful and thorough information regarding each lab preparation can be found in the Laboratory Resource Guide. Completely revised by Kenneth Kilborn, the guide is designed to help instructors make the laboratory experience a more meaningful one for the student. For handy reference, a list of suppliers is printed on the inside front cover. The Resource Guide is now divided into 3 parts. They are Laboratory Preparation and Instructions; Laboratory Exercises and Expected Results; and Answers to the Laboratory Review Questions.

## Transparencies and Lecture Enrichment Kit

A full set of transparency acetates also accompanies the text. These feature key illustrations from the text in both 2 and full color. They are accompanied by a Lecture Enrichment Kit, which is a set of lecture notes featuring additional high-interest information about the pictured process or concept and not presented in the text.

## Critical Thinking Case Study Workbook

Written by Robert Allen, this ancillary includes 30 additional critical thinking case studies of the type found in the text. Like the text case studies, they are designed to immerse students in the "process of science" and challenge them to solve problems in the same way biologists do. The case studies here are divided into 3 levels of difficulty (introductory, intermediate, and advanced) to afford instructors greater choice and flexibility. An answer key is printed in the Instructor's Manual.

## Extended Lecture Outline Software

This instructor software features extensive outlines of each text chapter with a brief synopsis of each subtopic to assist in lecture preparation. Written in ASCII files for maximum utility, it is available in IBM, Apple, or Mac formats. It is free to all adopters, upon request.

## Biology Art Masters

A set of 150 art masters consisting of one-color line art with labels can be used for additional transparencies or can be copied and used for student hand-outs.

The 7 packages include the following titles: Cell Biology, Genetics, Diversity, Plant Biology, Animal Biology, Ecology and Behavior, and Evolution.

Adopters of *Biology* can order one or all 7 packages for free.

## WCB TestPak with Enhanced QuizPak and GradePak

**WCB** TestPak, a computerized testing service, provides instructions with either a mail-in/call-in testing program or the complete test item file on diskette for use with the IBM PC, Apple, or Macintosh computer. **WCB** TestPak requires no programming experience.

**WCB** QuizPak, a part of TestPak, provides students with true/false, multiple choice, and matching questions for each chapter in the text. Using this portion of the program will help students to prepare for examinations. Also included with the QuizPak is an on-line testing option to allow professors to prepare tests for students to take on the computer. The computer will automatically grade the test and update a gradebook file.

**WCB** GradePak, also a part of TestPak, is a computerized grade management system for instructors. This program tracks student performance on examinations and assignments. It will compute each student's percentage and corresponding letter grade, as well as the class average. Printouts can be made utilizing both text and graphics.

## Other Titles of Related Interest from Wm. C. Brown Publishing

### You Can Make a Difference
*by Judith Getis*

This short, inexpensive supplement offers students practical guidelines for recycling, conserving energy, disposing of hazardous wastes, and other pollution controls. It can be shrink wrapped with the text, at minimal additional cost (ISBN 0-697-13923-9)

### How to Study Science
*by Fred Drewes, Suffolk County Community College*

This excellent new workbook offers students helpful suggestions for meeting the considerable challenges of a college science course. It offers tips on how to take notes; how to get the most out of laboratories; as well as how to overcome science anxiety. The book's unique design helps students develop critical thinking skills, while facilitating careful note-taking. (ISBN 0-697-14474-7)

### The Life Science Lexicon
*by William N. Marchuk, Red Deer College*

This portable, inexpensive reference helps introductory-level students quickly master the vocabulary of the life sciences. Not a dictionary, it carefully explains the rules of word construction and derivation, in addition to giving complete definitions of all important terms. (ISBN 0-697-12133-X)

### Biology Study Cards
*by Kent Van De Graaff, R. Ward Rhees, and Christopher H. Creek, Brigham Young University*

This boxed set of 300, 2-sided study cards provides a quick yet thorough visual synopsis of all key biological terms and concepts in the general biology curriculum. Each card features a masterful illustration, pronunciation guide, definition, and description in context. (ISBN 0-697-03069-5)

### The Gundy-Weber Knowledge Map of the Human Body
*by G. Craig Gundy, Weber State University*

This 13-disk, Mac-Hypercard program is for use by instructors and students alike. It features carefully prepared computer graphics, animations, labeling exercises, self-tests, and practice questions to help students examine the systems of the human body. Contact your local Wm. C. Brown representative or call 1-800-351-7671.

*The Knowledge Map Diagrams*
1. Introduction, Tissues, Integument System (0-697-13255-2)
2. Viruses, Bacteria, Eukaryotic Cells (0-697-13257-9)
3. Skeletal System (0-697-13258-7)
4. Muscle System (0-697-13259-5)
5. Nervous System (0-697-13260-9)
6. Special Senses (0-697-13261-7)
7. Endocrine System (0-697-13262-5)
8. Blood and the Lymphatic System (0-697-13263-3)
9. Cardiovascular System (0-697-13264-1)
10. Respiratory System (0-697-13265-X)
11. Digestive System (0-697-13266-8)
12. Urinary System (0-697-13267-6)
13. Reproductive System (0-697-13268-4)

Demo—(0-697-13256-0)
Complete Package—(0-697-13269-2)

### GenPak: A Computer Assisted Guide to Genetics
*by Tully Turney, Hampden-Sydney College*

This Mac-Hypercard program features numerous interactive/tutorial (problem-solving) exercises in Mendelian, molecular, and population genetics at the introductory level. (ISBN 0-697-13760-0)

## Acknowledgments

Many persons have contributed to the success of previous editions and helped me make this edition our very best effort. My editor, Kevin Kane, directed the efforts of all. Carol Mills, my developmental editor, served as a liaison between the editor, me, and many other people. I especially want to mention Susan Murray and Anne Packard of Plymouth State College, who carefully read the galleys and made many helpful suggestions.

The production team at Wm. C. Brown was ever-faithful to their duties. Anne Scroggin and Ann Fuerste were the production editors; Miriam Hoffman the art editor; Lori Gockel the photo researcher; Michelle Oberhoffer the photo researcher; and Mark Christianson the designer. My thanks to each of them for a job well done!

## The Contributors

With this edition, as with the last edition, there were several contributors to various sections of the book. They critically reviewed my revised chapters and told me how they could be improved. With their help, it was easier to make sure the content was sufficient, accurate, and up to date. The contributors were:

W. Dennis Clark (Ph.D. University of Texas), professor of botany at Arizona State University. Dr. Clark's expertise in the areas of chemistry, evolution, and physiology of plants was extremely helpful in the revision of the plant and energy chapters.

Thomas C. Emmel (Ph.D. Stanford University), professor of zoology at the University of Florida. He assisted invaluably by providing updated information in ecology, where he has published numerous articles and 2 books.

Robert M. Kitchin (Ph.D. University of California-Berkeley), professor of genetics at the University of Wyoming. He reviewed the genetics chapters and outlined new information and events taking place in this rapidly changing and expanding field.

## The Ancillary Contributors

The ancillary package that accompanies *Biology* also involved the efforts of many instructors currently teaching biology.

*Critical Thinking Case Study Book:* Robert Allen (Ph.D. University of California-Los Angeles) is vice-president for instruction at Victor Valley College, whose special interests include research on the development of critical thinking skills in science. He is a frequent presenter on the subject for the National Science Foundation Chatauqua series.

*Instructor's Manual-Test Item File:* Jean Helgeson, (M. A. The University of Texas Southwestern Medical Center) is professor and coordinator of biology and health science at Collin County Community College.

*Student Study Guide:* Jay Templin (Ed.D. Temple University) is an instructor at Widener University who has authored numerous student and instructor's ancillaries for several college biology textbooks. He is also a co-author of the *College Board Achievement Test* in biology.

*Laboratory Manual and Laboratory Resource Guide:* Kenneth Kilborn (M. A. San Jose State University) has been teaching general biology and general botany in addition to 20 years of teaching in the biology laboratory at Shasta College.

*QuizPak and The Lecture Enrichment Kit:* Jane Aloi (Ph.D. University of California, Davis) is an ecologist teaching courses in introductory biology, zoology, and human anatomy at Saddleback College.

## Reviewers

Many instructors of introductory biology courses around the country reviewed portions of the manuscript. Others assisted by taking the time to fill out survey forms that helped us make important decisions about content. With many thanks, we list their names here.

### *First Edition*

A. Lester Allen *Brigham Young University*
William E. Barstow *University of Georgia*
Lester Bazinet *Community College of Philadelphia*
Eugene C. Bovee *University of Kansas*
Larry C. Brown *Virginia State University*
L. Herbert Bruneau *Oklahoma State University*
Carol B. Crafts *Providence College*
John D. Cunningham *Keene State College*
Dean G. Dillery *Albion College*
H. W. Elmore *Marshall University*
David J. Fox *University of Tennessee*
Larry N. Gleason *Western Kentucky University*
E. Bruce Holmes *Western Illinois University*
Genevieve D. Johnson *University of Iowa*
Malcolm Jollie *Northern Illinois University*
Karen A. Koos *Rio Hondo College*
William H. Leonard *University of Nebraska—Lincoln*
A. David Scarfe *Texas A & M University*
Carl A. Scheel *Central Michigan University*
Donald R. Scoby *North Dakota State University*
John L. Zimmerman *Kansas State University*

### *Second Edition*

David Ashley *Missouri Western State College*
Jack Bennett *Northern Illinois University*
Oscar Carlson *University of Wisconsin-Stout*
Arthur Cohen *Massachusetts Bay Community College*
Rebecca McBride DeLiddo *Suffolk University*
Gary Donnermeyer *St. John's University (Minnesota)*
D. C. Freeman *Wayne State University*
Sally Frost *University of Kansas*
Maura Gage *Palomar College*
Betsy Gulotta *Nassau Community College*
W. M. Hess *Brigham Young University*
Richard J. Hoffmann *Iowa State University*
Trudy McKee
Brian Myres *Cypress College*
John M. Pleasants *Iowa State University*
Jay Templin *Widener University*

### *Third Edition*

Wayne P. Armstrong *Palomar College*
Mark S. Bergland *University of Wisconsin-River Falls*
Richard Blazier *Parkland College*

William F. Burke *University of Hawaii*
Donald L. Collins *Orange Coast College*
Ellen C. Cover *Manatee Community College*
John W. Crane *Washington State University*
Calvin A. Davenport *California State University-Fullerton*
Robert Ebert *Palomar College*
Darrell R. Falk *Pt. Loma Nazarene College*
Jerran T. Flinders *Brigham Young University*
Sally Frost *University of Kansas*
Elizabeth Gulotta *Nassau Community College*
Madeline M. Hall *Cleveland State University*
James G. Harris *Utah Valley Community College*
Kenneth S. Kilborn *Shasta College*
Donald R. Kirk *Shasta College*
Jon R. Maki *Eastern Kentucky University*
Ric Matthews *Miramar College*
Joyce B. Maxwell *California State University-Northridge*
Leroy McClenaghan *San Diego State University*
Leroy E. Olson *Southwestern College*
Barbara Yohai Pleasants *Iowa State University*
David M. Prescott *University of Colorado*
Robert R. Rinehart *San Diego State University*
Mary Beth Saffo *University of California—Santa Cruz*
Walter H. Sakai *Santa Monica College*
Frederick W. Spiegel *University of Arkansas-Fayetteville*
Gerald Summers *University of Missouri-Columbia*
Marshall Sundberg *Louisiana St. University*
Kathy S. Thompson *Louisiana State University*
Anna J. Wilson *Oklahoma City Community College*
Timothy S. Wood *Wright State University*

### Fourth Edition

Michael S. Gaines *University of Kansas*
Kerry S. Kilburn *West Virginia State College*
W. Sylvester Allred *Northern Arizona University*
Helen L. Grierson *Morehead State University*
Deborah K. Meinke *Oklahoma State University*
Robert Snetsinger *Queens University*
Gail Kingrey *Pueblo Community College*
Barbara Yohai Pleasants *Iowa State University*
Eugene Nester *University of Washington*

## Survey Respondents

Sister Julia Van Demack *Silver Lake College*
Ken Beatty *College of the Siskiyous*
John W. Metcalfe *Potsdam College at the State University of New York*
Brenda K. Johnson *Western Michigan University*
Tom Dale *Kirtland Community College*
Marlene Kayne *Trenton State College*
Peter M. Grant *Southwestern Oklahoma State University*
Janet Carter *Delaware Technical & Community College*
Frank F. Escobar *Holyoke Community College*
John A. Chisler *Glenville State College*

Nancy Goodyear *Bainbridge College*
John F. Lyon *University of Lowell*
Rob Snetsinger *Queens University*
Fred Stevens *Schreiner College*
Paula H. Dedmon *Gaston College*
Bruce G. Stewart *Murray State College*
Leo R. Finkenbinder *Southern Nazarene University*
Mary P. Greer *Macomb Community College*
Kenneth M. Allen *Schoolcraft College*
Robert Cerwonica *SUNY-Potsdam*
Darcy Williams *Cecil Community College*
Dave McShaffrey *Marietta College*
Elizabeth L. Nebel *Macomb Community College*
Jal S. Parakh *Western Washington University*
Dwight Kamback *Northampton Community College*
David L. Haas *Fayetteville State University*
A. Quinton White *Jacksonville University*
Paul J. Hummer, Jr. *Hood College*
Dee Forrest *Western Wyoming College*
Linden C. Haynes *Hinds Community College*
John De Banzie *Northeastern State University*
Anne T. Packard *Plymouth St. College*
L. H. Buff, Jr. *Spartanburg Methodist College*
Dennis Vrba *North Iowa Area Community College*
Donald L. Collins *Orange Coast College*
Larrie E. Stone *Dana College*
Brian K. Mitchell *Longview College*
Dennis M. Forsythe *The Citadel*
Bonnie S. Wood *University of Maine at Presque Isle*
Eunice R. Knouso *Spartanburg Methodist College*
Larry C. Brown *Virginia State University*
Clyde M. Senger *Western Washington University*
James V. Makowski *Messiah College*
James R. Coggins *University Wisconsin-Milwaukee*
Roland Vieira *Green River Community College*
Donna Barleen *Bethany College*
S. M. Cabrini Angelli *Regis College*
W. Brooke Yeager, III *Penn State-Wilkes-Barre*
Diana M. Colon *Northwest Technical College*
Carolyn K. Jones *Vincennes University*
Rita M. O'Clair *University of Alaska Southeast*
Fred M. Busroe *Morehead State University*
Edward R. Garrison *Navajo Community College*
Joseph V. Faryniarz *Maitatuck Community College*
K. Dale Smoak *Piedmont Technical College*
Donald A. Wheeler *Edinboro University of Pennsylvania*
W. Lee Williams *Alamance Community College*
Eugene A. Oshima *Central Missouri State University*
Richard C. Renner *Laredo Junior College*
Peggy Rae Doris *Henderson State University*
Charles H. Owens *Virginia Highlands Community College*
Jeanette Oliver *Flathead Valley Community College*
E. L. Beard *Loyola University*
Monica Macklin *Northeastern State University*

# 1

## A View of Life

Your study of this chapter will be complete when you can

1. list and explain 5 characteristics of life;
2. relate the concept of emergent properties to the increasingly complex levels of biological organization;
3. describe how populations of organisms interact within ecosystems;
4. explain how evolution accounts for both the unity and the diversity of life.

Bengal tiger and cub. Life comes only from life, and reproduction passes life's characteristics from one generation to another.

a.

b.

c.

**Figure 1.1**

Examples of life's diversity, rockhopper penguins are birds in good standing, but they cannot fly. Their stubby forelimbs are modified into flippers for fast swimming. Only a little over a half meter tall, these penguins are named for their skill in leaping from rock to rock. *a.* Male and female rockhoppers court on the rocks of their breeding ground. *b.* Rockhoppers greet each other by braying; the sound has been likened to the screech of a rusty wheelbarrow. *c.* Yellow feathers above the eyes give a rockhopper a striking pair of eyebrows.

ife on earth takes on a staggering variety of forms, often behaving in ways strange to us. For example, gastric-brooding frogs swallow their embryos and give birth to them later by throwing them up! Fetal sand sharks kill and eat their siblings while still inside their mother. Young black widow spiders spin a perfect web on their first try. Fringe-lipped bats swoop out of the black tropical night to attack mud-puddle frogs. Some *Ophrys* orchids look so much like female bees that male bees try to mate with them. Penguins are birds that cannot fly but can swim (fig. 1.1). Bacteria live out their entire lives in 15 minutes, while bristlecone pine trees outlive 10 generations of humans.

Categorizing the astounding array of life has always been difficult, and different types of classification systems have been and are in use today. Most modern biologists, however, use a system that divides organisms into 5 major groups, or **kingdoms:** protista, monera, fungi, plants, and animals. The reading in this chapter describes these kingdoms.

Despite life's diversity, all living things, be they pasteur roses, sidewinder snakes, or basketball players, are comprised of the same chemical elements that make up nonliving things and they obey the same laws of physics and chemistry as nonliving objects do. So, what is special about living things? Fortunately, all organisms share certain common characteristics. They give us insight into the nature of life and help us distinguish living things from nonliving things.

## Characteristics of Life

### Living Things Are Organized

The complex organization of living things (fig. 1.2) begins with the **cell,** the smallest, most basic unit of life. Most cells are microscopic specks, but some can be long and thin, as in the 3-foot-long fibers of human nerve cells. They can even be a blob the size of an apple, as with the ostrich's egg with its plentiful supply of yolk.

Some cells live by themselves as functioning organisms. Other cells sometimes cluster together in microscopic colonies. In multicellular organisms, similar cells combine to form a tissue, such as nerve tissue, for example. Tissues make up organs, as when various tissues combine to form the brain. Organs work together in systems; for example, the brain works with the spinal cord and a network of nerves to form the nervous system. Organ systems are joined within an individual (see fig. 3.1).

There are levels of biological organization that extend beyond the individual organism. All the members of one species in a particular area belong to a population. In your local forest there is a population of gray squirrels, for example. The populations of various animals and plants in the forest make up a community. The community of populations interacts with the physical environment and forms an ecosystem. How an ecosystem functions is discussed in more detail later in this chapter.

**Figure 1.2**
The high degree of organization of life shows up everywhere, from the patterns of veins on a dragonfly's wings to the arrangement of bundles of pine needles.

### *Emergent Properties*

We have seen that biological organization includes these levels of organization: cells, tissues, organs, organ systems, individuals, populations, communities, and ecosystems. Each level of organization is more complex than the level preceding it and has properties beyond those of the former level. For example, when cells are broken down into bits of membrane and oozing liquids, these parts themselves cannot carry out the business of living. Slice a lump of coal and fit the pieces together, and you still have a lump of coal. Slice a frog and arrange the slices, and you have a former frog. It can no longer grow and reproduce.

In the living world, the whole is indeed more than the sum of its parts. Each new level of biological organization has emergent properties that are due to interactions between the parts making up the whole. All properties, even emergent properties, are governed by the laws of physics and chemistry.

Living things have levels of organization from cells to ecosystems. Each level of organization has emergent properties that cannot be accounted for by simply adding up the properties of previous levels.

A golden-furred, rat-sized animal, discovered in Australia about 100 years ago, bores through desert sand like a mole but has a pouch like a kangaroo. Figuring out what to call such an oddity is the job of taxonomists. Taxonomy is the discipline of identifying and classifying organisms according to certain rules (*taxo*, to put into order; *nomy*, law or rule).

Taxonomists give each living thing a binomial (*bi*, two; *nomen*, name) in Latin, a custom started by the Swedish taxonomist Linnaeus. For example, 2 types of the mole mentioned are named *Notorcytes typhlops* and *Notorcytes caurinus*. The first word of the name is the genus and the second word is the specific epithet, which pinpoints a particular species within a genus.

Another aspect of taxonomy is the grouping of species into larger groups according to their shared characteristics. Taxonomists put similar species in the same genus, then group genera into families, families into orders, and so on as listed in table 1.A. All the species in a genus have fairly specific characteristics in common, whereas those organisms in the same kingdom have general characteristics in common. Most taxonomists now believe that the classification of an organism into progressively larger categories, defined by increasingly general characteristics, reflects the organism's evolutionary history. Species are defined by unique, recently evolved characteristics, but the traits that link all the organisms in a kingdom are ancient.

Taxonomists debate intensely over how many kingdoms to recognize and how to define them. Originally only 2 kingdoms were described—plants (Plantae) and animals (Animalia). Plants were literally organisms that were planted and immobile, while animals were animated and moved about. The discovery of photosynthesis gave an added characteristic to plants: they make their own food, whereas animals must eat. When microscopes were invented in the seventeenth century, scientists found many tiny creatures that weren't much like either plants or animals. What, for example, do you call a little cell that photosynthesizes like a plant but swims and eats like an animal?

These problems have inspired taxonomists to propose a number of classification systems. Some still retain just 2 kingdoms, and others use as many as 15 kingdoms. Such disagreements are good for progress in taxonomy but confusing to biology students. Therefore, we have elected to refer to just one classification system. Widely used today, this system divides living things into 5 kingdoms (table 1.B and fig. 1.A). Three of these kingdoms contain multicellular forms. These organisms differ in their mode of nutrition: plants make their own food through photosynthesis; animals ingest preformed food, and fungi live on preformed food but do not ingest it, secreting enzymes that break down food externally.

**Table 1.A**
Levels of Classification

| | Example: Human | Example: Corn |
|---|---|---|
| Kingdom | Animalia | Plantae |
| Phylum (Division) | Chordata | Anthophyta |
| Class | Mammalia | Monocotyledonae |
| Order | Primates | Commelinales |
| Family | Hominidae | Poaceae |
| Genus | *Homo* | *Zea* |
| Species | *Homo sapiens* | *Zea mays* |

## Living Things Metabolize

A bedroom can easily turn into a giant heap of laundry, notebooks, magazines, and random doughnut boxes unless the occupant puts in energy to maintain order. Living things have the same problem. Maintaining their high degree of organization takes energy. Living things also need raw materials from their surroundings (their environment). **Metabolism** is the term we use for all the chemical and energy transformations that occur in cells as they use energy and nutrients to grow and make repairs.

The ultimate source of energy for all life on earth is the sun. Plants and plantlike organisms capture a small fraction of the energy that reaches the earth from the sun. They photosynthesize, a process that lets them store solar energy by converting carbon dioxide and water into energy-rich sugars. Animals get energy by eating plants; then other animals eat those animals (fig. 1.3).

### Homeostasis

For metabolic processes to continue, living things need to keep themselves stable in temperature, moisture level, acidity, and other physiological factors. This is **homeostasis**—the maintenance of internal conditions within certain boundaries.

Living things can maintain homeostasis in a variety of ways. A chilly lizard may raise its internal temperature by basking in the sun on a hot rock. When the lizard starts to overheat, it scurries for cool shade. Biology students also need to maintain homeostasis. Consider a student who becomes so engrossed in her textbook that she forgets to eat lunch. In such an emergency, the liver responds by releasing stored sugar, giving her energy until she eats again.

An intake of materials and energy is needed if an organism's organization is to be maintained. The ultimate source of energy for life on earth is the sun.

## Living Things Respond

Living things find energy and raw materials by interacting with their surroundings. Even one-celled organisms can respond to their environment. Some can move toward or away from light or chemicals by beating microscopic hairs or by snapping whiplike tails. Multicellular organisms can manage more complex responses. A maple seed can respond to cold weather, followed by

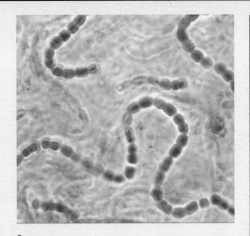

a.

**Figure 1.A**
Representatives from the 5 kingdoms recognized in this text: ***a.*** kingdom Monera: *Nostoc,* a cyanobacterium, magnification, X100; ***b.*** kingdom Protista: *Euglena,* a single-celled organism; ***c.*** kingdom Fungi: *Coprinus comatus,* a shaggy mane mushroom; ***d.*** kingdom Plantae: *Rosa* hybrid, a flowering plant; ***e.*** kingdom Animalia: *Felix caracal,* an African lynx (small cat).

**Table 1.B**
Kingdoms of Life

| Kingdom | Organization | Type of Nutrition | Representative Organisms |
|---|---|---|---|
| Monera (monerans) | Small, simple single cell (sometimes chains or mats) | Absorb food (some photosynthetic) | Bacteria including cyanobacteria |
| Protista (protists) | Large, complex single cell (sometimes chains or colonies) | Absorb, photosynthesize, or ingest food | Protozoans, algae of various types |
| Fungi | Multicellular and filamentous with specialized complex cells | Absorb food | Molds and mushrooms |
| Plantae (plants) | Multicellular with specialized complex cells | Photosynthesize food | Mosses, ferns, woody and nonwoody flowering plants |
| Animalia (animals) | Multicellular with specialized complex cells | Ingest food | Sponges, worms, insects, fish, amphibians, reptiles, birds, and mammals |

b.

c.

d.

e.

warm weather, by sprouting out of its seed coat—just plain warmth will not trigger this action. A vulture can smell meat a mile away and soar toward dinner. A monarch butterfly in Maryland can sense the approach of fall and begin its flight to Mexico for winter. And humans can detect the arrival of winter and head for the ski slopes, haul out sweaters, and buckle down to study for exams. All these responses make up what we call the **behavior** of an organism.

## Living Things Reproduce

Life only comes from life. Every type of living thing can **reproduce,** or make a copy like itself (fig. 1.4). Bacteria, protozoa, and other one-celled organisms simply split in 2. Most multicellular organisms begin the reproductive process by pairing a sperm from one partner with an egg from the other partner.

The united egg and sperm have a long way to go to become a sperm whale or a yellow daffodil or a biology professor. The embryos develop into such complex adults thanks to blueprints inherited from their parents. These instructions for organization and metabolism are encoded in **genes.** Genes are made of long molecules of DNA shaped somewhat like spiral staircases with thousands of steps. All the cells of an organism have a copy of the very same genes.

**a.**

**b.**

**Figure 1.3**
All organisms require an outside source of nutrient materials and energy. Many animals actively seek, kill, and then, like all other animals, ingest their food. *a.* A lioness stalking her prey, a group of zebras, moves closer and closer with muscles taut. *b.* She explodes into her charge joined by other lionesses. The zebras, sensing danger, bolt away. *c.* One zebra of the herd is trapped and cannot escape the claws and jaws of the lionesses.

**c.**

**Figure 1.4**

All organisms reproduce. **a.** A single-celled organism such as this *Amoeba proteus* simply divides in two. **b.** Multicellular plants produce seeds, which contain an embryo, an immature form that develops into the adult. This photograph shows the seed-containing fruit of a flowering dogwood. **c.** Multicellular animals go through developmental stages also. This fertilized queen wasp and workers are building a papery nest. Three stages in wasp development—eggs, larvae, and a pupa in its sealed cell—coexist in the nest. **d.** A hippopotamus and her young move into the water to cool off. Mammals give birth to live offspring that must be cared for.

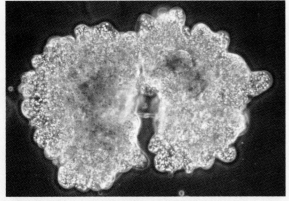

a.

b.

c.

d.

## Living Things Evolve

Occasionally cells make a mistake when they copy and pass on DNA. The change, or **mutation,** may cause an offspring to develop slightly differently. Most of the time, the change is for the worse, and the offspring dies or fails to reproduce well. The changed DNA disappears.

On rare occasions, however, the change in the DNA copy improves the offspring's chance of surviving and reproducing. For example, scientists in England noticed that a mutation was causing normally light-colored moths to be dark in color. In polluted areas, the mutant moths fared better because birds didn't eat them as often; dark bodies made them hard to spot against sooty trees.

**Table 1.1**
Characteristics of Living Things

| Characteristic | Result |
| --- | --- |
| 1. Are organized | Consist of cells |
| 2. Metabolize | Maintain their organization |
| 3. Respond to stimuli | Interact with the environment |
| 4. Reproduce | Have offspring |
| 5. Evolve | Adapt to the environment |

**Figure 1.5**
The feeding patterns of organisms in an ecosystem can be described in terms of food chains that indicate "who eats whom." The food chain always begins with a photosynthesizing producer population followed by one or more consumer populations. The decomposers break down dead organisms and return the chemicals to the producer population. As one population feeds on another, energy flows through the system and gradually dissipates as heat, but chemicals (see inorganic nutrient supply) cycle back to the producers once again.

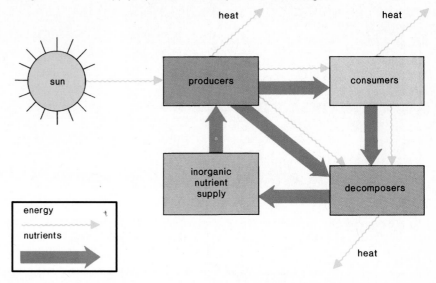

The process by which the characteristics of organisms change over time is called **evolution.** Evolutionary changes occur in groups of interbreeding individuals called **species.** Certain members of a species may inherit a genetic change that causes them to be better adapted or suited to a particular environment. These members can be expected to produce more surviving offspring, who also have the favorable characteristic. In this way the attributes of the species' members change over time. For example, the dark-colored moths just mentioned are better adapted to living in a polluted area than are the light-colored moths of the same species. The dark-colored moths will produce offspring in greater quantity than light-colored moths, and these offspring will also be dark in color.

An **adaptation** is any peculiarity of form or function or behavior that promotes the likelihood of a species' continued existence. If the environment should change, and members of a species no longer possess the genetic capability of adapting to the new environment, then extinction can follow.

Living things reproduce and pass on genes that determine the form and function of their offspring. Genetic changes account for the ability of a species to evolve and adapt to a particular environment.

All organisms are adapted to particular places, climates, food sources, and ways of life. Hence the mind-boggling diversity of life on earth. For example, penguins are adapted to an aquatic existence in the antarctic (see fig. 1.1). Most birds have forelimbs proportioned for flying, but a penguin has stubby, flattened wings suitable for swimming. Their feet and tails serve as rudders in the water, but the flat feet also allow them to walk on land. Rockhopper penguins have a bill adapted to eating small shell fish. Their eggs—one, at most 2—are carried on their feet, where they are protected by a pouch of skin. This allows the birds to huddle together for warmth while standing erect and incubating eggs.

Adaptations to different ways of life, including specific ways of reproducing and acquiring food, account for the great diversity of life forms.

Different as a penguin may be from a hummingbird, or from a house cat or a holly tree for that matter, they share basic characteristics (table 1.1).

## Ecosystems

Adaptations include those features that allow populations to interact with themselves and with the physical **environment** within an **ecosystem** (fig. 1.5). All the ecosystems taken together make up the **biosphere,** the network of life on earth. For a description of 2 ecosystems within this network, turn to figures 1.6 and 1.7. Note the many ways that organisms can live within the 2 environments. Their specific adaptations allow them to play particular roles in their ecosystem. The diversity of organisms is best understood in terms of the many different ways in which organisms carry on their life functions within the ecosystem where they live, acquire energy, and reproduce.

The interactions between **populations** in an ecosystem tend to keep the system relatively stable. Although a forest or pond changes—trees fall, ducks come and go, seeds sprout—each ecosystem remains recognizable year after year. We say it is in dynamic balance. In many cases, even the extinction of species, and their replacement by new species through evolution, still permits the dynamic balance of the system to be maintained.

If the ecosystem is big enough, it needs no raw materials from the outside. (Individual organisms never escape this need, but ecosystems can.) A big ecosystem just keeps cycling its raw materials, like water and nitrogen. The only input it needs is energy.

**Figure 1.6**

In a coral reef, populations of organisms interact among themselves and with their marine environment. The reef consists of the skeletons of stony corals, colonial animals that form deposits of calcium carbonate. Only the outer layer of the reef is alive; the rest forms an inert structure full of nooks and crannies, where fish can hide from their predators. Some types of fish venture out of hiding only at night, when they feed on plankton, the microscopic organisms that drift through the ocean. The coral animals themselves often feed at night also, taking in whatever comes within reach of their extended tentacles. Other fish are active during the day. These hover near the surface of the coral, grazing on algae and worms or on shrimp and crabs if their jaws are able to crunch through shells. Parrot fish even grind the stony coral skeletons themselves. All the smaller fishes are prey to large carnivores like moray eels, jack fish, and barracudas.

Palau coral reef

Tubastraea coral

moray eel

red grouper

lionfish

## Figure 1.7

A temperate deciduous forest is a terrestrial ecosystem found in the Northern Hemisphere where the climate is mild. Various types of insects and worms as well as bacteria and fungi live off the rich organic matter found in the soil. Some birds, like the robin and the tufted titmouse, feed on soil organisms. Other birds, like the red-eyed vireo and woodpeckers, prefer the insects that reside in the trees. Various small mammals, such as rabbits, squirrels, and white-footed mice, live mainly on vegetation, fruits, and insects. Wolves, bobcats, and gray foxes, which feed on these small animals, were more prevalent in days gone by; however, birds of prey like hawks and owls are still common and feed on these animals.

virgin forest at Porcupine Mountains State Park, Michigan

Eastern chipmunk

barn owl

millipede

marsh marigolds

bobcat

> Organisms are members of populations within ecosystems, which are units of the biosphere. Their various adaptations help keep the system in dynamic balance and assure its continued existence.

> The human population tends to modify ecosystems to suit human needs. We are now beginning to realize that this process can endanger us as well as other species.

## The Human Population

People modify ecosystems for their own purposes: clearing forests for farms, clustering houses into towns, and, finally, covering square miles with the asphalt and skyscrapers of big cities. With each step, more of an area's original organisms are driven out. Big cities are populated largely by humans and the few types of life forms, such as rats and pigeons and office plants, that can tolerate a hostile environment. With such sparse, disconnected groups of living things, natural cycles do not function well.

Unlike the original ecosystem, humans rely on manufactured goods of various types and they supplement solar energy with fossil fuel (coal and oil) energy to maintain farms, towns, and cities. These inputs are eventually converted to outputs like trash and carbon dioxide, which become pollutants when they build up and cannot be dispersed easily.

We are beginning to learn that the human population is a part of the biosphere and is dependent upon the natural cycles that occur within the biosphere. As more and more of the biosphere is converted to towns and cities, fewer of the natural cycles are able to function adequately to sustain the human population. For example, the air is cleaner over forests, where many trees absorb carbon dioxide, than over cities with few trees. An increasing human population tends to destroy the very cycles that sustain it.

## The Unity and Diversity of Life

Earlier we discussed the characteristics shared by living things. These common threads, the surprising unity of diverse living things, can be explained by descent from a common ancestor. Many kinds of evidence suggest that life began with single cells and that the present rainbow of organisms evolved from this common origin over hundreds of millions of years. In other words, the process of evolution explains the unity we observe in living things. The other striking thing about life on earth is its diversity. The same coral reef contains massive animals and mere pinpoints of life, rock-hard corals and jellyfish that are all frills and film. Yet, each body type suits a particular life-style. The process of evolution, which involves changes in the genetic material and then a sorting out of physical modifications suited to different environments, explains the diversity we observe in living things. This evolutionary process is the cornerstone of biology.

> Evolution from a common ancestor explains the unity of life. Adaptations to different ways of living explain the diversity of life.

---

## Summary

1. Although living things are diverse, they share certain characteristics in common.
2. All living things are organized; the cell is the smallest unit of life. Cells form tissues, tissues form organs, organs form systems in individuals, individuals are members of populations, and populations make up communities in ecosystems.
3. Each level of organization has emergent properties that cannot be accounted for by simply adding up the properties of the previous level.
4. Living things need an outside source of materials and energy. These are used during metabolism for repair and growth of the organism.
5. Organisms differ as to how they acquire materials and energy. Plants photosynthesize and make their own food; animals either eat plants or other animals. Therefore, there is a flow of energy from plants through all other living things.
6. Living things need a relatively constant internal environment and have various mechanisms and strategies for maintaining homeostasis.
7. Living things respond to external stimuli; taken together, these reactions constitute the behavior of the organism.
8. Living things reproduce and pass on genes, hereditary factors that control organization and metabolism. Changes in the hereditary material allow evolution to occur.
9. Organisms are adapted to various types of environments and to different ways of acquiring energy and reproducing.
10. Within an ecosystem, populations interact with each other and with the physical environment. Materials cycle through an ecosystem. Energy does not cycle.
11. Two examples of ecosystems, the coral reef and the temperate deciduous forest, show that the adaptations of organisms allow them to play particular roles within an ecosystem.
12. The human population is a part of the biosphere, and the diversity of the biosphere should be preserved in order to ensure the continuance of the human population.
13. The process of evolution explains both the unity and the diversity of life. Descent from a common ancestor explains why all organisms share the same characteristics, and adaptation to various ways of life explains the diversity of life forms.

### Writing Across the Curriculum

In order to practice writing skills, students should write out the answers to any or all of the study questions and the critical thinking questions. The study questions are sequenced in the same order as the text. Suggested answers to the critical thinking questions are in appendix D.

# Study Questions

1. What are the 5 common characteristics of life listed in the text?
2. What evidence can you cite to show that living things are organized?
3. Why do living things require an outside source of materials and energy?
4. What is passed from generation to generation when organisms reproduce? What has to happen to the hereditary material in order for evolution to occur?
5. Choose an organism from a temperate forest or coral reef and tell how it is adapted to its way of life.
6. What is an ecosystem, and why should human beings preserve ecosystems?
7. How does evolution explain both the unity and diversity of life?
8. What kingdoms are used in the classification system adopted by this text? What types of organisms are found in each kingdom?

# Objective Questions

For questions 1–5, match the statements in the key with the sentences below.

*Key:*

   a. Living things are organized.
   b. Living things metabolize.
   c. Living things respond.
   d. Living things reproduce.
   e. Living things evolve.

1. Genes made up of DNA are passed from parent to child.
2. Zebras run away from approaching lions.
3. All organisms are composed of cells.
4. Cells use materials and energy for growth and repair.
5. There are many different kinds of living things.
6. Evolution from the first cell(s) best explains why
   a. ecosystems have populations of organisms.
   b. photosynthesizers produce food.
   c. diverse organisms share common characteristics.
   d. All of these.

7. Adaptation to a way of life best explains why living things
   a. display homeostasis.
   b. are diverse.
   c. began as single cells.
   d. are classified into 5 kingdoms.
8. Into which kingdom would you place a multicellular land organism that carries on photosynthesis?

   a. kingdom Monera
   b. kingdom Protista
   c. kingdom Plantae
   d. kingdom Animalia

9. Place the following labels on this diagram of chemical cycling and energy flow in an ecosystem: plants, animals, death and decay, nutrients (e.g., water, fertilizer) for plants.

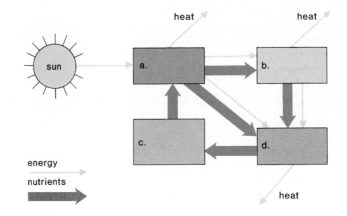

# Concepts and Critical Thinking

1. *All living things evolved from a common ancestor.*

   How does this concept explain the unity of life forms?

2. *Organisms are adapted to particular ways of life.*

   How does this concept explain the diversity of life?

3. *Each level of biological organization has emergent properties.*

   How does this concept help explain the difference between the living and the nonliving?

# Selected Key Terms

**kingdom** (king'dum) 3
**cell** (sel) 3
**metabolism** (me-tab'o-lizm) 4
**homeostasis** (ho"me-o-sta'sis) 4
**behavior** (be-hāv'yor) 5

**reproduce** (re"pro-dūs') 5
**gene** (jēn) 5
**mutation** (mu-ta'shun) 7
**evolution** (ev"o-lu'shun) 8
**species** (spe'shēz) 8

**adaptation** (ad"ap-ta'shun) 8
**environment** (en-vi'ron-ment) 8
**ecosystem** (ek"o-sis'tem) 8
**biosphere** (bi'o-sfēr) 8
**population** (pop"u-la'shun) 8

# 2

# The Scientific Method

Your study of this chapter will be complete when you can

1. outline the basic steps of the scientific method;
2. explain how inductive and deductive reasoning are used during the steps of the scientific method;
3. explain why experimental results can falsify a hypothesis but cannot prove a hypothesis true;
4. give an example of an investigation that follows the steps of the scientific method, and identify the experimental variable and the dependent variable;
5. tell why investigators often report mathematical data and publish their results in scientific journals;
6. design a controlled experiment;
7. show that descriptive research also follows the scientific method;
8. explain what is meant by a "theory" in science, and give examples of scientific theories.

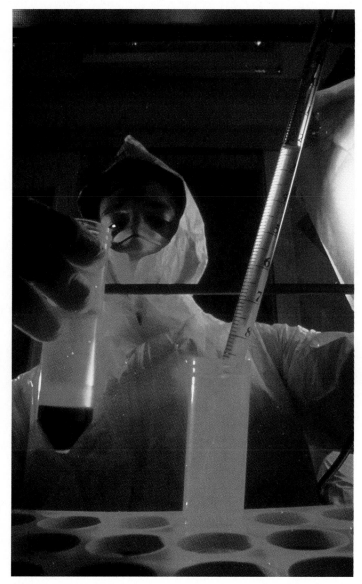

Many biological experiments are carried out in the laboratory where conditions can be rigidly controlled. This technician, who is testing blood for the presence of the HIV virus, is wearing protective clothing.

cience helps human beings understand the natural world and is concerned solely with information gained by observing and testing that world. Scientists, therefore, ask questions only about events in the natural world and expect that the natural world in turn will provide all the information needed to understand events. It is the aim of science, too, to be objective rather than subjective, even though it is very difficult to make objective observations and to come to objective conclusions because human beings are often influenced by their particular prejudices. Still, anything less than a completely objective observation or conclusion is not considered scientific. Finally, the conclusions of science are subject to change. Quite often in science, new studies, which might utilize new techniques and equipment, indicate that previous conclusions need to be modified or changed entirely.

> Scientists ask questions and carry on investigations that pertain to the natural world. The conclusions of these investigations are tentative and subject to change.

## Scientific Method

Scientists, including biologists, employ an approach for gathering information that is known as the **scientific method.** Although the approach is as varied as scientists themselves, there are still certain processes that can be identified as typical of the scientific method. Figure 2.1 outlines the primary steps in the scientific method. On the basis of **data** (factual information), which may have been collected by someone previously, a scientist formulates a tentative statement, called a **hypothesis,** that can be used to guide further observations and experimentation. It is often said that this portion of the scientific method involves **inductive reasoning;** that is, a scientist uses isolated facts to arrive at a general idea that may explain a phenomenon. For example, scientists not only observed the prevalence of dark-colored peppered moths on trees in polluted areas, they also observed the prevalence of light-colored peppered moths on trees in nonpolluted areas. This caused them to formulate the hypothesis that predatory birds were responsible for the unequal distribution of these moths because the birds feed on the moths they can see (see fig. 20.10).

Once the hypothesis has been stated, **deductive reasoning** comes into play. Deductive reasoning begins with a general statement that infers a specific conclusion. It often takes the form of an "if ... then" statement: *If* predatory birds are responsible for

**Figure 2.1**

Steps involved in the scientific method. Inductive reasoning is used to formulate a hypothesis from accumulated scientific data, and deductive reasoning is used to decide what observations and experiments are appropriate in order to test the hypothesis.

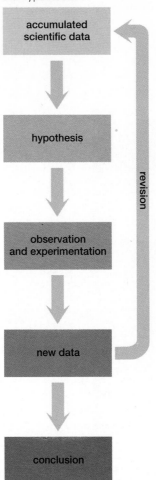

the unequal distribution of dark- and light-colored moths, *then* it will be possible to get evidence of birds feeding primarily on light-colored moths in polluted areas and on dark-colored moths in nonpolluted areas.

To see if this deduction was correct, the British scientist H.B.D. Kettlewell performed an experiment. He released equal numbers of the 2 types of moths in the 2 different areas. He observed that birds captured more dark-colored moths in nonpolluted areas and more light-colored moths in polluted areas. From the unpolluted area, Kettlewell recaptured 13.7% of the light-colored moths and only 4.7% of the dark form. From the polluted area, he recaptured 27.5% of the dark form and only 13% of the light-colored moths. Mathematical data like these are preferred because they are objective and cannot be influenced by the scientist's subjective feelings. These particular data supported the hypothesis (fig. 2.2).

Sometimes, as in this case, the data support the hypothesis; however, at other times, the data do not support the hypothesis, and it has to be rejected. Knowing that some future observation and/or experiment might falsify the hypothesis, a scientist never says that the data "prove" the hypothesis. In other words, a hypothesis cannot be proven true, but it can be shown to be false. Because of this feature, some think of science as what is left after alternative hypotheses have been rejected.

> The scientific method consists of making observations, formulating a hypothesis, testing, and coming to a conclusion. Hypotheses can be supported but not proven true. Hypotheses can be shown to be false, and then they are rejected.

### Reporting the Experiment

It is customary for an investigator to report findings in a scientific journal so that the design and results of an experiment are available to all. This is especially desirable because all experimental results must be repeatable; that is, other scientists using the same procedures should get the same results. If they do not, the original data cannot be considered to support the hypothesis.

Often, too, the authors of a report suggest other experiments that could be done to clarify or broaden the understanding of the matter under study.

> The results of observations and experiments are published in scientific journals, where they can be examined by all. These results must be repeatable, that is, they should be obtainable by anyone doing the exact same procedure.

## Figure 2.2

Analysis of experiment performed by Kettlewell.

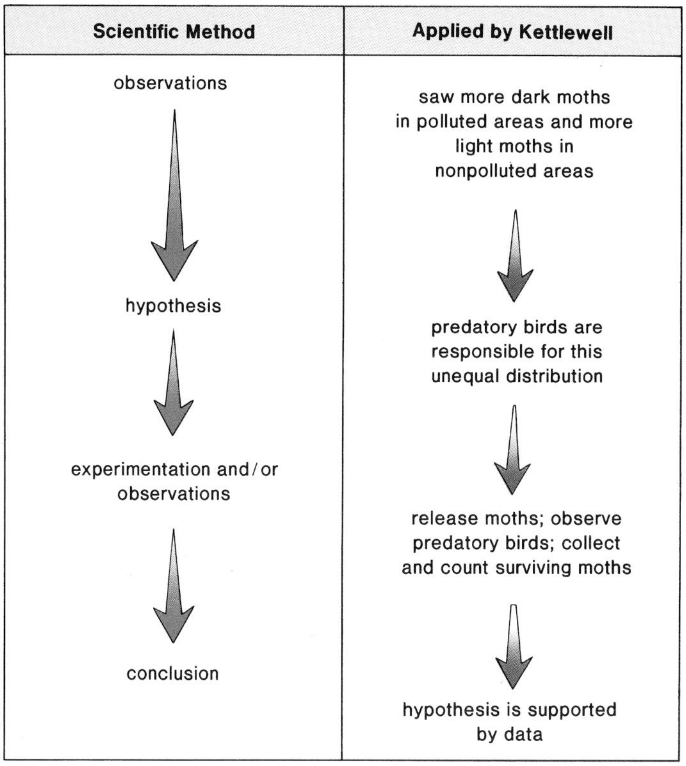

| Scientific Method | Applied by Kettlewell |
|---|---|
| observations | saw more dark moths in polluted areas and more light moths in nonpolluted areas |
| ↓ | ↓ |
| hypothesis | predatory birds are responsible for this unequal distribution |
| ↓ | ↓ |
| experimentation and/or observations | release moths; observe predatory birds; collect and count surviving moths |
| ↓ | ↓ |
| conclusion | hypothesis is supported by data |

## Revision

Science is an ongoing endeavor. Because hypotheses are never proven true, they are reviewed constantly. Particularly when new procedures and instruments make it possible, the process of revision begins. Previous data are used as a starting point, and then new data is collected. The new data may require a new conclusion and the formulation of revised hypotheses, which will then be tested.

## Controlled Experiment

Kettlewell performed his experiment in the field, but scientists often prefer to work in the laboratory, where they can oversee all elements of the experiment. These are controlled experiments.

| Experimental Variable | Dependent Variable |
|---|---|
| Element of the experiment being tested | Result or change that occurs due to the experimental variable |

### Elements of an Experiment

When conducting a controlled experiment, all conditions should be held constant except the one being tested. This one element is called the **experimental variable;** in other words, the investigator deliberately varies this portion of the experiment. Then the investigator observes the effects of the experiment; in other words, he or she observes the **dependent variable:** To be absolutely sure that the results are due to the experimental variable and not due to some unknown factor, it is customary to have a separate sample called the control group. A **control group** undergoes all the steps of the experiment except the one being tested.

### Design of an Experiment

As an example of a controlled experiment, suppose physiologists wish to determine if sweetener S is a safe additive for foods. On the basis of available information, they hypothesize that sweetener S has no effect on health at a low concentration but causes bladder cancer at a high concentration. They then decide to feed sweetener S to groups of mice at ever-greater percentages of the total intake of food:

Group 1: no sweetener S in food (control group)
Group 2: sweetener S in 5% of food
Group 3: sweetener S in 10% of food
↓
Group 11: sweetener S in 50% of food

The experimenters first place a certain number of randomly chosen, inbred (genetically identical) mice into the various groups--say 10 mice per group. It is hoped that if any of the mice are different from the others, random selection will distribute them evenly among the groups. The experimenters also make sure that all environmental conditions, such as availability of water, cage conditions, and temperature of surroundings are the same for all groups. The food for each group is exactly the same except for the amount of sweetener S. (The amount of sweetener S is the experimental variable.) After a designated period of time, the animals are killed and dissected to examine their bladders for evidence of cancer. (Evidence of bladder cancer is the dependent variable.) Figure 2.3 outlines the design of this experiment.

### Results of an Experiment

Usually the data from experiments such as these are presented in the form of a table or graph (fig. 2.4). A statistical test may be run to determine if the difference in the number of cases of bladder cancer among the various groups is significant. After all, if a significant number of mice in the control group develop cancer, the results may be invalid. On the basis of the results, the experimenters try to develop a recommendation concerning the safety of adding sweetener S to the food of humans. They may determine that ever-larger amounts of the sweetener over 10% of food intake are expected to cause a progressively increased incidence of bladder cancer, for example.

Many scientists work in laboratories in which they carry out controlled experiments.

## Descriptive Research

Scientists don't always gather data by experimenting. Much of the data they gather is purely observational, but even so, the steps described for the scientific method are still applicable: observations are made, a hypothesis is formulated, predictions are made on

**Figure 2.3**

Design of a controlled experiment. Genetically identical mice are randomly divided into a control group and the experimental groups. All groups are exposed to the same environmental conditions, such as housing, temperature, and water supply. Only the experimental groups are subjected to the test (experimental variable)--in this case the presence of sweetener S in the food. At the end of the experiments, all mice are examined for evidence of bladder cancer.

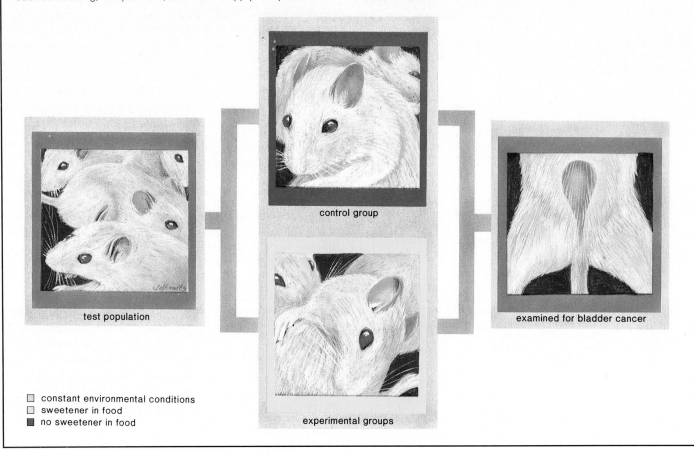

test population

control group

examined for bladder cancer

experimental groups

☐ constant environmental conditions
☐ sweetener in food
■ no sweetener in food

the basis of the hypothesis, and data are collected that support or disprove the hypothesis. For example, Kenneth E. Glander[1] was interested in determining the feeding pattern of mantled howling monkeys, which are herbivores that feed on plant products (fig. 2.5). He believed 2 of the hypotheses he found in the scientific literature could apply to the monkeys.

1. Most large herbivores try to obtain an optimal mix of nutrients.
2. Mammalian herbivores tend to avoid plant foods that contain toxins.

Glander felt that there was insufficient data to decide if one or both of these statements were pertinent to the feeding behavior of the howling monkeys. So he and his associates made new observations.

1. They marked off 2 test sites and identified every type of tree in each site. Altogether they identified 1,699 trees.

[1]Glander, K. E. 1981. Feeding patterns in mantled howling monkeys. In *Foraging behavior: ecological, ethological and psychological approaches,* ed. A. C. Kamil and T. D. Sargent, 231-57. New York: Garland STPM Press.

2. They observed the feeding behavior of a group of monkeys consisting of 2 adult males, 6 adult females, and 5 juveniles. A total of 1,853 hours of animal observation was accumulated for a period of one year. The average observation day lasted 12 hours and 11 minutes.
3. They performed chemical analyses of fresh plant samples from about 50 trees to determine their nutritional (amino acid) and toxin contents.

Glander found that the study group of monkeys obtained a majority of their food from 62 relatively rare tree species. During the 12-month study period the monkeys obtained all of their food from 331 different kinds of trees, or 19.5% of all trees available to them. Further, they tended to choose mature leaves and fruits in the wet season and new leaves and flowers in the dry season. The chemical analyses of the plant products showed that the howlers were maximizing their intake of water, digestible protein, total protein, and amino acids and minimizing their intake of fiber, ash, and toxic plant compounds. Glander concluded that the howlers chose those foods that gave the greatest net nutrient return. Glander stated, "A tropical

## Figure 2.4

Scientists often acquire mathematical data, which they report in the form of a table or a graph. Mathematical data are more decisive and objective than visual observations. The data in this instance suggest that there is an ever greater chance of bladder cancer development if the food contains more than 10% sweetener S. The same experiment is repeated many times to test these results, and statistical analyses are conducted to see if the results are significant or simply due to chance alone.

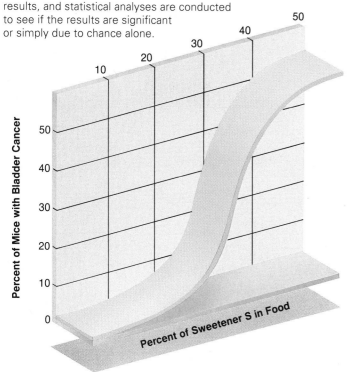

## Figure 2.5

Mantled howling monkeys are herbivores that feed on plant products in South American forests. They tend to choose mature leaves and fruit in the wet season and new leaves and flowers in the dry season.

forest should not be viewed as a well-stocked larder waiting to be exploited, but rather as a spatially and temporally changing mosaic of items of varying value and availability."

This proposed model of the nature of the food sources for forest-dwelling herbivores is one of the most important outcomes of Glander's observations. A scientific **model** is a suggested explanation that can help direct future research. Again, the steps in the scientific method can be applied (fig. 2.6).

> A large proportion of scientific information is based on purely observational (descriptive) data, but the steps previously suggested for the scientific method still apply.

## Theories and Principles

The ultimate goal of science is to understand the natural world in terms of **theories,** which are interpretations that take into account the results of many experiments and observations. In a movie, a detective may claim to have a theory about the crime. Or you may say that you have a theory about the won-lost record of your favorite baseball team. But in science, the word *theory* is reserved for a conceptual scheme that is supported by a large number of observations and has not yet been found lacking.

## Figure 2.6

Analysis of Glander's study of howling monkeys.

| Scientific Method | Applied by Glander |
|---|---|
| observations | studied scientific literature |
| ↓ | ↓ |
| hypothesis | feeding behavior may be influenced by nutrient value of food including avoidance of toxins |
| ↓ | ↓ |
| experimentation and/or observations | observe monkeys feeding and perform chemical analyses of food selected |
| ↓ | ↓ |
| conclusion | hypothesis is supported by data |

**Table 2.1**
Unifying Theories of Biology

| Name of Theory | Concept | Investigator |
|---|---|---|
| Cell | All organisms are composed of cells | Matthias Schleiden, 1838 Theodor Schwann, 1839 |
| Biogenesis | Life comes only from life | Louis Pasteur, 1865 |
| Homeostasis | The internal environment remains within a normal range | Claude Bernard, 1851 |
| Evolution | All living things have a common ancestor and are adapted to a particular way of life | Charles Darwin, 1858 |
| Gene | Organisms contain coded information that dictates their form, function, and behavior | Gregor Mendel, 1866 James Watson and Francis Crick, 1953 |

Table 2.1 lists some of the unifying theories of biology. You can see that in general the theories pertain to the characteristics of life we discussed in chapter 1. The theory of evolution enables scientists to understand the history of life, the variety of living things, and the anatomy, physiology, development, and even behavior of organisms. Because the theory of evolution has been supported by so many observations and experiments for over 100 years, some biologists refer to the "principle" of evolution, suggesting that this is the appropriate term for theories that are generally accepted as valid by an overwhelming number of scientists.

Sometimes scientists do experiments that they *expect* to be explainable by a certain theory:

*The role of an experiment is illustrative. It allows a scientist to demonstrate the power of his theory, not as a collection of truths, but as a set of ideas. When an experiment succeeds, this shows that a certain way of describing the world has proved itself useful. When an experiment fails it shows that one's concepts were inadequate or confused.*[2]

For example, in 1975 David P. Barash[3] conducted a study of 2 mountain bluebird mating pairs, seemingly with the intention of using the theory of evolution to explain their behavior. He posted a model of a male bluebird 1 m from each nest while the resident male was foraging. Upon returning to its nest, each resident encountered the model in proximity to his female. This was done during 3 different stages of the nesting cycle. The behavior of the male toward the model and toward his female during the first 10 minutes of his discovery of the model was observed and noted. As a control, a model male robin was also posted 4 days after the initial

[2]Harre, R. 1981. *Great scientific experiments.* London: Phaidon Press Limited.

**Figure 2.7**
Results of an experiment with nesting mountain bluebirds. The graph shows that resident males are more likely to aggressively approach a male model and female mate before the nest is completed, less likely after the first egg is laid, and least likely after the eggs have hatched. The experimenter gives an evolutionary interpretation to the results: it is more adaptive for a male bluebird to be aggressive before mating in order to be sure the offspring will be his and less adaptive at the other stages mentioned.

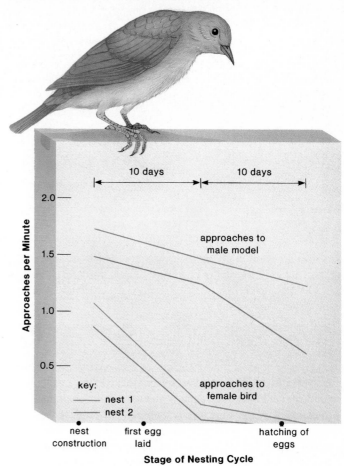

presentation of the bluebird model. The robin model elicited no response. The number of aggressive approaches toward the bluebird male model and female mate is given in graph form in figure 2.7. Aggression was most severe when the bluebird male model was presented before the first egg was laid, less severe once the first egg was laid, and least severe after the eggs had hatched.

In his report of this experiment, Barash explains how evolutionary theory can account for these results. It is adaptive for males to be especially aggressive before the eggs are laid because there is a limited number of nesting sites available to the birds, and he wants to be sure he is the one mating and passing on his genes. It is adaptive for males to be less aggressive after the first egg is laid; the reproductive process has begun and the male is more sure the

[3]Barash, D. P. 1976. The male response to apparent female adultery in the mountain bluebird, *Sialia currucoides:* An evolutionary interpretation. *American Naturalist* 110:1097-101.

offspring will be his. Finally, it would be maladaptive for the male to spend time and energy being too aggressive toward a rival and his mate after hatching because his offspring are already present.

Barash's discussion of his results seems to support the suggestion that scientists sometimes do experiments with the intention of explaining their results on the basis of a particular theory. He shows how his results can be given an evolutionary interpretation. It is adaptive for a male bluebird to be aggressive toward rivals, especially before mating, because it helps ensure that he will be the one to pass on genes. Barash shows that this is an observed characteristic among these birds. In doing so, he also shows that in addition to anatomy and physiology, behavior is subject to an evolutionary interpretation.

> It is the aim of science to formulate theories and principles that encompass a number of hypotheses concerning the natural world. The power of a theory is evident when it can be used to explain diverse observations and experiments.

## Science and Social Responsibility

There are many ways in which science has improved life. The most obvious examples are in the field of medicine. The discovery of antibiotics, such as penicillin, and the vaccines for polio, measles, and mumps have increased the life span by decades. Cell biology research may someday help us understand the mechanisms that cause cancer. Genetic research has produced new strains of agricultural plants that have eased the burden of feeding our burgeoning population.

But science has also fostered technologies of grave concern to some of us. Even though we would not like a cancer patient to be denied the benefits of radiation therapy, we are concerned about the nuclear power industry and do not want nuclear wastes stored near our homes. Then, too, there may be undesirable side effects even from those technologies of which we approve. Not one of us wishes to go back to the days in which large numbers of people died of measles or polio. Yet as science has conquered one disease after another, the world's death rate has fallen and the human population has exploded. Few of us are willing to give up technology's gift of the private automobile, but we are concerned about the amount of air pollution that automobiles generate.

All too often we blame science for these side effects and say that our lives are infinitely more threatened than they were in "the good old days." We think that scientists should label research "good" or "bad" and be able to predict whether any resulting technology would primarily benefit or harm humanity. Yet science, by its very nature, is impartial and simply attempts to study natural phenomena. Science does not make ethical or moral decisions. When we wish to make value judgments, we must go to other fields of study to find the means to make those judgments. And these judgments must be made by all people. The responsibility for how we use the fruits of science, including a given technology, must rest with people from all walks of life, not upon scientists alone.

Scientists should provide the public with as much information as possible when such issues as nuclear power or recombinant DNA technology are being debated. Then, they, along with other citizens, can help make decisions about the future role of these technologies in our society. All men and women have a responsibility to decide how to use scientific knowledge so that it benefits the human species and all living things.

## Summary

1. Scientific investigations always pertain to the natural world, and the conclusions of the investigations are always subject to change.
2. In general, scientists use the scientific method. Based on their own observations or on ones found in the scientific literature, inductive reasoning is used to formulate a hypothesis. Using deductive reasoning, a plan of action is devised and carried out. The results either support or fail to support the hypothesis.
3. Most scientists do controlled experiments. Particularly, in the laboratory (1) the environmental conditions are held constant, (2) there is a control group that goes through all the steps of the experiment except the one being tested, and (3) one or more experimental groups are involved.

4. In a controlled experiment, the experimental variable is that portion of the experiment that is being tested and the dependent variable is the results of the experiment.
5. Much of biological investigation is descriptive rather than experimental. Descriptive research requires that the investigator make careful observations. Descriptive research still follows the steps in the scientific method.
6. The goal of science is to understand the natural world in terms of theories, which can be used to explain diverse observations and experiments. Some theories are so well supported that many suggest they should be called principles.

7. Investigators sometimes make observations or do experiments with the purpose of explaining their results on the basis of a particular theory.

### Writing Across the Curriculum

In order to practice writing skills, students should write out the answers to any or all of the study questions and the critical thinking questions. The study questions are sequenced in the same order as the text. Suggested answers to the critical thinking questions are in appendix D.

## Study Questions

1. There is no standard scientific method but still it is possible to identify certain steps that are commonly accepted as comprising this method. What are these steps?
2. Apply these steps to Kettlewell's experiment regarding unequal distribution of dark- and light-colored moths.
3. Which of the steps identified in question 1 requires the use of inductive reasoning? Which requires the use of deductive reasoning?
4. Why do scientists prefer mathematical data and use charts and graphs to report such data?
5. For what purposes are scientific journals useful?
6. Design a controlled experiment using an example other than the one given in the text.
7. What is descriptive research, and how does it differ from experimental research?

8. What is the difference between a hypothesis and a theory? List 5 biological theories that pertain to all living things, and name some of the investigators that helped formulate these theories.
9. An investigator spills dye on a culture plate and then notices that the bacteria live despite exposure to sunlight. He decides to expose 2 culture plates to ultraviolet light--one plate contains bacteria and dye; the other plate contains only bacteria. The bacteria on both plates die. Fill in the right-hand portion of this diagram:

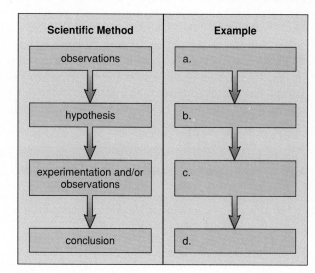

## Objective Questions

1. Which of these is the second step in the scientific method?
   a. experimentation
   b. conclusion
   c. researching the scientific literature
   d. formulating the hypothesis
2. Which of these is the third step in the scientific method?
   a. experimentation
   b. conclusion
   c. researching the scientific literature
   d. formulating the hypothesis
3. With which of these steps in the scientific method do you associate inductive reasoning?
   a. experimentation
   b. conclusion
   c. researching the scientific literature
   d. formulating the hypothesis
4. With which of these steps in the scientific method would you associate deductive reasoning?
   a. experimentation
   b. conclusion
   c. researching the scientific literature
   d. formulating the hypothesis

5. Which is the experimental variable in this experiment?
   a. Conditions like temperature and housing are the same for all groups.
   b. The amount of sweetener S in food is varied.
   c. Two percent of the group who were fed food that was 10% sweetener S got bladder cancer, and 90% of the group who were fed food that was 50% sweetener S got bladder cancer.
   d. The data were presented as a graph.
6. Which is the control group in this experiment?
   a. All mice in this group died because a lab assistant forgot to give them water.
   b. All mice in this group received food that was 10% sweetener S.
   c. All mice in this group received no sweetener S in food.
   d. Some mice in all groups got bladder cancer; therefore, there was no control group.

7. Which is an example of descriptive research?
   a. Jones put broken egg shells in the nest of some gulls and watched their behavior.
   b. Smith measured the length of the twigs of all trees in the marked off area.
   c. Green put pesticide into one jar of amoebas but not into the other jar.
   d. Kettlewell counted how many moths the birds did not eat.
8. Which of these statements is correct?
   a. Because scientists speak of the "theory of evolution" they believe that evolution does not have much merit.
   b. A theory is simply a hypothesis that needs further experimentation and observation.
   c. Theories are hypotheses that have failed to be supported by experimentation and observation.
   d. The term theory in science is reserved for those hypotheses that have proven to have the greatest explanatory power.

## Concepts and Critical Thinking

1. *Scientists use the scientific method to gather information about the natural world.*

   When utilizing the scientific method, hypotheses can be proven false but not true. Explain.

2. *There are several underlying theories in biology.*

   What is the difference between the use of the word theory in the everyday sense and in the scientific sense?

3. *All persons, not just scientists, should decide on the role of various technologies in society.*

   Science is not obligated to make policy decisions. Why not?

## Selected Key Terms

**scientific method** (si″en-tif′ik meth′ud) 14
**data** (da′tah) 14
**hypothesis** (hi-poth′e-sis) 14
**inductive reasoning** (in-duk′tiv re′zun-ing) 14
**deductive reasoning** (de-duk′tiv re′zun-ing) 14

**experimental variable** (ek-sper″ ı-men′tal va′re-ah-b′l) 15
**dependent variable** (de-pen′dent va′re-ah-b′l) 15

**control group** (kon-trōl′ grōōp) 15
**model** (mod′el) 17
**theory** (the′o-re) 17

## Suggested Readings

Baker, J., and Allen, G. 1971. *Hypothesis, prediction, and implication in biology.* Reading, Mass.: Addison-Wesley.

Barash, D. P. 1976. The male response to apparent female adultery in the mountain bluebird, *Sialia currucoides:* An evolutionary interpretation. *American Naturalist.*

Glander, K. E. 1981. Feeding patterns in mantled howling monkeys. A. C. Kamil and T. D. Sargent, eds. *Foraging behavior: ecological, ethological and psychological approaches.* New York: Garland STPM Press.

*Life: Origin and evolution: Readings from Scientific American.* Intro. by C. E. Folsome. 1979. San Francisco: W. H. Freeman.

Scientific American Editors. 1978. *Evolution: A Scientific American book.* San Francisco: W. H. Freeman.

Volpe, E. P. 1985. *Understanding evolution.* 5th ed. Dubuque, Iowa: Wm. C. Brown Publishers.

# The Cell

All organisms, including these wandering albatrosses, have the same levels of organization. In this part, we examine the structure and function of atoms, molecules, and cells—the first few levels of organization. Organization determines what an organism is like and how it acquires energy and carries on life's functions. These birds are performing a courtship ritual prior to reproducing.

# 3

## Basic Chemistry

A hippopotamus keeps cool by remaining in water. Water has many biological functions, both without and within organisms. Cells are largely composed of this inorganic molecule, which facilitates chemical reactions and helps maintain a normal body temperature because it is slow to heat.

Your study of this chapter will be complete when you can

1. name and describe the subatomic particles of an atom, indicating which one accounts for the occurrence of isotopes;
2. describe and discuss the energy levels (electron shells) of an atom, including the orbitals of the first 2 levels;
3. draw a simplified atomic structure of any atom with an atomic number less than 20;
4. distinguish between ionic and covalent reactions, and draw representative atomic structures for ionic and covalent molecules;
5. tell which atom has been reduced and which has been oxidized in a particular reaction;
6. describe the chemical properties of water, and explain their importance for living things;
7. define an acid and a base; describe the pH scale, and state the significance of buffers.

ometimes it is difficult to realize that life has a chemical basis. Maybe this is because we usually deal with whole objects, such as trees, dogs, and people, and do not often consider their minute composition. If we did consider it, though, we would soon learn that an organism is composed of organ systems with organs made up of tissues, that tissues contain cells made up of molecules, and that the molecules themselves are atoms bonded together (fig. 3.1). This shows that fundamentally, living things are made up of chemicals.

The structure and the function of living things are dependent upon chemicals. The success of the cheetah that darts across the plain to capture the gazelle is dependent upon the chemical reactions that occur throughout its body, such as those that allow its eyes to see and its muscles to contract. Also, a bittersweet vine can climb a fence only because of the chemical reactions that permit growth to occur.

As recently as the nineteenth century, scientists believed that only nonliving things, like rocks and metals, consisted of chemicals alone. They thought that living things, such as the cheetah and the bittersweet vine, were different and had a special force, called a vital force, necessary to being alive. Scientific investigation, however, has repeatedly shown that both nonliving and living things have only a chemical and physical basis. Therefore, it is appropriate for us to study chemical principles as an introduction to our study of life. In this chapter, we will consider the basic facts of chemistry, and in the next chapter we will take a look at some complex molecules that are especially associated with

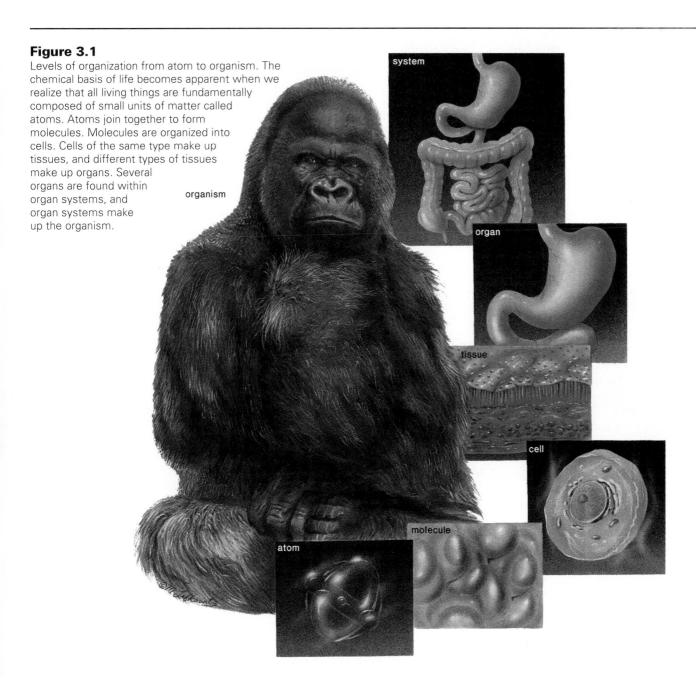

**Figure 3.1**

Levels of organization from atom to organism. The chemical basis of life becomes apparent when we realize that all living things are fundamentally composed of small units of matter called atoms. Atoms join together to form molecules. Molecules are organized into cells. Cells of the same type make up tissues, and different types of tissues make up organs. Several organs are found within organ systems, and organ systems make up the organism.

## Table 3.1
Common Elements in Living Things

| Element* | Atomic Symbol | Atomic Number | Atomic Weight$^d$ | Comment |
|---|---|---|---|---|
| Hydrogen | H | 1 | 1 | These elements make up most biological molecules. |
| Carbon | C | 6 | 12 | |
| Nitrogen | N | 7 | 14 | |
| Oxygen | O | 8 | 16 | |
| Phosphorus | P | 15 | 31 | |
| Sulfur | S | 16 | 32 | |
| Sodium | Na | 11 | 23 | These elements occur mainly as dissolved salts. |
| Magnesium | Mg | 12 | 24 | |
| Chlorine | Cl | 17 | 35 | |
| Potassium | K | 19 | 39 | |
| Calcium | Ca | 20 | 40 | |
| Iron | Fe | 26 | 56 | These elements also play vital roles in organisms. |
| Copper | Cu | 29 | 64 | |
| Zinc | Zn | 30 | 65 | |

*A periodic table of the elements appears in appendix B.
$^d$Average of most common isotopes.
Note: The atomic number gives the number of protons (and electrons in electrically neutral atoms). The number of neutrons is equal to the atomic weight minus the atomic number. There are no neutrons in a hydrogen atom.

living things. Then we will discuss how complex molecules are organized into the structures found within cells. In this way, not only the sameness but also the uniqueness of living things will begin to be apparent.

---

Living things are composed of chemicals organized to provide the structure and to perform the functions necessary to life.

---

Only 6 elements—carbon, hydrogen, nitrogen, oxygen, phosphorus, and sulfur—make up most (about 98%) of the body weight of organisms. The acronym CHNOPS helps us remember these 6 elements. The other elements noted in table 3.1 also play important roles in cells.

---

All living and nonliving things are matter composed of elements. Six elements in particular are commonly found in living things.

---

## Matter

**Matter** refers to anything that takes up space and has weight.[1] It is helpful to remember that matter can exist as a solid, a liquid, or a gas. Then we can realize that not only are we matter but so too are the water we drink and the air we breathe.

[1]The term *atomic weight* will be used as if it was the same as atomic mass. Weight, however, is dependent on gravity, and therefore atomic weight (but not mass) would be different on the moon, for example, compared to the earth.

All matter, both nonliving and living, is composed of certain basic substances called **elements.** Most matter contains more than one type of element, but some matter contains only one type of element. For example, air contains more than one type of element, but the oxygen we respire contains only the element of that name. It is quite remarkable that there are only 92 naturally occurring elements (see appendix B). We know these are elements because any particular element cannot be broken down to substances with different properties (a property is a chemical or physical characteristic, such as denseness, smell, taste, and reactivity).

## Atoms

In the early 1800s, the English scientist John Dalton proposed that elements actually contain tiny particles called **atoms.** He also deduced that there is only one type of atom in each type of element. You can see why, then, the name assigned to each element is the same as the name assigned to the type of atom it contains. Some of the names we use for the elements (atoms) are derived from English and some are derived from Latin. One or 2 letters create the *atomic symbol,* which stands for this name. For example, the symbol H stands for a hydrogen atom, and the symbol Na (for *natrium* in Latin) stands for a sodium atom. Table 3.1 gives the atomic symbols for the other elements (atoms) commonly found in living things.

### Atoms Have Weight

From our discussion of elements, we would expect atoms to have a certain weight. The weight of an atom is in turn dependent upon the presence of certain subatomic particles. Although physicists have identified a number of subatomic particles, we will consider only the most stable of these: **protons, neutrons,** and **electrons.** Protons and neutrons are located within the nucleus of an atom, and electrons move about the nucleus. Figure 3.2 shows the arrangement of the subatomic particles in helium, an atom that has only 2 electrons.

Our concept of an atom has changed greatly since Dalton's day. If we could draw an atom the size of a football field, the nucleus would be like a gumball in the center of the field and the electrons would be tiny specks whirling about in the upper stands. Most of an atom is empty space. We should also realize that we can only indicate where the electrons are expected to be most of the time. In our analogy, the electrons may even stray outside the stadium at times.

---

Atoms contain protons, neutrons, and electrons that are arranged in a definite manner.

---

The subatomic particles are so light that their weight is indicated by special units called atomic mass units (table 3.2). Protons and neutrons each have only one atomic unit of weight. In comparison electrons have almost no weight; an electron weighs about 1/1,800 that of a proton or neutron. Therefore, it is customary to disregard the combined weight of the electrons when calculating the total weight, called the *atomic weight,* of an atom.

The Cell

## Figure 3.2

Model of helium (He), a simple atom. Atoms contain subatomic particles which are located where shown. Protons and neutrons are found within the nucleus, and electrons are outside the nucleus. *a.* The stippling shows the probable location of the electrons in the helium atom. *b.* The average location of electrons is sometimes represented by a circle, and each electron is represented by a small sphere.

= proton

= neutron

= electron

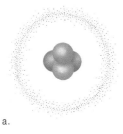

a.

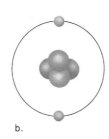

b.

## Figure 3.3

All atoms of a particular element have the same atomic number, which tells the number of protons. This number is often written as a subscript in the lower left of the atomic symbol. The atomic weight (mass) is often written as a superscript at the upper left of the atomic symbol. For example, the carbon atom can be noted in this way:

atomic weight ———— $^{12}_{6}C$ ———— atomic symbol
atomic number ————

Isotopes are atoms of the same element that differ in weight. The number of protons stays the same, but the number of neutrons can vary. The above configuration is the most common isotope of carbon; other isotopes are

$^{13}_{6}C$  $^{14}_{6}C$  $^{15}_{6}C$

## Table 3.2
Subatomic Particles

| Name | Charge | Weight |
|------|--------|--------|
| Electron | One negative unit | Almost none |
| Proton | One positive unit | One atomic unit |
| Neutron | No charge | One atomic unit |

## Figure 3.4

The thyroid gland is a butterfly-shaped organ located in front of the throat region of humans. It produces the hormone thyroxine, which contains iodine. Therefore, if a patient is administered radioactive iodine (atomic number 53) a scan of the thyroid viewed sometime later will show the distribution of the iodine throughout the tissue.

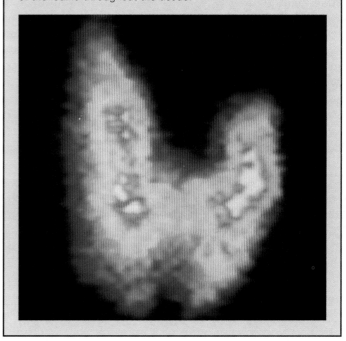

All atoms of an element have the same number of protons. This is called the atom's *atomic number*. In table 3.1, atoms are listed according to increasing atomic number, as they are in the periodic table of the elements found in appendix B. This indicates that it is the number of protons (i.e., the atomic number) that makes an atom unique. The number of protons not only contributes to the physical properties (weight) of an atom but also *indirectly* determines the chemical properties, which we will discuss in a later section.

### Isotopes

All atoms of an element have the same number of protons but may differ in the number of neutrons. Table 3.1 allows you to calculate only the usual number of neutrons for hydrogen, carbon, and oxygen. But any specific carbon atom, for example, may have a few more or a few less neutrons than the number indicated. Atoms that have the same atomic number and differ only by the number of neutrons are called **isotopes** (fig. 3.3). Most isotopes are stable, but a few are unstable and tend to break down to more stable forms. They emit radiation as they break down. These radioactive isotopes can be detected by physical or photographic means and are used by biologists as "labels" in biochemical experiments. For example, researchers used radioactive carbon dioxide ($^{14}CO_2$) to detect the sequential biochemical steps occurring during pho-

tosynthesis. Radioactive isotopes are sometimes used in medical diagnostic procedures also. For instance, because the thyroid gland uses iodine, it is possible to administer radioactive iodine and then scan the gland to determine the location of the iodine isotope (fig. 3.4).

**Figure 3.5**

Electron energy levels. *a.* Electrons possess stored energy to different degrees. Those with the least amount of energy are found closest to the nucleus, and those with more energy are found at a greater distance from the nucleus. These energy levels are also called electron shells. *b.* An electron can absorb energy from an outside source, such as sunlight, and then move to a higher energy level. *c.* An electron can later give up the energy absorbed and drop back to its former energy level. This feature of electron behavior is important during photosynthesis studied in chapter 8.

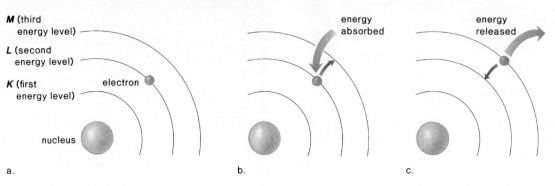

### Carbon[14]

Most carbon atoms have 6 protons and 6 neutrons (shown as $^{12}C$, carbon 12) and are not radioactive; however, $^{14}C$ (carbon 14) has 6 protons and 8 neutrons and is radioactive. With the release of radioactivity, $^{14}C$ spontaneously breaks down over a period of time into $^{14}N$ (nitrogen 14). A small amount of $^{14}C$ is present in all living things. Scientists can determine the age of a fossil because the rate at which $^{14}C$ breaks down is known. If the fossil still contains organic matter, the relative amounts of $^{14}C$ and $^{14}N$ are measured and the age is calculated. This method is unreliable for measuring fossil ages in excess of 20,000 years, however, because the radio-activity is so slight in such fossils that it is difficult to make an accurate determination.

## Atoms Have Chemical Properties

Protons and electrons carry a charge; protons have a positive (+) electrical charge and electrons have a negative (-) electrical charge. When an atom is electrically neutral, the number of protons equals the number of electrons. Therefore, in such atoms, the atomic number tells you the number of electrons. For example, a carbon atom has an atomic number of 6 (table 3.1) and when electrically neutral, it has 6 electrons. Atoms have different chemical proper-ties (the way they react with other atoms) because they each have a different number of electrons arranged about the nucleus.

> The chemical properties of atoms differ because the number and the arrangement of their electrons are different.

### Atomic Energy Levels

All electrons have the same weight and charge, but they vary in energy content. **Energy** is defined as the ability to do work. We recognize this in everyday life by suggesting that those who are energetic do more work than those who are not. Electrons differ in the amount of their *potential energy,* that is, stored energy ready to do work. The relative amount of stored energy is designated by placing electrons at *energy levels,* also called *electron shells,* about the nucleus (fig. 3.5). Electrons with the least amount of potential energy are located in the shell closest to the nucleus, called the *K* shell. Electrons in the next higher shell, called the *L* shell, have more energy, and so forth as we proceed from shell to shell outside the nucleus. An analogy may help you appreciate that the farther electrons are from the nucleus, the more potential energy they possess. Falling water has potential energy, as witnessed by how it can turn a waterwheel connected to a shaft that transfers the energy to machinery for grinding grain, cutting wood, or weaving cloth, for example. The higher the waterfall, the greater the amount of energy released per unit amount of water.

Our analogy is not exact because the potential energy possessed by electrons is not due to gravity; it is due to the attraction between the positively charged protons and the nega-tively charged electrons. It takes energy to keep an electron farther away from the nucleus as opposed to closer to the nucleus. You have often heard that sunlight provides the energy for photosynthe-sis in green plants, but it may come as a surprise to learn that when a pigment such as chlorophyll absorbs the energy of the sun, electrons move to a higher energy level about the nucleus. When these electrons return to their previous level, the energy released is used to make food molecules (fig. 3.5*b* and *c*).

### Electrons Occupy Orbitals

It is convenient to imagine electrons occupying concentric electron shells about a nucleus, as in figure 3.5; however, such depictions should be regarded as diagrams of convenience. In actuality, it is not possible to determine where rapidly moving electrons are from moment to moment. It is possible, however, to describe the pattern of their motion. These volumes of space where electrons are most often found are called **orbitals.**

At the first energy level, there is only a single spherical orbital, where at most 2 electrons are found about the nucleus (fig. 3.6). The space is spherical-shaped because the most likely loca-tion for each electron is a fixed distance in all directions from the

The Cell

## Figure 3.6

Electron orbitals. Each electron energy level (see fig. 3.5) contains one or more orbitals, a volume of space in which the rapidly moving electrons are most likely found. The nucleus is at the intersection of axes *x*, *y*, and *z*. **a.** The first energy level (electron shell) contains only one orbital, which has a spherical shape. Two electrons can occupy this orbital. **b.** The second energy level contains one spherical-shaped orbital and 3 dumbbell-shaped orbitals at right angles to each other. Since 2 electrons can occupy each orbital, there are a total of 8 electrons in the second electron shell. Each orbital is drawn separately here, but actually the second spherical orbital surrounds the first spherical orbital and the dumbbell-shaped orbitals pass through the spherical ones.

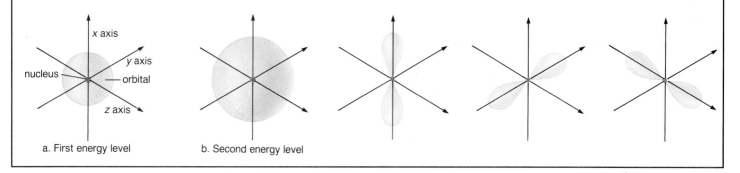

a. First energy level     b. Second energy level

nucleus. At the second energy level, there are 4 orbitals; one of these is spherical-shaped but the other 3 are dumbbell-shaped. This is the shape that allows the electrons to be most distant from one another. Since each orbital can hold 2 electrons, there are a maximum of 8 electrons in this shell. Higher shells can be more complex and contain more orbitals if they are inside another shell. If such a shell is the outer shell, it too has only 4 orbitals and a maximum number of 8 electrons.

### Electron Configurations

Although it is a simplification to depict atoms by indicating only electron shells, it is a convenient way to keep track of the number of electrons in the outer shell (fig. 3.7). This is important because the number of electrons in the outer shell determines whether or not an atom reacts with another atom as well as the manner in which the atom reacts. If an atom has only one shell, this outer shell is complete when it has 2 electrons. Otherwise the *octet rule*, which states that the outer shell is complete when it has 8 electrons, holds. It is observed that atoms with 8 electrons in the outer shell do not react at all. They are said to be inert. Atoms that have fewer than eight electrons in the outer shell react with other atoms in such a way that after the reaction, each has a completed outer shell. Atoms can give up, accept, or share electrons in order to have a completed outer shell.

> The manner in which atoms react with one another is determined by the number of electrons in the outer shell.

## Compounds and Molecules

When 2 or more atoms of different elements react or bond together, a **compound** results. A **molecule** is the smallest part of a compound that still has the properties of that compound. Molecules can also form when 2 or more atoms of the same element react with one another.

## Figure 3.7

Since the manner in which atoms react with one another is dependent upon the number of electrons in the outer shell, it is beneficial to have a convenient way to determine and to show this number. These examples of electrically neutral atoms will help you learn to do this.

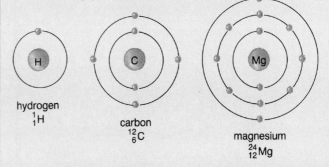

hydrogen
$^{1}_{1}H$

carbon
$^{12}_{6}C$

magnesium
$^{24}_{12}Mg$

Hydrogen, the smallest atom, has only one proton and one electron; therefore, the outer shell has only one electron. Carbon, on the other hand, has an atomic number of 6. It has 6 protons in the nucleus and 6 electrons in the shells (2 electrons in the first shell and 4 electrons in the second, or outer, shell). Each lower energy level is filled with electrons before the next higher energy level contains any electrons. Magnesium, with an atomic number of 12, has 3 shells (two electrons in the first shell, 8 electrons in the second shell, and 2 electrons in the third, or outer, shell).

### Chemical Bonds

Electrons possess energy; therefore, the bonds that exist between atoms in molecules are energy relationships. Energy is required for a bond to form, and energy is released when a bond is broken. The provision of energy for bond formation and the utilization of energy released when a bond is broken are top priorities for organisms.

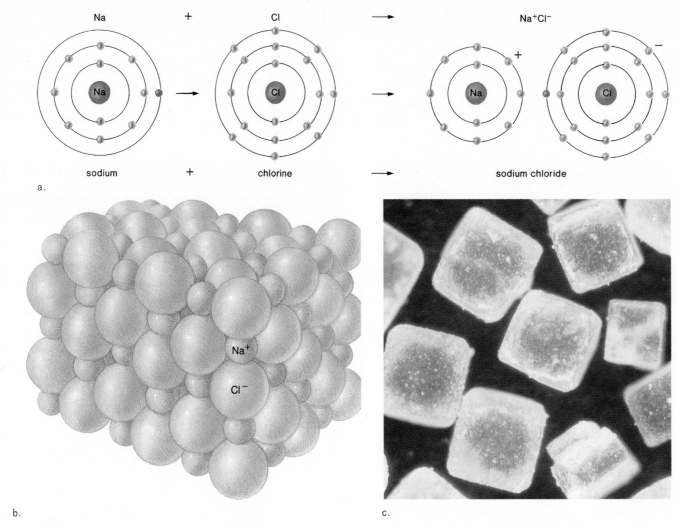

**Figure 3.8**

During an ionic reaction, an electron is transferred from one atom to another. ***a.*** Formation of sodium chloride, an ionic compound. Here a single electron passes from sodium (Na) to chlorine (Cl). At the completion of the reaction, each atom has 8 electrons in the outer shell, but each also carries a charge. Sodium now has a net charge of +1, symbolized as Na⁺, and chlorine has a net charge of -1, symbolized as Cl⁻. An ionic bond is the attraction between ions, atoms that carry a charge. ***b.*** Ionic bonding between Na⁺ and Cl⁻ causes the atoms to assume a 3-dimensional lattice in which each sodium ion is surrounded by 6 chlorine ions and each chlorine ion is surrounded by 6 sodium ions. ***c.*** The crystals of table salt can vary in size but their structure still is due to the lattice formation shown here. Magnification, ×30.

## Ionic Bonding

Ionic bonds form when electrons are transferred from one atom to another. For example, sodium (Na), with only one electron in its third shell, tends to be an electron *donor* (fig. 3.8). Once it gives up this electron, the second shell, with 8 electrons, becomes its outer shell. Chlorine (Cl), on the other hand, tends to be an electron *acceptor*. Its outer shell has 7 electrons, so it needs only one more electron to have a completed outer shell. When a sodium atom and a chlorine atom come together, an electron is transferred from the sodium atom to the chlorine atom. Now both atoms have 8 electrons in their outer shells.

This electron transfer, however, causes a charge imbalance in each atom. The sodium atom has one more proton than it has electrons; therefore, it has a net charge of +1 (symbolized by Na⁺). The chlorine atom has one more electron than it has protons; therefore, it has a net charge of -1 (symbolized by Cl⁻). Such charged atoms are called **ions.** Sodium (Na⁺) and chlorine (Cl⁻) are not the only biologically important ions. Some, such as potassium (K⁺), are formed by the transfer of a single electron to another atom; others, such as calcium (Ca⁺⁺) and magnesium (Mg⁺⁺), are formed by the transfer of 2 electrons.

## Figure 3.9

Killer whale emerges from the sea. Seawater and the blood of vertebrates are strikingly similar in the kinds of salts present and in the relative concentrations of these salts. Life is believed to have originated in the sea, and the first organisms were adapted to a seawater environment. Since then, life forms have also become adapted to the freshwater and land environments. Even so, the blood of these animals retains something of the same pattern of salts as the sea.

## Figure 3.10

Covalently bonded molecules. In a covalent bond, atoms share electrons. Sharing is represented as overlapping outer shells, and the shared electrons are counted as belonging to each atom. When this is done, each atom has a completed outer shell. *a.* A molecule of hydrogen ($H_2$) contains 2 hydrogen atoms sharing a pair of electrons. This single covalent bond can be represented in any of the 3 ways shown. *b.* A molecule of oxygen ($O_2$) contains 2 oxygen atoms sharing 2 pairs of electrons. This results in a double covalent bond. *c.* A molecule of methane ($CH_4$) contains one carbon atom bonded to 4 hydrogen atoms. Each hydrogen atom has a completed outer shell because the first shell only holds 2 electrons. By sharing with 4 hydrogens, the carbon atom has a completed outer shell containing 8 electrons.

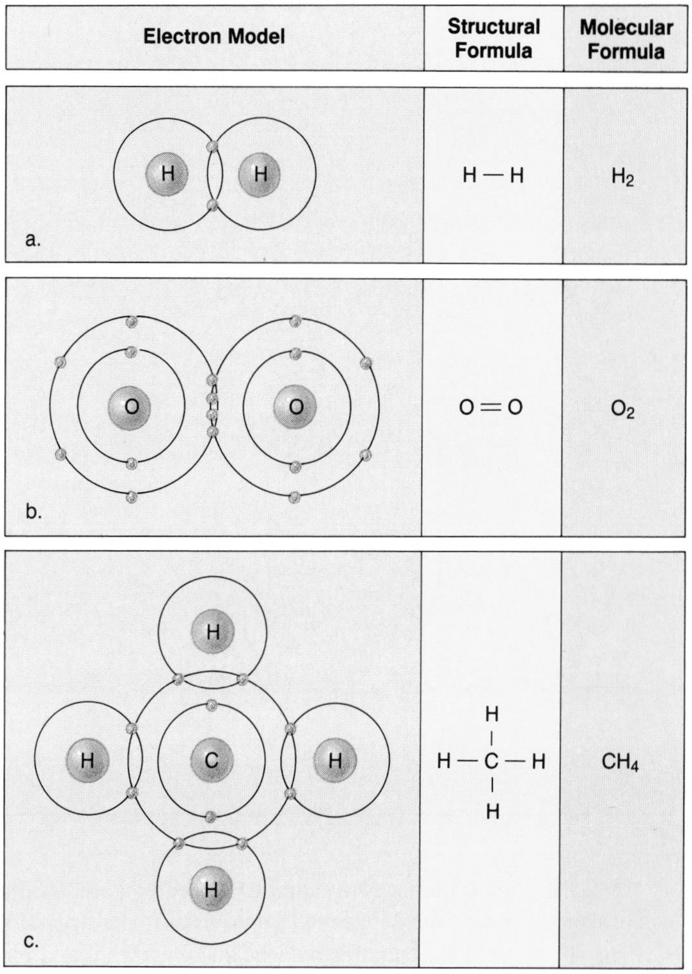

| Electron Model | Structural Formula | Molecular Formula |
|---|---|---|
| a. | H — H | $H_2$ |
| b. | O = O | $O_2$ |
| c. | H—C—H (with H above and H below) | $CH_4$ |

Ionic compounds are held together by an attraction between the charged ions called an **ionic bond.** For example, when sodium reacts with chlorine, an ionic compound called sodium chloride ($Na^+Cl^-$) results, and the reaction is called an ionic reaction. Sodium chloride is a salt commonly known as table salt because it is used at the table to season our food (fig. 3.8*c*). Salts can exist as dry solids, but when such a compound is placed in water, the ions separate as the salt dissolves. For example, $Na^+Cl^-$ separates into $Na^+$ and $Cl^-$. Ionic compounds are most commonly found in this dissociated (ionized) form in biological systems because these systems are 70%–90% water (fig. 3.9).

The transfer of electron(s) between atoms results in ions that are held together by an ionic bond, the attraction of negative and positive charges.

## Covalent Bonding

A **covalent bond** results when 2 atoms *share* electrons in such a way that each atom has a completed outer shell. Consider the hydrogen atom, which has one electron in the first shell, a shell that is complete when it contains 2 electrons. If hydrogen is in the presence of a strong electron acceptor, it gives up its electron to become a hydrogen ion ($H^+$). But if this is not possible, hydrogen

can share with another atom and thereby have a completed outer shell. For example, a hydrogen atom can share with another hydrogen atom. In this case, the 2 orbitals overlap and the electrons are shared between them (fig. 3.10*a*). Because they share the electron pair, each atom has a completed outer shell. When a reaction results in a covalent molecule, it is called a covalent reaction.

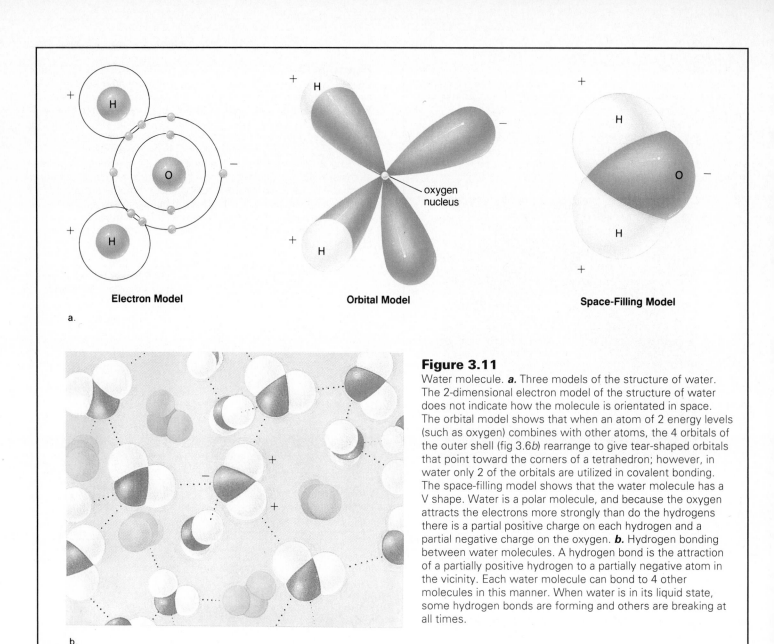

**Electron Model**

**Orbital Model**

oxygen
nucleus

**Space-Filling Model**

a.

b.

**Figure 3.11**

Water molecule. ***a.*** Three models of the structure of water. The 2-dimensional electron model of the structure of water does not indicate how the molecule is orientated in space. The orbital model shows that when an atom of 2 energy levels (such as oxygen) combines with other atoms, the 4 orbitals of the outer shell (fig 3.6*b*) rearrange to give tear-shaped orbitals that point toward the corners of a tetrahedron; however, in water only 2 of the orbitals are utilized in covalent bonding. The space-filling model shows that the water molecule has a V shape. Water is a polar molecule, and because the oxygen attracts the electrons more strongly than do the hydrogens there is a partial positive charge on each hydrogen and a partial negative charge on the oxygen. ***b.*** Hydrogen bonding between water molecules. A hydrogen bond is the attraction of a partially positive hydrogen to a partially negative atom in the vicinity. Each water molecule can bond to 4 other molecules in this manner. When water is in its liquid state, some hydrogen bonds are forming and others are breaking at all times.

A more common way to symbolize that atoms are sharing electrons is to draw a line between the 2 atoms as in the structural formula H—H. In a molecular formula the line is even omitted and the molecule is simply written as $H_2$.

**Double and Triple Covalent Bonds** Like a single bond between 2 hydrogen atoms, a double bond can also allow 2 atoms to complete their octets. In a double covalent bond, 2 atoms share 2 pairs of electrons (fig. 3.10*b*). In order to show that oxygen gas ($O_2$) contains a double bond, the molecule can be written as O=O.

It is even possible for atoms to form triple covalent bonds as in nitrogen gas ($N_2$), which can be written as N≡N. Single covalent bonds between atoms are quite strong, but double and triple bonds are even stronger.

In a covalent molecule, atoms share electrons; not only single bonds but also double and even triple bonds are possible.

**Polar Covalent Bonds**

Normally, the sharing of electrons between 2 atoms is fairly equal, and the covalent bond is nonpolar. All the molecules in figure 3.10, including methane ($CH_4$), are nonpolar. In the case of water ($H_2O$), however, the sharing of electrons between oxygen and each hydrogen is not completely equal. The larger oxygen atom, with the greater number of protons, dominates the $H_2O$ association and attracts the electron pair to a greater extent. This causes the oxygen atom to assume a slightly negative charge, showing that it is electronegative in relation to the hydrogen atoms. Each hydrogen atom assumes a slightly positive charge, showing that it is electropositive in relation to oxygen. The unequal sharing of electrons in a covalent bond creates a **polar covalent bond,** and the molecule itself is a polar molecule (fig. 3.11).

> The water molecule has an electronegative end and an electropositive end indicating that it is a polar molecule.

### Hydrogen Bonding

Polarity within a water molecule causes the hydrogen atoms in one molecule to be attracted to the oxygen atoms in other molecules (fig. 3.11*b*). This attractive force creates a weak bond called a **hydrogen bond.** This bond is often represented by a dotted line because a hydrogen bond is easily broken. Hydrogen bonding is not unique to water. It occurs whenever an electropositive hydrogen atom is attracted to an electronegative atom in another molecule or parts of the same molecule.

Although a hydrogen bond is more easily broken than a covalent bond, many hydrogen bonds taken together are quite strong. Hydrogen bonds between parts of cellular molecules help maintain their proper structure and function. We will see that some of the important properties of water are also due to hydrogen bonding.

> A hydrogen bond occurs between a slightly positive hydrogen atom of one molecule and a slightly negative atom of another molecule or between parts of the same molecule.

## Chemical Reactions

Chemical reactions are often indicated by a chemical equation. The atoms or molecules to the left of an arrow are the *reactants,* the substances that react with one another to give the product(s). The *product(s)* is placed to the right of the arrow. The reactants and the products are designated by formulas that give the number of atoms and possibly an indication of how they are arranged within the compound or molecule. For example, the equation for the formation of water from hydrogen gas and oxygen gas is written as follows:

$$2H_2 \ + \ O_2 \longrightarrow 2H_2O$$
$$\text{hydrogen} \quad \text{oxygen} \qquad \text{water}$$

Notice that all atoms present in the reactants are accounted for in the products; there are 4 hydrogen atoms and 2 oxygen atoms on both sides of the equation. The equation must be balanced in this way because mass can never be created and can never disappear.

### Oxidation-Reduction Reactions

Oxidation-reduction reactions are an important type of reaction in cells, but the terminology was derived from studying reactions outside of cells. When oxygen combines with a metal, oxygen receives electrons and becomes negatively charged and the metal loses electrons and becomes positively charged.

Today, the terms **oxidation** and **reduction** are applied to many ionic reactions, whether or not oxygen is involved. Very simply, *oxidation refers to the loss of electrons, and reduction refers to the gain of electrons.* In our previous ionic reaction, $Na + Cl \rightarrow Na^+Cl^-$, the sodium has been oxidized (loss of electron) and the chlorine has been reduced (gain of electron).

The terms oxidation and reduction are also applied to certain covalent reactions. In this case, however, oxidation is the loss of hydrogen atoms, and reduction is the gain of hydrogen atoms. A hydrogen atom contains one proton and one electron; therefore, when a molecule loses a hydrogen atom, it has lost an electron, and when a molecule gains a hydrogen atom, it has gained an electron. We will have occasion to refer to this form of oxidation-reduction reaction again in chapter 5.

> When oxidation occurs, an atom is oxidized (loses electrons). When reduction occurs, an atom is reduced (gains electrons). These 2 processes occur concurrently in oxidation-reduction reactions.

## Water

### Properties of Water

Life evolved in water, and living things are 70%-90% water. The chemical properties of water are absolutely essential to the continuance of life (fig. 3.12). Water is a polar molecule—the oxygen end of the molecule is electronegative, and the hydrogen end is electropositive. Water molecules are hydrogen bonded one to the other (fig. 3.11). A hydrogen bond is much weaker than a covalent bond within a water molecule, but taken together, hydrogen bonds cause water molecules to cling together. Without hydrogen bonding between molecules, water would boil at -80° C and freeze at -100° C, making life as we know it impossible. But because of hydrogen bonding, water is a liquid at temperatures suitable for life. It boils at 100° C and freezes at 0° C. Water also has other important properties (table 3.3).

1.  *Water is the universal solvent and facilitates chemical reactions both outside of and within living systems.* When a salt, such as sodium chloride ($Na^+Cl^-$), is put into water, the electronegative ends of the water molecules are attracted to the sodium ions, and the electropositive ends of the water molecules are attracted to the chlorine ions. This causes the sodium ions and the chlorine ions to separate and to dissolve in water:

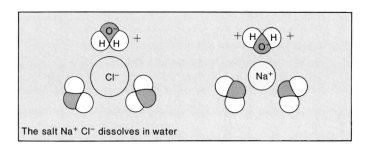

The salt $Na^+ Cl^-$ dissolves in water

**Figure 3.12**

*a.* The temperature of water changes slowly and a great deal of heat is needed for vaporization to occur. *b.* Ice is less dense than liquid water and therefore bodies of water freeze from the top down, making ice fishing possible after a hole is drilled through the ice. *c.* The heat needed to vaporize water can help animals in a hot climate to maintain internal temperatures.

a.

b.

c.

Water is also a solvent for larger molecules that contain ionized atoms or are polar molecules:

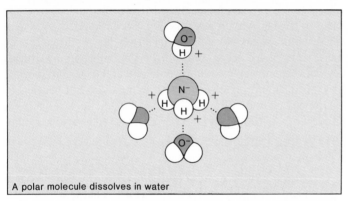

A polar molecule dissolves in water

When ions and molecules disperse in water, they move about and collide, allowing reactions to occur. Those molecules that can attract water are said to be hydrophilic, meaning "water loving." Nonionized and nonpolar molecules that cannot attract water are said to be hydrophobic, or "water hating."

2. *Water molecules are cohesive.* Water flows freely, yet water molecules do not break apart. They cling together because of hydrogen bonding. Water molecules also adhere to surfaces, particularly polar surfaces. Therefore, water can fill a tubular vessel and still flow so that dissolved and suspended molecules are evenly distributed throughout a system. For these reasons, water is an excellent transport system both outside of and within living organisms. One-celled organisms rely on external water to transport nutrient and waste molecules, but multicellular organisms often contain internal vessels in which water serves to transport nutrients and wastes.

3. *The temperature of liquid water rises and falls more slowly than that of most other liquids.* A calorie of heat energy is needed to raise the temperature of 1 g of water 1°C.[2] This is about twice the amount of heat required for other covalently bonded liquids. The many hydrogen bonds that link water molecules help water absorb heat without a change in temperature. Because water holds heat so effectively, its temperature falls more slowly. This property of water is important not only for aquatic organisms but also for all living things. Water protects organisms from rapid temperature changes and helps them maintain their normal internal temperatures.

4. *Water tends to remain a liquid; it does not readily change to ice or steam.* Converting 1 g of liquid water to ice requires the loss of 80 calories of heat energy. Converting 1 g of water to steam requires an input of 540 calories of heat energy (fig. 3.13). Hydrogen bonds must be broken to change water to steam; this accounts for the very large amount of heat needed for evaporation. This

[2]A calorie (cal) is the amount of heat required to raise the temperature of 1 g of water 1°C. A kilocalorie (kcal) equals 1,000 calories and is the unit used to express the energy value of food.

The Cell

## Table 3.3
Water

| Properties | Chemistry | Result |
|---|---|---|
| Universal solvent | Polarity | Facilitates chemical reactions |
| Adheres and is cohesive | Polarity; hydrogen bonding | Serves as transport medium |
| Resists changes in temperature | Hydrogen bonding | Helps keep body temperatures constant |
| Resists change of state (from liquid to ice and from liquid to steam) | Hydrogen bonding | Moderates earth's temperature Evaporation helps bodies remain cool |
| Less dense as ice than as liquid water | Hydrogen bonding changes | Ice floats on water |

## Figure 3.13

Water is the only common molecule that can be a solid, liquid, or gas according to the environmental temperature. At ordinary temperatures and pressure, water is a liquid and it takes a large input of heat to change it to steam. (When we perspire the heat of our bodies is causing water to vaporize, and therefore, sweating causes us to cool off.) In contrast, water gives off heat when it freezes and this heat will keep the environmental temperature higher than expected. Can you see why there are less severe changes in temperature along the coasts?

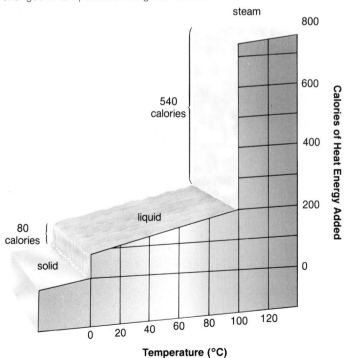

## Figure 3.14

Most substances contract when they solidify, but water expands. Water is most dense at 4° C and becomes less dense as it either cools or warms. At 0° C water freezes, forming a lattice structure in which the hydrogen bonds are fixed. The water molecules in the lattice are further apart than in liquid water; therefore ice is less dense than liquid water. This means that ice can float on liquid water. When water vaporizes at 100° C, all the hydrogen bonds are broken, and the molecules move away from one another.

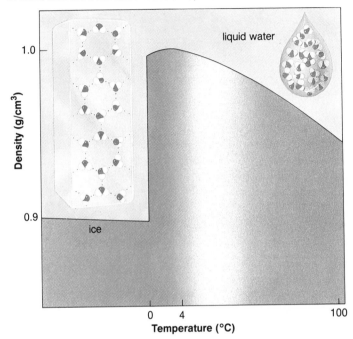

property of water helps moderate the earth's temperature so that it promotes the continuance of life. It also gives animals in a hot environment an efficient way to release excess body heat. When an animal sweats, body heat is used to vaporize the sweat, thus cooling the animal.

5. *Frozen water is less dense than liquid water.* As water cools, the molecules come closer together. They are densest at 4° C, but they are still moving about (fig. 3.14). At temperatures below 4° C, the water molecules cease moving, and hydrogen bonding becomes more rigid but also more open. This makes ice less dense than liquid water which is why ice floats on liquid water. Bodies of water always freeze from the top down. When a body of water freezes on the surface, the ice acts as an insulator to prevent the water below it from freezing. This protects many aquatic organisms so that they can survive the winter.

Water has unique properties that allow cellular activities to occur and that make life on earth possible.

## Ionization of Water

Water has another property that has important consequences for living things. It tends to ionize or dissociate, releasing an equal number of hydrogen ions ($H^+$) and hydroxyl ions ($OH^-$):

$$H—O—H \rightleftharpoons H^+ + OH^-$$
water — hydrogen ion — hydroxyl ion

This equation is written with a double arrow because the reaction is reversible. Just as water can release hydrogen and hydroxyl ions, these same ions can join together to form water. In fact, only a few water molecules at a time are actually dissociated.

There are many other substances that also dissociate and contribute ions to a given solution, which is a liquid containing a dissolved substance. **Acids** are molecules that release hydrogen ions when they dissociate. A strong acid, such as hydrochloric acid (HCl), dissociates almost completely when put into water:

$$HCl \longrightarrow H^+ + Cl^-$$
hydrochloric acid — hydrogen ion — chloride ion

In contrast, **bases** are molecules that can take up hydrogen ions. When sodium hydroxide (NaOH) dissociates, the hydroxyl ion can combine with a hydrogen ion to form water. Sodium hydroxide is termed a strong base because it dissociates almost completely when put into water:

$$NaOH \longrightarrow Na^+ + OH^-$$
sodium hydroxide — sodium ion — hydroxyl ion

### pH Scale

Acids and bases affect the relative concentration of hydrogen ions and hydroxyl ions. Biologists use the **pH scale** because it indicates the relative concentrations of $H^+$ and $OH^-$ in a solution (fig. 3.15). The pH scale range is from 0 to 14. The pH of pure water is 7; this is neutral pH. Acids lower the pH; any pH below 7 indicates that there are more hydrogen ions than hydroxyl ions. Bases increase the pH; any pH above 7 indicates that there are fewer hydrogen ions than hydroxyl ions. A pH change of one unit involves a 10-fold change in hydrogen ion concentration. Therefore, pH 6 has 10 times the [$H^+$] of pH 7, and pH 5 has 100 times the [$H^+$] of pH 7.

### Buffers

Most organisms maintain a pH of about 7; for example, human blood has a pH of about 7.4. A much higher pH or much lower pH causes illness. Normally, pH stability is possible because organisms have built-in mechanisms to prevent pH changes. Buffers are the most important of these mechanisms. A **buffer** is a chemical or

**Figure 3.15**
The pH scale. The proportionate amount of hydrogen ions to hydroxide ions is indicated by the diagonal line. Any pH above 7 is basic, while any pH below 7 is acidic.

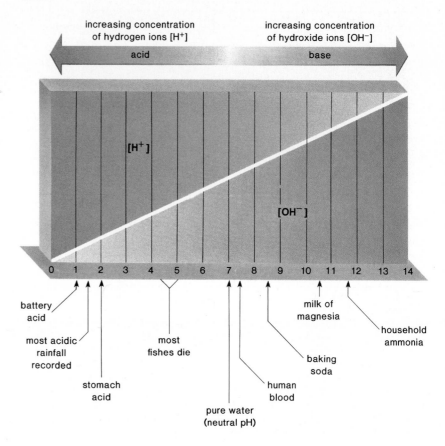

a combination of chemicals that can both take up and release hydrogen ions. Carbonic acid ($H_2CO_3$) helps buffer human blood because it is a weak acid that does not totally dissociate:

$$H_2CO_3 \rightleftharpoons H^+ + HCO_3^-$$
carbonic acid — hydrogen ion — bicarbonate ion

When excess hydrogen ions are present in blood, the reaction goes to the left, and carbonic acid forms to maintain the pH. If hydroxyl ions are added to blood, water will form, effectively removing hydrogen ions. Other types of bases can directly combine with hydrogen ions. When a base removes hydrogen ions, the reaction goes to the right, and the pH is maintained.

It is possible to overcome an organism's buffering ability. Usually, however, buffers keep the pH within normal limits despite the many biochemical reactions that either release or take up hydrogen or hydroxyl ions.

Acids have a pH that is less than 7, and bases have a pH that is greater than 7. Organisms contain buffers that help maintain the pH within a normal range.

# Acid Deposition

ormally rainwater has a pH of about 5.6 because the carbon dioxide in the air combines with water to form a weak solution of carbonic acid. Rain falling in the northeastern United States and southeastern Canada has a pH between 5.0 and 4.0. You have to remember that a pH of 4 is 10 times more acidic than a pH of 5 to appreciate the increase in acidity this represents.

There is very strong evidence now that this observed increase in rainwater acidity is a result of the burning of fossil fuels like coal and oil, as well as the gasoline derived from oil. When these substances are burned, sulfur oxides and nitrogen oxides are produced. These combine with water vapor in the atmosphere to form acids. These acids fall out of the atmosphere as rain or snow, in a process properly called wet deposition but more often called acid rain. Also, dry particles of

sulfate salts and nitrate salts can fall out of the atmosphere, a process called dry deposition. The net result is that the burning of fossil fuels and related products leads to the presence of acids in the air, and these return to the earth sometime later.

These air pollutants can be carried in the wind to places far distant from their origin. Acid deposition in Canada and the northeastern United States is due to the burning of fossil fuels in factories and power plants located in the Midwest. Similarly, the Scandinavian countries are the recipients of air pollutants from England and other northern European countries. Unfortunately, regulations that require the use of tall smokestacks to reduce local air pollution only result in the pollutants being carried farther away.

Acid deposition has created a very serious situation in certain areas of the world. In the United States, vulnerable ar-

eas include not only the Northeast but also the Great Smokey Mountains, the lakes of Wisconsin and Minnesota, the Pacific Northwest, and the Colorado Rockies. The soil in these areas is thin and lacks limestone (calcium carbonate, $CaCO_3$), which can buffer acid deposition. The first cry of alarm came from Sweden, which reported as early as the 1960s that its lakes were dying. Today dying lakes are found across northeastern North America, Western Europe, and Eastern Europe. The acid rain leaches aluminum from the soil and carries this element into the lakes. The acid also allows mercury deposits in lake bottom sediments to be converted to soluble methyl mercury. Not only do the lakes become more acidic, they also show an accumulation of substances that are toxic to living things. Sweden now reports 15,000 of its lakes are too acidic to support any higher forms of aquatic life (fig. 3.A).

a. Factories, power plants, and automobiles put out sulfur dioxide ($SO_2$), nitrogen oxides ($NO_x$), and other pollutants.

b. In the atmosphere, $SO_2$ is converted to sulfuric acid. $NO_x$ is converted to nitric acid. Some of the pollutants quickly fall to the ground, helping to cause local environmental problems.

c. Many of the acids, however, are carried hundreds of miles by the prevailing winds.

d. Far from their source, the acids fall from the sky either in the form of dry particles or as wet deposition. They cause lakes to become sterile and statues and buildings to crumble and can seriously harm forests.

**Figure 3.A**
The cause and effect of acid deposition.

*Continued on next page*

In the early 1980s, evidence began to accumulate that acid deposition was also damaging to forests. As of mid-1986, some 19 countries in Europe reported damage to their woodlands ranging from roughly 5%-15% of the forested area in Yugoslavia and Sweden to 50% or more in the Netherlands, Switzerland, and Germany. More than one-fifth of Europe's forests are now damaged.

These are not the only effects of acid deposition. Other effects include reduction of agricultural yields, damage to marble and limestone monuments and buildings, and even an increase in the illnesses of humans. Acid deposition is believed to be implicated in the increased incidence of lung cancer and possibly colon cancer along the East Coast of the United States. Tom McMillan, Canadian Minister of the Environment, says that acid deposition "is destroying our lakes, killing our fish, undermining our tourism, retarding our forests, harming our agriculture, devastating our heritage and threatening our health."

There are, of course, things that can be done.

1. Use alternative energy sources, such as solar, wind, hydropower, and geothermal energy whenever possible.
2. Use low-sulfur coal or remove the sulfur impurities from coal before it is burned.
3. Require the use of scrubbers to remove sulfur from factory and power plant emissions.
4. Require people to use mass transit rather than their own automobiles.
5. Reduce our energy needs through other means of energy conservation.

These measures and possibly others could be taken immediately. It is only necessary for us to determine that they are worthwhile.

This reading is based on these references: G. Tyler Miller, *Living in the Environment,* Wadsworth Publishing Company, Belmont, CA, 1988 and Lester R. Brown, et al., *State of the World,* W. W. Norton & Company, New York, NY, 1991.

## Summary

1. Both living and nonliving things are composed of matter consisting of elements. Each element contains atoms of just one type. The acronym CHNOPS recalls the most common elements (atoms) found in living things.
2. Atoms contain subatomic particles. Protons and neutrons in the nucleus determine the weight of an atom. Electrons are outside the nucleus and have almost no weight.
3. The atomic number indicates the number of protons and the number of electrons in electrically neutral atoms. Protons have a positive charge and electrons have a negative charge. Before atoms react, the charges are equal.
4. Isotopes are atoms of the same type that differ by the number of neutrons. Radioactive isotopes are used as tracers in biological experiments and medical procedures.
5. Electrons occupy energy levels (electron shells) at discrete distances from the nucleus. When electrons absorb energy from an outside source, they move to a higher level, and when they drop back, they release energy. Certain biological processes are powered by this release of energy.
6. Electron shells contain orbitals. The first shell contains a single spherical-shaped orbital. The second shell contains 4 orbitals: the first is spherical-shaped and the others are dumb-bell-shaped.
7. The number of electrons in the outer shell determines the reactivity of an atom. The first shell is complete when it has 2 electrons; the second shell is complete when it has 8 electrons; other shells can hold more electrons, but if they are an outer shell, they are also complete with 8 electrons. This is the octet rule.
8. Most atoms, including those common to living things, do not have completed outer shells. This causes them to react with one another to form compounds and/or molecules. Following the reaction, the atoms have completed outer shells.
9. When an ionic reaction occurs, one or more electrons are transferred from one atom to another. The ionic bond is an attraction between the resulting ions.
10. When a covalent reaction occurs, atoms share electrons. A covalent bond is the sharing of these electrons. There are single, double, and triple covalent bonds.
11. In polar covalent bonds, the sharing of electrons is not equal; one end of the bond (molecule) is electronegative, and the other end is electropositive. A hydrogen bond is a weak bond that occurs between an electropositive hydrogen atom and an electronegative atom present in another molecule or another part of the same molecule. Hydrogen bonds help maintain the shape of cellular molecules.
12. Chemical equations are used to symbolize chemical reactions. An atom that has lost electrons (or hydrogen atoms) has been oxidized, and an atom that has gained electrons (or hydrogen atoms) has been reduced.
13. Water is a polar molecule, and hydrogen bonding occurs between water molecules. These 2 features account for the unique properties of water, which are summarized in table 3.3. These features allow cellular activities to occur and make life on earth possible.
14. Water dissociates to give an equal number of hydrogen ions and hydroxyl ions. This is termed neutral pH. In acidic solutions, there are more hydrogen ions than hydroxyl ions; these solutions have a pH less than 7. In basic solutions, there are more hydroxyl ions than hydrogen ions; these solutions have a pH greater than 7. Cells are sensitive to pH changes, and biological systems tend to have buffers that help keep the pH within a normal range.

## Study Questions

1. Name the kinds of subatomic particles studied; tell their weight, charge, and location in an atom. Which of these varies in isotopes?
2. Define energy level and orbital. What is their relationship?
3. Draw a simplified atomic structure for a carbon atom that has 6 protons and 6 neutrons.

4. Draw an atomic representation for the molecule $Mg^{++}Cl_2^-$. Using the octet rule, explain the structure of the compound.
5. Tell whether $CO_2$ (O—C—O) is an ionic or covalent compound. Why does this arrangement satisfy all atoms involved?
6. Explain why water is a polar molecule. What is the relationship between polarity of the molecule and hydrogen bonding between water molecules?

7. Define oxidation and reduction and tell which has been oxidized and which reduced in this equation:
$4 \, Fe + 3 \, O_2 \rightarrow 2 \, Fe_2O_3$
Satisfy yourself that this equation is balanced.
8. Name 5 properties of water and relate them to the structure of water including its polarity and hydrogen bonding between molecules.
9. Define an acid and a base. On the pH scale, which numbers indicate an acid, a base, and a neutral pH?

## Objective Questions

1. An atom that has 2 electrons in the outer shell would most likely
   a. share to acquire a completed outer shell.
   b. lose these 2 electrons and become a negatively charged ion.
   c. lose these 2 electrons and become a positively charged ion.
   d. form hydrogen bonds only.
2. The atomic number tells you
   a. the number of neutrons in the nucleus.
   b. the number of protons in the atom.
   c. the weight of the atom.
   d. to what element the atom belongs.
3. An orbital is
   a. the same volume and charge as an electron shell.
   b. the same space and energy content as the energy level.
   c. the volume of space most likely occupied by an electron.
   d. what causes an atom to react with another atom.
4. A covalent bond is indicated by
   a. plus and minus charges attached to atoms.

   b. dotted lines between hydrogen atoms.
   c. concentric circles about a nucleus.
   d. overlapping electron shells or a straight line between atomic symbols.
5. An atom has been oxidized when it
   a. combines with oxygen.
   b. gains an electron.
   c. loses an electron.
   d. Both a and c.
6. In which of these are the electrons shared unequally?
   a. double covalent bond
   b. triple covalent bond
   c. hydrogen bond
   d. polar covalent bond
7. In the molecule

   a. all atoms have 8 electrons in the outer shell.
   b. all atoms are sharing electrons.
   c. carbon could accept more hydrogen atoms.
   d. All of these.

8. Which of these properties of water is not due to hydrogen bonding?
   a. stabilizes temperature inside and outside cell
   b. molecules are cohesive
   c. universal solvent
   d. ice floats on water
9. Acids
   a. release hydrogen ions.
   b. have a pH value above 7.
   c. take up hydroxyl ions.
   d. Both a and b.
10. Complete this diagram of a nitrogen atom by placing the correct number of protons and neutrons in the nucleus and electrons in the shells. Explain why the correct formula for ammonia is $NH_3$ and not $NH_4$.

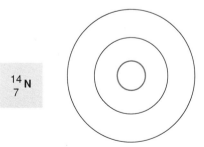

## Concepts and Critical Thinking

**1.** *Life has a chemical and physical basis.*

Give an example from your knowledge of nutrition, medicine, or environmental effects to show that this concept has an everyday application.

**2.** *The organization of life begins with atoms and continues from molecules to cells to tissues to organism.*

Why does figure 3.1 suggest that the first 2 levels of organization in living things are atoms and molecules? Why aren't cells the first level of organization?

**3.** *Life is dependent upon the special properties of water.*

Show that living things are absolutely dependent upon a particular property of water by telling what would happen if water didn't have this property.

## Selected Key Terms

**atom** (at'om) 26
**electron** (e-lek'tron) 26
**isotope** (i'so-tōp) 27
**orbital** (or'bı-tal) 28
**compound** (kom'pownd) 29

**molecule** (mol'e-kūl) 29
**ionic bond** (i-on'ik bond) 31
**covalent bond** (ko-va'lent bond) 31
**polar covalent bond** (po'lar ko-va'lent bond) 33
**hydrogen bond** (hi'dro-jen bond) 33

**oxidation** (ok"sı-da'shun) 33
**reduction** (re-duk'shun) 33
**acid** (as'id) 36
**base** (bās) 36
**pH scale** (pe āch skāl) 36
**buffer** (buf'er) 36

# 4

# The Chemistry of Life

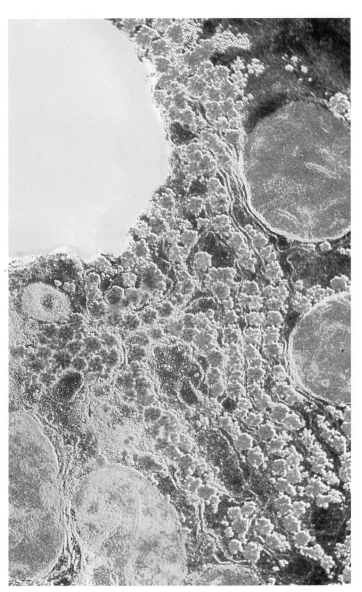

This electron micrograph of a liver cell has been colored to reveal its contents, which includes many types of organic molecules. Energy is stored in glycogen granules (red) and in a fat vacuole (yellow). Magnification, ×9,000.

Your study of this chapter will be complete when you can

1. explain and demonstrate the bonding patterns of carbon;
2. recognize the various functional groups found in cellular molecules;
3. distinguish between condensation of monomers and hydrolysis of polymers;
4. give examples of monosaccharides, disaccharides, and polysaccharides, and state their functions;
5. recognize the molecular and structural formulas for glucose;
6. give examples of various lipids, and state their functions;
7. recognize the structural formula for a saturated fatty acid and an unsaturated fatty acid and the structural formula for a fat;
8. give examples of proteins, and state their functions;
9. recognize an amino acid, and demonstrate how a peptide bond is formed;
10. relate the 4 levels of structure of a protein to the bonding patterns observed at each level;
11. give examples of nucleotides and nucleic acids, and state their functions;
12. state the components of a nucleotide, and tell how these monomers are joined to form a nucleic acid;
13. compare the structures of DNA and RNA.

iving things are highly organized; therefore, it is easy for us to differentiate between the plant, the animal, and the bacterium in figure 4.1. We might even be inclined to think that these organisms are so different that they must contain entirely different types of chemicals. But this is not the case. Instead, they and all living things contain the same classes of primary molecules: carbohydrates, proteins, lipids, and nucleic acids. But within these classes there is molecular diversity among organisms. For example, the plant, the animal, and the bacterium all utilize a carbohydrate molecule for structural purposes, but the exact carbohydrate is different in each of them.

The most common elements in living things are remembered by the acronym CHNOPS (see table 3.1), and of these, carbon, hydrogen, nitrogen, and oxygen constitute about 95% of your body weight. But we will see that both the sameness and the diversity of life are dependent upon the chemical characteristics of carbon, an atom whose chemistry is essential to living things.

The bonding of hydrogen, oxygen, nitrogen, and other atoms to carbon creates the molecules known as **organic** molecules. For a molecule to be organic, it must contain carbon and hydrogen. This is why carbon dioxide is termed an inorganic molecule. It is the organic molecules that characterize the structure and the function of living things, like the primrose, the blueshell crab, and the bacterium in figure 4.1. **Inorganic** molecules constitute nonliving matter, but even so, inorganic molecules, like salts (e.g., $Na^+Cl^-$), also play important roles in living things.

Of the elements most common to living things, the chemistry of carbon allows the formation of varied organic molecules. This accounts for both the sameness and the diversity of living things.

# Chemistry of the Cell and the Organism

Carbon has 4 electrons in its outer shell, and this allows it to bond with as many as 4 other atoms. Moreover, because these bonds are covalent, they are quite strong. Usually carbon bonds to hydrogen, oxygen, nitrogen, or another carbon atom. The ability of carbon to bond to itself makes carbon chains of various lengths and shapes possible. Long chains containing 50 or more carbon atoms are not unusual in living systems. Carbon can also share 2 pairs of electrons with another atom, giving a double covalent bond. Carbon-to-carbon bonding sometimes results in ring compounds of biological significance.

## Small Organic Molecules

Carbon chains make up the skeleton or backbone of organic molecules. The small organic molecules in living things—sugars, fatty acids, amino acids, and nucleotides—all have a carbon backbone, but in addition they have characteristic functional groups (fig. 4.2). *Functional groups* are clusters of atoms with a certain pattern and that always behave in a certain way. Therefore, functional groups help determine the characteristics of organic molecules.

## Figure 4.1

One thing these organisms (primrose, blueshell crab, and bacterium) have in common is the use of carbohydrate to provide structure. **a.** The *primrose* is held erect partly by the incorporation of the polysaccharide cellulose into the cell walls, which surround each cell. **b.** The shell of the *crab* contains chitin, a different type of polysaccharide. **c.** Most *bacterial* cells, like plant cells, are enclosed by a cell wall, but in this case the wall is strengthened by another type of polysaccharide known as peptidoglycan. Magnification, ✕38,000.

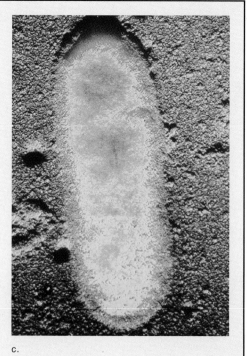

a.                                    b.                                                              c.

One characteristic of an organic molecule of some importance is whether it is hydrophilic or hydrophobic. You will recall in chapter 3 that we showed how a polar molecule with plus and minus charges is **hydrophilic** and attracted to water because water is also a polar molecule. A hydrocarbon (molecule with only carbon and hydrogen) is always **hydrophobic,** but if there is a functional group that can ionize the molecule becomes hydrophilic:

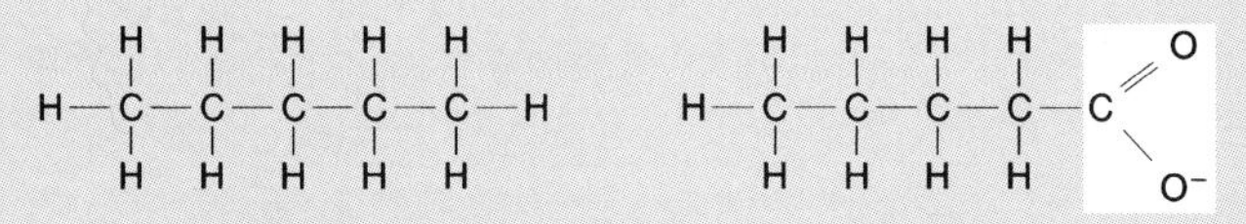

hydrocarbon
nonpolar
(hydrophobic)

acid in ionized form
polar
(hydrophilic)

The carboxyl (acid) functional group, —COOH, can give up a hydrogen ion ($H^+$) and become ionized as shown. Other functional groups (—OH, —CO, and —NH$_2$) are also polar even though they do not ionize. Polar molecules have nonbonding electrons that associate closely with water molecules. Since cells are 70%–90% water, the ability or inability to interact with water profoundly affects the function of organic molecules in cells.

Functional groups add diversity to organic molecules. Another factor contributing to the diversity of organic molecules is the existence of isomers. **Isomers** are molecules that have identical molecular formulas because they contain the same numbers and kinds of atoms, yet they are different molecules because the atoms in each isomer are arranged differently. For example, figure 4.3 shows 2 compounds with the molecular formula $C_3H_6O_3$. Each of these molecules is assigned its own chemical name because the molecules differ structurally.

## Large Organic Molecules

Each of the small organic molecules already mentioned can be a subunit of a large organic molecule, often called a *macromolecule.* A subunit is called a *monomer,* and the macromolecule is called a **polymer.** Simple sugars (monosaccharides) are the monomers within polysaccharides; fatty acids and glycerol are found in fat, a lipid; amino acids join to form proteins; and nucleotides are the subunits of nucleic acids:

| Macromolecule | Monomer |
|---|---|
| polysaccharide | monosaccharide |
| lipid (e.g., fat) | glycerol and fatty acid |
| protein | amino acid |
| nucleic acid | nucleotide |

We will see that macromolecules in the same class can vary because of differences in monomer type or the way the monomers are joined. Also, different types of functional groups can be bonded to the same monomer backbone. This is why, for example, carbohydrates can have different characteristics and therefore play different roles in cells.

**Figure 4.2**
Molecules with the exact same type of backbone can still differ according to the type of functional group attached to the backbone. Many of these functional groups are polar, helping to make the molecule soluble in water. In this illustration, the remainder of the molecule (aside from the functional groups) is represented by an R.

| Functional Groups | | |
|---|---|---|
| **Name** | **Structure** | **Found in** |
| hydroxyl (alcohol) | R — OH | sugars |
| carboxyl (acid) | R — C (double bond O, OH) | sugars fats amino acids |
| ketone | R — C (double bond O) — R | sugars |
| aldehyde | R — C (double bond O, H) | sugars |
| amine (amino) | R — N (H, H) | amino acids proteins |
| sulfhydryl | R — SH | amino acids proteins |
| phosphate | R — O — P (double bond O, OH, OH) | phospholipids nucleotides nucleic acids |

R = remainder of molecule

**Figure 4.3**
Isomers have the same molecular formula but different configurations. Both of these compounds have the formula $C_3H_6O_3$. *a.* In glyceraldehyde, oxygen is double-bonded to an end carbon. *b.* In dihydroxyacetone, oxygen is double-bonded to the middle carbon.

a. Glyceraldehyde

b. Dihydroxyacetone

There are only a few types of molecules in cells, but these molecules still have great variety and therefore play different roles in cells.

## Figure 4.4

*a.* In cells, synthesis often occurs when monomers are joined together by condensation (removal of $H_2O$). *b.* Breakdown occurs when the monomers in a polymer are separated by hydrolysis (addition of $H_2O$).

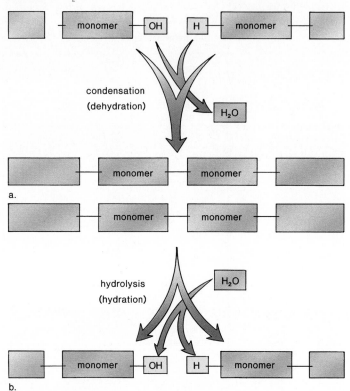

## Figure 4.5

*a.* Glucose is a 6-carbon sugar that can exist as an open chain or as a ring compound; the ring is more common in cells. The shape of glucose is indicated in the space-filling model. Notice that the carbon atoms in glucose are numbered and that the colors allow you to decipher the manner in which the open chain becomes the ring. *b.* Fructose is an isomer of glucose. It too has the molecular formula $C_6H_{12}O_6$. Notice, though, that the atoms are bonded in a slightly different manner in fructose. This causes the open chain to form a slightly different ring and therefore also contributes to a different shape of the molecule.

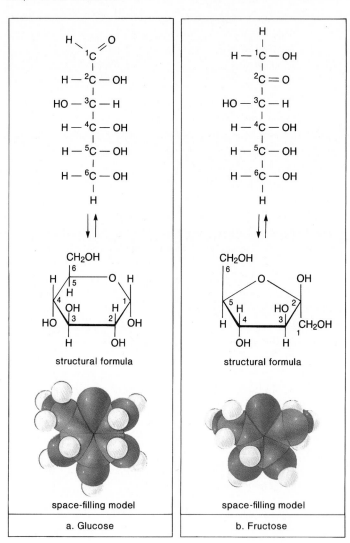

## Condensation and Hydrolysis

All macromolecules are made within a cell in essentially the same way. Two monomers join together when a hydroxyl (—OH) group is removed from one monomer and a hydrogen (—H) is removed from the other (fig. 4.4*a*). This **condensation** of monomers is a *dehydration synthesis* because water is removed (dehydration) and a bond is made (synthesis). Condensation does not take place unless the proper enzyme (a molecule that speeds up a chemical reaction in cells) is present and energy is expended. When a bond is formed, energy is needed.

Polymers are broken down by **hydrolysis,** which is essentially the reverse of condensation; an —OH group from water attaches to one monomer and a —H attaches to the other (fig. 4.4*b*). This is a hydrolysis reaction because water (*hydro*) is used to break (*lyse*) a bond. When a bond is broken, energy is released, or made available.

Macromolecules are formed and are broken down in cells as each cell renews its many parts. Also, polysaccharide breakdown makes energy available to cells.

---

Condensation (the removal of water) joins monomers together to form macromolecules, and hydrolysis (the addition of water) breaks macromolecules apart into monomers.

---

## Simple Sugars to Polysaccharides

Sugars and polysaccharides belong to a class of compounds called **carbohydrates.** Carbohydrates usually contain carbon and the components of water—hydrogen and oxygen—in a 1:2:1 ratio. The general formula for any carbohydrate is $(CH_2O)_n$; the *n* subscript stands for whatever number of these groups there are. For example, a carbohydrate with the molecular formula $C_6H_{12}O_6 = (CH_2O)_6$.

## Figure 4.6

Sucrose is a disaccharide that contains glucose and fructose units. During condensation, a bond forms between the glucose and fructose molecules as the components of water are removed. During hydrolysis, the components of water are added as the bond is broken.

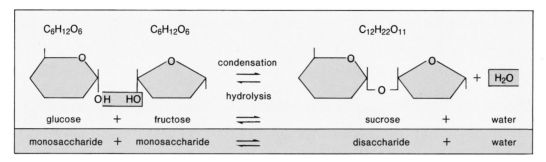

Simple sugars are *monosaccharides* (one sugar) with a carbon backbone of 3 to 7 carbon atoms. The best-known sugars are those that have 6 carbons (*hexoses*). **Glucose** is in the blood of animals, and *fructose* is frequently found in fruits. These sugars are isomers of one another. They all have the molecular formula $C_6H_{12}O_6$, but they differ in structure (fig. 4.5). Structural differences cause molecules to vary in shape, which is best seen by using space-filling models. Shape is very important in determining how molecules interact with one another.

There are 2 5-carbon sugars (*pentoses*) of significance. They are *ribose* and *deoxyribose*, which are found in the nucleic acids RNA and DNA, respectively (RNA and DNA are discussed later in the chapter).

A *disaccharide* contains 2 monosaccharides that have joined together by condensation. *Sucrose* is a disaccharide that contains glucose and fructose (fig. 4.6). Sugar is transported within the body of a plant in the form of sucrose, and this is the sugar we use at the table to sweeten our food. We acquire this sugar from plants, such as sugarcane and sugarbeets.

*Lactose* is a disaccharide that contains galactose and glucose and is found in milk. *Maltose* (composed of 2 glucose molecules) is a disaccharide of interest because it is found in our digestive tract as a result of starch digestion.

The most common *polysaccharides* in living things are starch, glycogen, and cellulose. Each of these is a polymer of glucose.

### Starch and Glycogen

The structures of starch and glycogen differ only slightly (fig. 4.7). **Glycogen** is characterized by many side branches, which are chains of glucose that go off from the main chain. **Starch** has few of these chains.

Plant cells store extra carbohydrate as complex sugars or starches. When leaf cells are actively producing sugar by photosynthesis, they store some of this sugar within the cell in the form of starch granules (fig. 4.7b). Roots also store sugars as starch and so do seeds. Grains (e.g., the seeds of wheat, corn, and rice plants) are used by humans as a source of nourishment.

Animal cells store extra carbohydrate as glycogen, sometimes called "animal starch." After a human eats, the liver stores glucose as glycogen (fig. 4.7c). Between meals, the liver releases glucose to keep the blood concentration of glucose near the normal 0.1%.

> Cells use sugars, especially glucose, as an immediate energy source. Glucose is stored as starch in plants and as glycogen in animals.

### Cellulose and Chitin

**Cellulose** contains glucose molecules that are joined together differently than they are in starch and glycogen. The orientation of the bonds in starch and glycogen allows these polymers to form compact spirals, making these polymers suitable as storage compounds. The orientation of the bond in cellulose causes the polymer to be straight and fibrous, making it suitable as a structural compound.

As shown in figure 4.8, the long, unbranched polymers of cellulose are held together by hydrogen bonding to form microfibrils. Several microfibrils, in turn, make up a fibril. Layers of cellulose fibrils make up plant cell walls. The cellulose fibrils are parallel within each layer, but the layers themselves lie at angles to one another to give added strength.

Humans have found many uses for cellulose. Cotton fibers are almost pure cellulose, and we often wear cotton clothing. Furniture and buildings are made from wood, which contains a high percentage of cellulose. Most animals, however, including humans, cannot digest cellulose because the enzymes that digest starch are unable to break the linkage between the glucose molecules in cellulose. Even so, cellulose is a recommended part of our diet because it provides bulk (also called fiber or roughage) that helps the body maintain regularity of elimination. As mentioned, grains are a source of starch, but they are also a source of fiber.

Cattle and sheep receive nutrients from grass because they have a special stomach chamber, the rumen, where bacteria that can digest cellulose live. Unfortunately, cattle are often

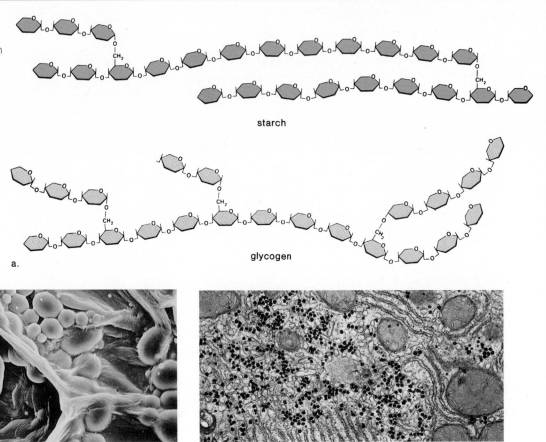

**Figure 4.7**
*a.* Starch and glycogen are polymers of glucose. *b.* Starch is the storage form of glucose in plant cells. This is a scanning electron micrograph of starch grains in a plant cell in which starch granules are highly visible. Magnification, X2,600. *c.* Glycogen is the storage form of glucose in animal cells. This is an electron micrograph of a liver cell in which glycogen granules are visible. Magnification, X24,000.

starch

glycogen

a.

b.

c.

## Wheat

H umans commonly use only about 12 species of plants as food sources. Three of these are cereals: rice, corn, and wheat. A grain of wheat contains a seed and is composed of 3 parts: the embryonic plant, the endosperm (stored food), and the seed coat along with other protective layers (fig. 4.A). The embryonic plant is wheat germ, which is high in vitamins. The starchy endosperm is used to produce the flour from which our white breads are made. The seed coat is bran, composed mostly of cellulose, which cannot be digested by humans. For this reason, bran is an excellent roughage substance that adds bulk to the diet.

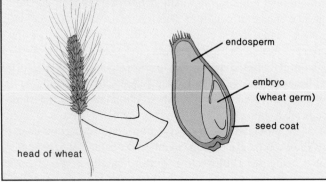

head of wheat

endosperm

embryo
(wheat germ)

seed coat

kept in feedlots, where they are fed grains instead of grass. This practice wastes fossil fuel energy because it takes energy to grow the grain, to process it, to transport it, and finally to feed it to the cattle. Also, range-fed cattle move about and produce a leaner meat than do cattle raised in a feedlot. There is growing evidence that lean meat is healthier for humans than fatty meat.

**Chitin,** which is found in the exoskeleton of crabs and related animals like lobsters and insects, is also a polymer of glucose. Each glucose, however, has an amino group attached to it. The linkage between the glucose molecules is like that found in cellulose; therefore, chitin is not a digestible material. Recently, scientists have discovered how to turn chitin into thread that can be used as suture material. They hope to find other uses for treated chitin. If so, all those discarded crab shells that pile up beside crabmeat processing plants will not go to waste.

## Fatty Acids to Lipids

A variety of organic compounds are classified as **lipids.** Many of these are insoluble in water because they lack any polar groups. The most familiar lipids are those found in fats and oils. Organisms use these molecules as long-term energy storage

**Figure 4.8**
Cellulose fibrils are present in plant cell walls. Magnification, X30,000. Each fibril contains several microfibrils, and each microfibril contains many chains of glucose hydrogen-bonded together. Finally, we have a close-up view of 3 cellulose polymers, each made up of glucose molecules.

cell wall

fibril

plant cell

microfibril

cellulose molecules

hydrogen bond

cellulose polymer

compounds. Phospholipids and steroids are also important lipids found in living things. For example, they are components of plasma membranes.

---

Lipids are quite varied in structure, but they tend to be insoluble in water.

---

## Fats and Oils

**Fats** and **oils** contain 2 types of unit molecules: fatty acids and glycerol.

Each *fatty acid* consists of a long hydrocarbon chain with a carboxyl (acid) group at one end. Because the carboxyl group is a polar group, fatty acids are soluble in water. Most of the fatty acids in cells contain 16–18 carbon atoms per molecule, although smaller ones are also found. Fatty acids are either saturated or unsaturated (fig. 4.9). *Saturated* fatty acids have no double bonds between the carbon atoms. The carbon chain is saturated, so to

speak, with all the hydrogens that can be held. *Unsaturated* fatty acids have double bonds in the carbon chain wherever the number of hydrogens is less than 2 per carbon atom:

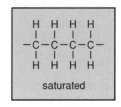

saturated          unsaturated

*Glycerol* is a compound with 3 hydroxyl groups (fig. 4.10). Hydroxyl groups are polar and therefore glycerol is soluble in water. When fat is formed, the acid portions of 3 fatty acids react with these hydroxyl groups so that fat and 3 molecules of water form. Again, the larger fat molecule is formed by condensation, and a fat can be hydrolyzed to its components. Since there are 3 fatty acids per one glycerol molecule, the fat molecule is sometimes called a triglyceride. Triglycerides are often referred to as neutral fats because they, unlike their component parts, lack polar groups that can hydrogen bond with water (p. 33). Therefore, for example, cooking oils do not mix with water even though they are both liquids. Even after shaking, the oil simply separates out.

Triglycerides containing unsaturated hydrocarbon chains melt at a lower temperature than those containing saturated chains. This is because a double bond makes a kink that prevents close packing among the chains (fig. 4.9). We can reason, then, that butter, which is a solid at room temperature, must contain saturated hydrocarbon chains and corn oil, which is a liquid even when placed in the refrigerator, must contain unsaturated chains. This difference is made use of by living things. For example, the feet of reindeer and penguins contain unsaturated triglycerides, and this helps protect these exposed parts from freezing.

Nearly all organisms use fats for long-term energy storage. Fats have mostly C—H bonds, making them a richer supply of chemical energy than carbohydrates, which have many C—OH bonds. Molecule per molecule, animal fat contains over twice as much energy as glycogen; gram per gram though, fat stores 6 times as much energy as glycogen. This is because fat droplets, being nonpolar, do not contain water. Small birds, like the broad-tailed hummingbird (fig. 4.11), store a great deal of fat before they start their long spring and fall migratory flights. About 0.15 g of fat per gram of body weight is accumulated each day. If the same amount of energy were stored as glycogen, a bird would be so heavy it would not be able to fly.

> Fats and oils are triglycerides (one glycerol plus 3 fatty acids). They are used as long-term energy storage compounds in plants and animals.

## Waxes

In *waxes,* a long-chain fatty acid bonds with a long-chain alcohol. Waxes are also solid at normal temperatures because they have a high melting point. Being hydrophobic, they are also waterproof and resistant to degradation. In many plants, waxes form a protective covering that retards the loss of water for all exposed parts. In animals, waxes are involved in skin and fur maintenance. In humans, wax is produced by glands in the outer ear canal. Here its function is to trap dust and dirt, preventing them from reaching the eardrum.

## Phospholipids

**Phospholipids,** as implied by their name, contain a phosphate group. A phosphate group is a polar group that can ionize:

$$O \quad\quad\quad\quad\quad O$$
$$\|\quad\quad\quad\quad\quad\quad\|$$
$$R—P—OH \quad\quad\quad R—P—O^-$$
$$|\quad\quad\quad\quad\quad\quad\quad|$$
$$OH \quad\quad\quad\quad\quad\quad O^-$$

nonionized phosphate        ionized phosphate

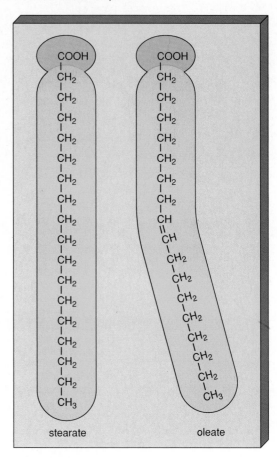

**Figure 4.9**
Fatty acid structure. Fatty acids are long hydrocarbon chains ending in a carboxylic (acid) group. Stearate is a saturated fatty acid, and oleate is an unsaturated fatty acid.

Essentially, phospholipids are constructed like neutral fats, except that in place of the third fatty acid, there is a phosphate group or a grouping that contains both phosphate and nitrogen. This group becomes the polar head of the molecule, while the hydrocarbon chains of the fatty acids become the nonpolar tails (fig. 4.12). When phospholipid molecules are placed in water, they form a sheet in which the polar heads face outward and the nonpolar tails face each other. This property of phospholipids means that they can form an interface or separation between 2 solutions such as the interior and exterior of a cell. The plasma membrane of cells is basically a phospholipid bilayer.

> Phospholipids have a polar head and nonpolar tails. They arrange themselves in a double layer in the presence of water, a property that makes them suitable as the plasma membrane of cells.

## Steroids

**Steroids** are lipids that have an entirely different structure from neutral fats. Each steroid has a backbone of 4 fused carbon rings and varies from other steroids primarily by the type of functional

## Figure 4.10

Formation of a neutral fat. Three fatty acids plus glycerol react to produce a fat molecule and 3 water molecules. A fat molecule plus 3 water molecules react to produce 3 fatty acids and glycerol.

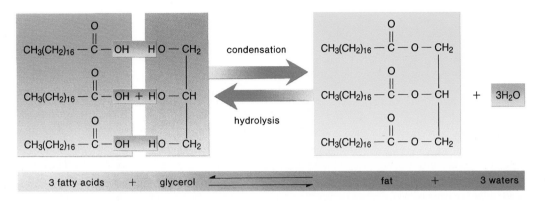

## Figure 4.11

A broad-tailed hummingbird, like other small birds that migrate, stores a lot of energy as fat in order to have enough fuel for its long journey. Fat is more energy rich than glycogen, and it is found in concentrated droplets; therefore, it is an efficient way to store energy.

## Figure 4.12

Structurally, phospholipids have a polar head and nonpolar tails (*left and center*). In water (*right*), the molecules arrange themselves as shown. The polar heads are attracted to water; the nonpolar tails are not.

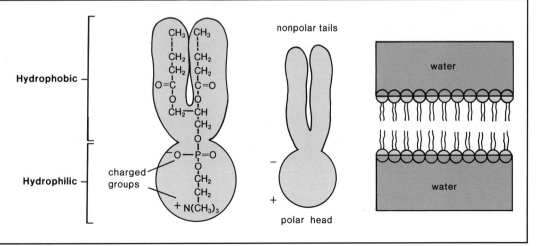

group attached to the ring (fig. 4.13). Cholesterol is the precursor of several other steroids, including the vertebrate hormones such as aldosterone, a hormone that helps regulate the sodium content of the blood, and the sex hormones, which help maintain male and female characteristics. Their functions vary due primarily to the different attached groups.

For many years, nutritionists have gathered evidence that a diet high in saturated fats and cholesterol can lead to reduced blood flow caused by the deposit of fatty materials on the linings of blood vessels. This problem is discussed at length in a reading in chapter 34.

> Steroids are ring compounds that have a similar backbone but vary according to the attached group. This causes them to have different functions in the bodies of humans and other animals.

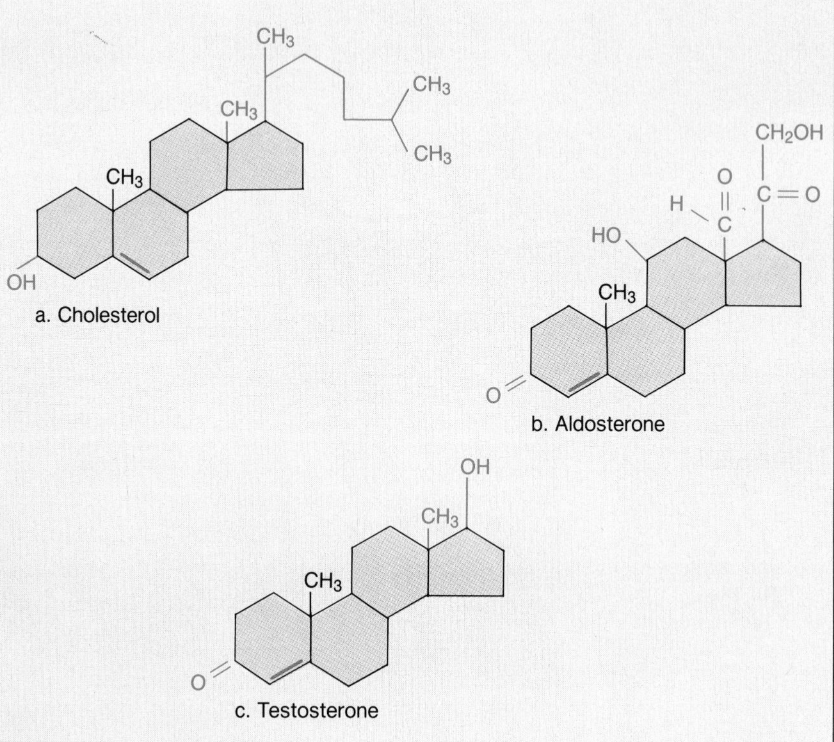

## Amino Acids to Proteins

**Amino acids** are the monomers that condense to form **proteins,** which are very large molecules that have both structural and metabolic functions. For example, in animals the proteins myosin and actin are the contractile components of muscle; insulin is a hormone that regulates the sugar content of the blood; hemoglobin transports oxygen in the blood; and collagen fibers support many organs. Proteins are present in the membrane that surrounds each cell, and they also exist within the cell. The cellular proteins, called **enzymes,** are organic catalysts that speed up chemical reactions within cells.

In this section we will point out that the role a protein plays in cells is dependent on its biological properties, which in turn are dependent on its structure.

### Amino Acids and Peptides

There are 20 different amino acids commonly found in cells, and figure 4.14 gives several examples. All amino acids contain 2 important functional groups: a carboxyl (acid), —COOH, group and an amino, —$NH_2$, group, both of which ionize at normal body pH in this manner:

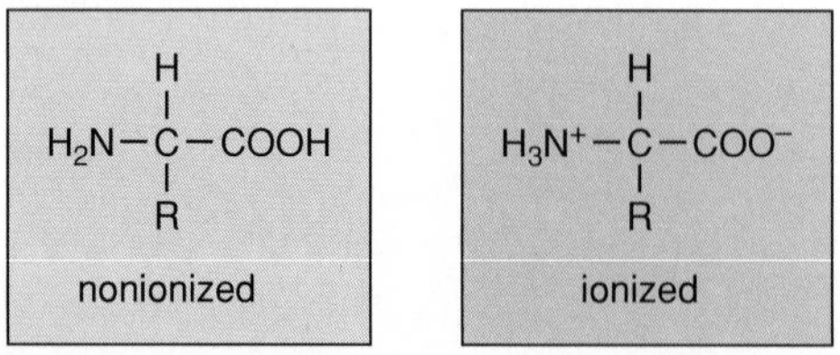

| | |
|---|---|
| H<br>&#124;<br>$H_2N$ — C — COOH<br>&#124;<br>R<br>**nonionized** | H<br>&#124;<br>$H_3N^+$ — C — $COO^-$<br>&#124;<br>R<br>**ionized** |

Amino acids differ in the nature of the R group (*R*emainder of the molecule), which ranges in complexity from a single hydrogen to complicated ring compounds. The unique chemical properties of

**Figure 4.13**
Steroid diversity. Like cholesterol in **a,** steroid molecules have 4 adjacent rings, but their effects on the body largely depend on the attached groups indicated in red. **b.** Aldosterone is involved in the regulation of sodium blood levels. **c.** Testosterone is the male sex hormone.

a. Cholesterol

b. Aldosterone

c. Testosterone

an amino acid depend on those of the R group. For example, some R groups are polar and some are not. Also, the amino acid cysteine has an R group that ends with a sulfhydryl, —SH, group that often serves to connect one chain of amino acids to another by a disulfide bond, —S—S.

A **peptide** is 2 or more amino acids joined together, and a *polypeptide* is a chain of many amino acids joined by peptide bonds. Since a protein can contain more than one polypeptide chain, you can see why a protein could have a very large number of amino acids.

Figure 4.15 shows how 2 amino acids are joined by a condensation reaction between the carboxyl group of one and the amino group of another. The resulting covalent bond between 2 amino acids is called a **peptide bond.** The atoms associated with the peptide bond share the electrons unevenly because both oxygen and nitrogen atoms are electronegative. The hydrogen attached to the nitrogen is electropositive. The polarity of the peptide bond means that hydrogen bonding is possible between parts of a polypeptide.

> Amino acids are joined by peptide bonds in polypeptides and proteins. Proteins have structural and metabolic functions in cells. Some proteins are enzymes that speed up chemical reactions.

## Figure 4.14

Representative amino acids; the R groups are in the contrasting color. Some R groups are nonpolar and hydrophobic, some are polar and hydrophilic, and some are ionized and hydrophilic.

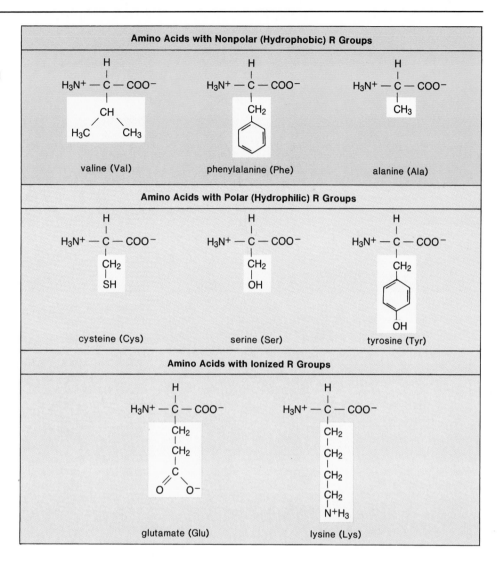

**Amino Acids with Nonpolar (Hydrophobic) R Groups**

valine (Val)    phenylalanine (Phe)    alanine (Ala)

**Amino Acids with Polar (Hydrophilic) R Groups**

cysteine (Cys)    serine (Ser)    tyrosine (Tyr)

**Amino Acids with Ionized R Groups**

glutamate (Glu)    lysine (Lys)

## Levels of Structure

The final shape of a protein in large measure determines the function of a protein in the cell or body of an organism. An analysis of protein shape shows that proteins can have up to 4 levels of structure (table 4.1 and fig. 4.16).

The *primary structure* of a protein is the sequence of the amino acids joined together by peptide bonds. In 1953, Frederick Sanger determined the amino acid sequence of the hormone insulin, the first protein to be sequenced. He did it by first breaking insulin into fragments and then determining the amino acid sequence of the fragments before determining the sequence of the fragments themselves. It was a laborious 10-year task, but it established for the first time that a particular protein has a particular sequence of amino acids. Today, there are automated sequencers that tell scientists the sequence of amino acids in a protein within a few hours. Notice that since each amino acid differs from another by its R group, it is correct to say that proteins differ from one another by a particular sequence of the R groups. The fact that some of these are polar and some are not influences the final shape of the polypeptide.

The *secondary structure* of a protein comes about when the polypeptide takes a particular orientation in space. Linus Pauling and Robert Corey, who began studying the structure of amino acids in the late 1930s, concluded about 20 years later that polypeptides must have particular orientations in space. They said that the α (alpha) helix and the ß (beta) sheet were 2 possible patterns of amino acids within a polypeptide. They called it the "α" helix because it was the first pattern they discovered and the "ß" sheet because it was the second pattern they discovered.

Hydrogen bonding between the oxygen atoms and nitrogen atoms of amino acid monomers, in particular, stabilizes the α helix. In the ß sheet, or pleated sheet, the polypeptide turns back upon itself and hydrogen bonding occurs between these extended lengths of the polypeptide. Fibrous proteins are so-called because they are found in tough fibers. Keratin found in

### Figure 4.15

Synthesis of a peptide. The peptide bond forms as the components of water are removed from the carboxyl group of one amino acid and the amino group of the other amino acid. There is a partial negative charge on the oxygen and nitrogen and a partial positive charge on the hydrogen within the peptide bond. These charges are not shown.

## Figure 4.16

Levels of protein structure for a globular protein. **a.** The primary structure is the sequence of amino acids. **b.** The secondary structure contains α helix segments and ß sheet segments. **c.** The tertiary structure is a twisting and turning of the polypeptide molecule. **d.** The quaternary structure contains more than one polypeptide each with its own levels of structure.

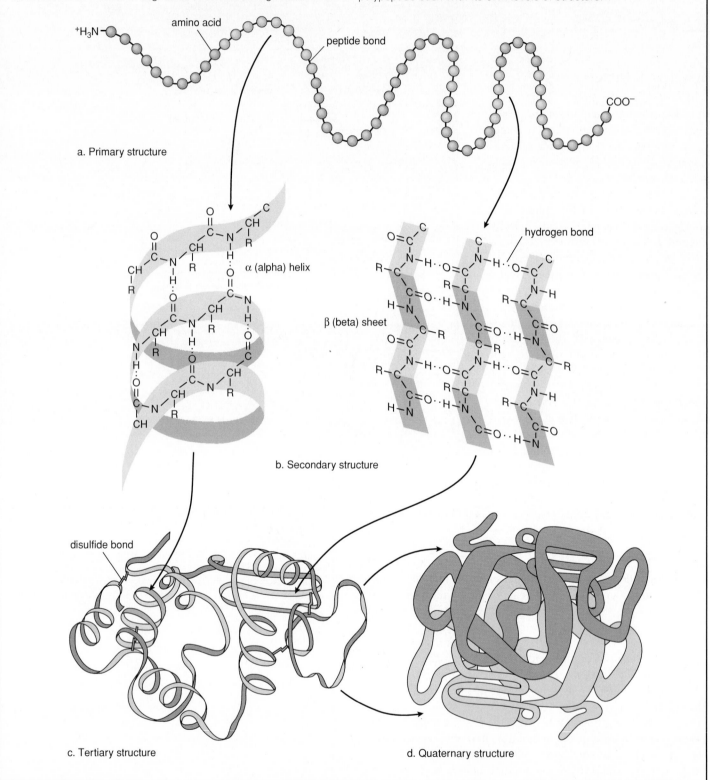

a. Primary structure

α (alpha) helix

β (beta) sheet

b. Secondary structure

c. Tertiary structure

d. Quaternary structure

hair and wool is an example of a fibrous protein in which the polypeptides are an α helix; silk is an example of a fibrous protein in which the polypeptides are ß sheets.

Globular proteins, so-called because they have a globular shape, have regions of both alpha helix and beta sheet arrangements depending upon the particular amino acids in the primary structure. The polypeptide of a globular protein folds and twists into a *tertiary shape* that is maintained by various types of bonding between the R groups. Indeed, the folding and twisting is determined by those R groups that can bond with one another, giving stability to the shape of the molecule. Hydrogen bonds, ionic bonds, and covalent bonds are seen; when cysteines are brought close to one another, a disulfide, S—S, bond serves as a bridge between 2 amino acid monomers. The hydrophobic R groups, which do not react with other R groups, tend to be pushed together into a common inner region, where they are not exposed to water. (These are called hydrophobic interactions.)

The various enzymes of a cell are globular proteins that usually have only a tertiary level structure. Some proteins have more than one polypeptide chain, each with its own primary, secondary, and tertiary structures. Within the protein, these separate chains are arranged to give a fourth level of structure termed the *quaternary structure*. Various types of interactions between the polypeptide chains are observed. Hemoglobin is a much studied globular protein that has a quaternary structure of 4 polypeptides. Each polypeptide has a heme group associated with it that carries oxygen reversibly (fig. 37.10c).

> A protein has up to 4 levels of structure that account for its final 3-dimensional shape. The shape of a protein determines its function in cells.

## Denaturation of Proteins

Both temperature and pH can bring about a change in protein shape. For example, we are all aware that the addition of acid to milk causes curdling; heating causes egg white, a protein called albumin, to congeal, or coagulate. When a protein loses its normal configuration, it is said to be denatured. Denaturation occurs because the normal bonding patterns between parts of a molecule have been disturbed. Once a protein loses its normal shape, it is no longer able to perform its usual function.

If the conditions that caused denaturation were not too severe, and if these are removed, some proteins regain their normal shape and biological activity. This shows that the primary structure of a polypeptide forecasts its final shape (fig. 4.17).

> The final shape of a protein is dependent upon the primary structure—the sequence of amino acids in the polypeptide(s).

**Table 4.1**
Levels of Protein Structure

| Level of Structure | Description | Type of Bond |
|---|---|---|
| Primary | Sequence of amino acids | Covalent (peptide) bond between amino acids |
| Secondary | Alpha helix and beta sheet | Hydrogen bond between members of peptide bond |
| Tertiary | Folding and twisting of polypeptide | Hydrogen, ionic, covalent (S-S), hydrophobic interactions* between R groups |
| Quaternary | Several polypeptides | Hydrogen, ionic between polypeptide chains |

*Strictly speaking, these are not bonds, but they are very important in creating and stabilizing tertiary structure.

**Figure 4.17**
Denaturation and reactivation of the enzyme ribonuclease. When a protein is denatured, it loses its normal shape and activity. If denaturation is gentle and if the conditions are removed, some proteins regain their normal shape. This shows that the normal conformation of the molecule is due to the various interactions between a set sequence of amino acids. Each type of protein has a particular sequence of amino acids.

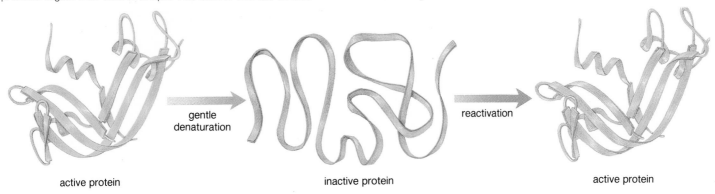

active protein     gentle denaturation     inactive protein     reactivation     active protein

## Table 4.2
DNA Structure Compared to RNA Structure

|  | DNA | RNA |
|---|---|---|
| Sugar | Deoxyribose | Ribose |
| Bases | Adenine, guanine, thymine, cytosine | Adenine, guanine, uracil, cytosine |
| Strands | Double-stranded with base pairing | Single-stranded |
| Helix | Yes | No |

## Nucleotides to Nucleic Acids

Every **nucleotide** is a molecular complex of 3 types of unit molecules: phosphate (phosphoric acid), a pentose sugar, and a nitrogen-containing base:

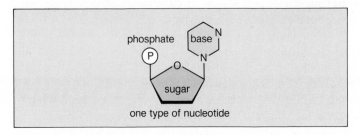

one type of nucleotide

Nucleotides have metabolic functions in cells. For example, some are components of coenzymes, which perform functions facilitating enzymatic reactions. ATP (adenosine triphosphate) is a nucleotide used in cells to supply energy for synthetic reactions and for various other energy-requiring processes. Nucleotides are also monomers in nucleic acids.

**Nucleic acids** are huge polymers of nucleotides with very specific functions in cells; for example, **DNA** (deoxyribonucleic acid) is the genetic material that stores information regarding its own replication and the order in which amino acids are to be joined to make a protein. Another important nucleic acid, **RNA** (ribonucleic acid), works in conjunction with DNA to bring about protein synthesis.

The nucleic acids DNA and RNA are polymers of nucleotides. DNA makes up genes and along with RNA controls protein synthesis within the cell.

In DNA the sugar is deoxyribose, and in RNA the sugar is ribose; the difference accounts for their respective names (table 4.2). There are 4 different types of nucleotides in DNA and RNA. Figure 4.18 shows the types of nucleotides that are present in DNA. The base can be one of the purines, adenine or guanine, which have a double ring, or one of the pyrimidines, thymine or cytosine, which have a single ring. These structures are called bases because their presence raises the pH of a solution. In RNA, the base uracil occurs instead of the base thymine.

## Figure 4.18
All nucleotides found in DNA contain phosphate, the pentose sugar deoxyribose, and a nitrogen-containing organic base. **a.** The bases adenine and guanine are double-ringed purine bases. **b.** The bases thymine and cytosine are single-ringed pyrimidine bases.

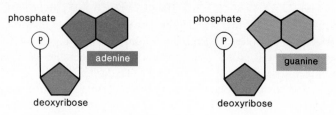

a. DNA nucleotides with purine bases

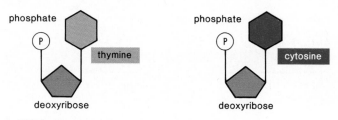

b. DNA nucleotides with pyrimidine bases

## Figure 4.19
RNA is a single-stranded polymer of nucleotides. When the nucleotides join, the phosphate group Ⓟ of one is bonded to the sugar of the next. The bases project to the side of the resulting sugar-phosphate backbone.

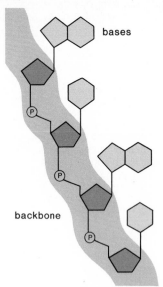

Nucleotides join together in a definite sequence when DNA and RNA form by condensation. The nucleotides form a linear molecule called a strand in which the backbone is made up of phosphate-sugar-phosphate-sugar, with the bases projecting to one side of the backbone. Since the nucleotides occur in a definite order, so do the bases.

# How Do You Use Radioactive Isotopes to Detect Macromolecule Synthesis?

**R**adioactive isotopes are often used today to trace the synthesis of macromolecules in cells and tissues. The first step is to expose the cells to a radioactively labeled molecule. For example, to detect the synthesis of DNA, $^3$H- or $^{14}$C-thymidine* can be used, and to detect the synthesis of RNA, $^3$H- or $^{14}$C-uridine* can be used. Other labeled molecules are available for protein and polysaccharide synthesis. After time has elapsed and synthesis has occurred, the labeled macromolecule is extracted from the cell by chemical means and placed in a vial that contains a special solution. The solution produces a light flash every time the labeled macromolecule gives off a radioactive particle. The number of flashes (called scintillations) are detected by placing the vial in a scintillation counter, an instrument that counts the flashes. The greater the amount of synthesis, the greater the radioactivity in the sample, and the greater the counts per minute (cpm). Figure 4.B shows the results of an experiment in which the amount of RNA synthesis was determined as a function of cell culture age. As cell cultures age, the amount of RNA synthesis decreases.

Autoradiography is a technique that is used to determine the placement of labeled molecules in cells. In this case, autoradiography would indicate where RNA synthesis is occurring. After exposure of cells (or tissues) to, say, radioactively labeled uridine, the cells are washed clean of any excess radioactivity before they are sectioned and placed on a slide. The slide is coated with an emulsion layer that contains silver bromide crystals. The radioactively labeled molecules in the cell (or tissues) interact with the silver bromide crystals in the emulsion, causing black grains to appear when the slide is photographically developed. It takes microscopic examination to see the location of the black grains in the emulsion and thereby tell which part of the cell took up the radioactively labeled molecules (fig. 4.C).

*These molecules are nucleosides, that is, they lack the phosphate component but do contain the base and sugar of a nucleotide.

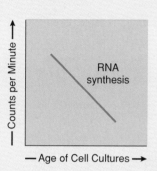

**Figure 4.B**
A scintillation chamber determines the relative amount of radioactivity in a number of samples, and a graph can be used to display the results.

scintillation chamber

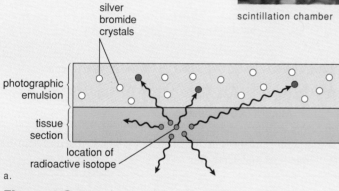

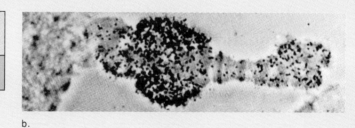

**Figure 4.C**
**a.** Radioactive isotopes in uridine molecules within a tissue interact with silver bromide crystals in a photographic emulsion. **b.** When the emulsion is developed, black grains show up that indicate the location of RNA synthesis.

---

RNA is single-stranded (fig. 4.19), but DNA is double-stranded. The 2 strands of DNA twist about one another in the form of a double helix (fig. 4.20). The 2 strands are held together by hydrogen bonds between purine and pyrimidine bases. Thymine (T) in one strand is always paired with adenine (A) in the opposite strand, and guanine (G) is always paired with cytosine (C). This is called **complementary base pairing.**

If the DNA helix is unwound, it resembles a ladder (fig. 4.20c). The sides of the ladder are made entirely of phosphate and sugar molecules, and the rungs of the ladder are made only of the complementary paired bases. The bases can be in any order within a strand, but between strands, A is always paired with T and G is always paired with C, and vice versa. Therefore,

## Figure 4.20

Overview of DNA structure. **a.** Double helix. **b.** Double helix showing complementary base pairing between bases. **c.** When the helix unwinds, the ladder structure of DNA is more evident. Sugar- phosphate molecules make up the sides of the ladder, and the bases, held together by hydrogen bonding, make up the rungs. One nucleotide has been boxed in the diagram.

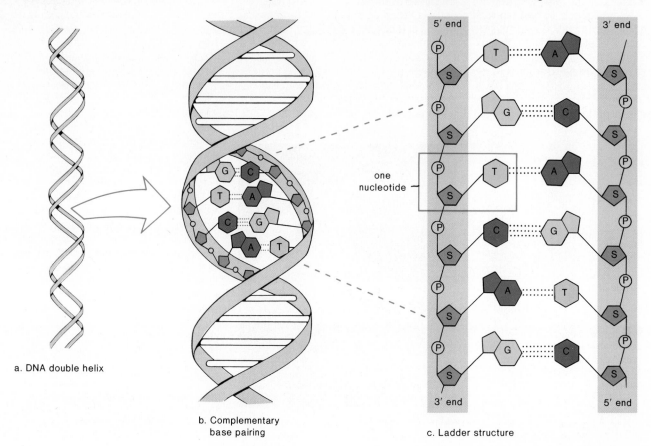

a. DNA double helix

b. Complementary base pairing

c. Ladder structure

regardless of the order or the quantity of any particular base pair, the number of purine bases always equals the number of pyrimidine bases.

DNA has a structure like a twisted ladder: sugar molecules and phosphate molecules make up the sides of the ladder, and hydrogen-bonded bases make up the rungs of the ladder. RNA differs from DNA in several respects.

The structure and the function of nucleic acids are discussed further in chapters 15 and 16 of this text.

## Summary

1. The chemistry of carbon accounts for the diversity of organic molecules found in living things. Carbon can bond with as many as 4 other atoms and can also bond with itself to form both chains and rings. The former are called the backbone of the molecule.

2. Each type of small molecule in living things—simple sugars, fatty acids, amino acids, and nucleotides—has a carbon backbone and is characterized by the presence of functional groups.

Some functional groups are hydrophobic, and some are hydrophilic.

3. Differences in the backbone and attached functional groups cause molecules to have different shapes, and shape helps determine how a molecule interacts with other molecules in cells.

4. The large organic molecules in cells are macromolecules, polymers formed by the joining together of monomers. Enzymes carry out both condensation

(building up) and hydrolysis (breaking down) of macromolecules. For each bond formed, a molecule of water is removed, and for each bond broken, a molecule of water is added.

5. Monosaccharides, disaccharides, and polysaccharides are all carbohydrates. Therefore, the term *carbohydrate* includes both the monomers (e.g., glucose) and the polymers (e.g., starch, glycogen, and cellulose). Starch and glycogen are energy-storage

The Cell

compounds, but cellulose has a structural function in plants.

6. Lipids include a wide variety of compounds that are insoluble in water. Fats and oils allow long-term energy storage and contain one glycerol and 3 fatty acids. Fats tend to contain saturated fatty acids, and oils tend to contain unsaturated fatty acids. In a phospholipid, one of the fatty acids is replaced by a phosphate group. In the presence of water phospholipids form a double layer because the head of the molecule is polarized and the tails are not. Waxes and steroids are also lipids.

7. Proteins are polymers of amino acids. Some proteins are enzymes, and proteins also have structural roles in cells and organisms.

8. A polypeptide is a long chain of amino acids joined by peptide bonds. There are 20 different amino acids in cells that differ only by their R groups. Polarity and nonpolarity are important aspects of the R groups.

9. A polypeptide has 3 levels of structure: the primary level is the sequence of the amino acids; the secondary level contains alpha (α) helices and beta (ß) sheets held in place by hydrogen bonding between peptide bonds; and the tertiary level is the final folding and twisting of the polypeptide that is held in place by bonding and hydrophobic interactions between R groups. Proteins that contain more than one polypeptide have a quaternary level of structure.

10. The shape of a polypeptide influences its biological activity. A polypeptide can be denatured and lose its normal shape and activity but, if reactivation conditions are appropriate, it assumes both of these again. This shows that the final shape of a polypeptide is dependent upon the primary structure.

11. Nucleic acids are polymers of nucleotides. Each nucleotide has 3 components: a sugar, a base, and phosphate (phosphoric acid). DNA, which contains the sugar deoxyribose, is the genetic material that stores information for its own replication and for the order in which amino acids are to be sequenced in proteins. DNA, with the help of RNA, controls protein synthesis.

12. Table 4.3 also summarizes our coverage of organic molecules in cells.

**Table 4.3**
Organic Molecules in Cells

| | Categories | Examples | Functions |
|---|---|---|---|
| **Carbohydrates** | Monosaccharides 6-carbon sugar | Glucose | Immediate energy source |
| | Disaccharides 12-carbon sugar | Sucrose | Transport sugar in plants |
| | Polysaccharides polymer of glucose | Starch, glycogen Cellulose | Energy storage Plant cell wall structure |
| **Lipids** | Triglycerides 1 glycerol + 3 fatty acids | Fats, oils | Long-term energy storage |
| | Waxes fatty acid + alcohol | Cuticle | Protective covering for parts of plants |
| | | Ear wax | Protective barrier |
| | Phospholipids like triglyceride except one fatty acid is replaced by a phosphate group | —— | Plasma membrane component |
| | Steroids backbone of 4 fused rings | Cholesterol Testosterone | Plasma membrane component Male sex hormone |
| **Proteins** | Polypeptides polymer of amino acids; 3 levels of structure— some proteins have 4 levels | Enzymes Myosin and actin Insulin Hemoglobin Collagen | Speed up cellular reactions Muscle cell components Regulates sugar content of blood Oxygen carrier in blood Fibrous support of body parts |
| **Nucleic Acids** | Nucleic acids polymer of nucleotides | DNA RNA | Genetic material Protein synthesis |
| | Nucleotides | ATP Coenzymes | Energy carrier Assist enzymes |

# Study Questions

1. How are the chemical characteristics of carbon reflected in the characteristics of organic molecules?

2. Give examples of functional groups, and discuss the importance of their being hydrophobic or hydrophilic.

3. What molecules are monomers of the polymers studied in this chapter? How are monomers joined to give polymers, and how are polymers broken down to monomers?

4. Name several monosaccharides, disaccharides, and polysaccharides, and give a function of each. How are these molecules structurally distinguishable?

5. Name the different types of lipids, and give a function for each type. What is the difference between a saturated and unsaturated fatty acid? Explain the structure of a fat molecule by stating its components and how they are joined together.

**6.** How does the structure of a phospholipid differ from that of a fat? How do phospholipids form a double layer in the presence of water?

**7.** Draw the structure of an amino acid and of a dipeptide, pointing out the peptide bond.

**8.** Discuss the 4 levels of structure of a protein and relate each level to particular bonding patterns.

**9.** How is the tertiary structure of a polypeptide related to its primary structure? Mention denaturation as evidence of this relationship.

**10.** How are nucleotides joined to form nucleic acids? Name several differences between the structure of DNA and the structure of RNA.

---

## Objective Questions

**1.** Which of these is not a characteristic of carbon?
  **a.** forms 4 covalent bonds
  **b.** bonds with itself
  **c.** is sometimes ionic
  **d.** forms long chains

**2.** The functional group COOH is
  **a.** acidic.
  **b.** basic.
  **c.** never ionized.
  **d.** All of these.

**3.** A hydrophilic group is
  **a.** attracted to water.
  **b.** a polar or ionized group.
  **c.** found in fatty acids.
  **d.** All of these.

**4.** Which of these is an example of hydrolysis?
  **a.** amino acid + amino acid → dipeptide + $H_2O$
  **b.** dipeptide + $H_2O$ → amino acid + amino acid
  **c.** Both of these.
  **d.** Neither of these.

**5.** Which of these makes cellulose nondigestible?
  **a.** a polymer of glucose subunits
  **b.** a fibrous protein
  **c.** the linkage between the glucose molecules
  **d.** the peptide linkage between the amino acid molecules

**6.** A fatty acid is unsaturated if it
  **a.** contains hydrogen.
  **b.** contains double bonds.
  **c.** contains an acidic group.
  **d.** bonds to glycogen.

**7.** Which of these is not a lipid?
  **a.** steroid
  **b.** fat
  **c.** polysaccharide
  **d.** wax

**8.** The difference between one amino acid and another is found in the
  **a.** amino group.
  **b.** carboxyl group.
  **c.** R group.
  **d.** peptide bond.

**9.** The shape of a polypeptide is
  **a.** maintained by bonding between parts of the polypeptide.
  **b.** important to its function.
  **c.** ultimately dependent upon the primary structure.
  **d.** All of these.

**10.** Which of these is the peptide bond?

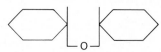

  a.

  $\begin{matrix} & O \\ & || \\ -\!\!&C-O- \end{matrix}$    $\begin{matrix} & O & H \\ & || & | \\ -\!\!&C-N- \end{matrix}$
  b.                    c.

**11.** Nucleotides
  **a.** contain a sugar, a nitrogen-containing base, and a phosphate molecule.
  **b.** are the monomers for fats and polysaccharides.
  **c.** join together by covalent bonding between the bases.
  **d.** All of these.

**12.** DNA
  **a.** has a sugar-phosphate backbone.
  **b.** is single-stranded.
  **c.** has a certain sequence of amino acids.
  **d.** All of these.

**13.** Label the following diagram using the terms monomer, hydrolysis, condensation, and polymer, and explain the diagram:

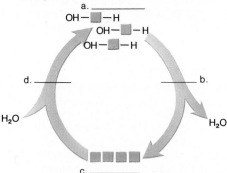

**14.** Label the levels of protein structure in this diagram of hemoglobin:

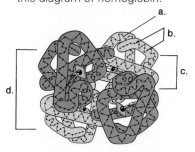

**15.** Complete the other strand of a DNA molecule.

| T | C | A | A | G | C | C | G | T | A | C | G |

## Concepts and Critical Thinking

1. *The unity and the diversity of life begin at the molecular level of organization.*

   How do the organisms in figure 4.1 demonstrate this biological concept?

2. *The atoms and bonds within a molecule determine its chemical and physical properties.*

   Compare fats that largely contain either saturated or unsaturated fatty acid to demonstrate this concept.

3. *The properties of a molecule determine the role that the molecule plays in the cells or body of an organism.*

   Choose either a phospholipid or the protein keratin to demonstrate this biological concept.

## Selected Key Terms

**organic** (or-gan′ik) 42
**hydrophilic** (hi″dro-fil′ik) 43
**hydrophobic** (hi″dro-fo′bik) 43
**isomer** (i′so-mer) 43
**polymer** (päl′e-mer) 43
**condensation** (kän-den-sa′shen) 44
**hydrolysis** (hi-drol′i-sis) 44

**carbohydrate** (kar″bo-hi′drāt) 44
**lipid** (lip′id) 46
**phospholipid** (fos″fo-lip′id) 48
**steroid** (ste′roid) 48
**amino acid** (ah-me′no as′id) 50
**protein** (pro′te-in) 50

**enzyme** (en′zīm) 50
**peptide** (pep′tīd) 50
**nucleotide** (nu′kle-o-tīd) 54
**nucleic acid** (nu-kle′ik as′id) 54
**DNA** 54
**RNA** 54

# 5

# Cell Structure and Function

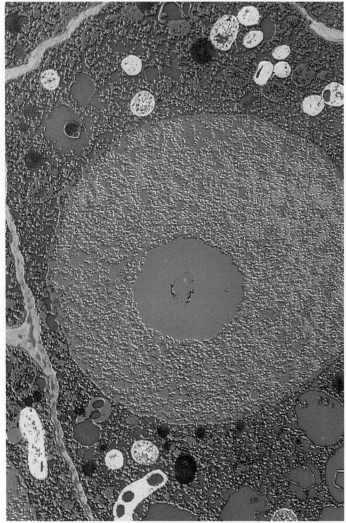

A cell in the root tip of a corn plant. The nucleus dominates the cell, signifying its importance in the life of the cell. The red granular material contains DNA, and the concentration of red contains RNA. These 2 nucleic acids are involved in protein synthesis within the cytoplasm, which lies outside the nucleus. The plant cell has many vacuoles; some appear yellow and others appear blue. The white outline of the cell is the cell wall, which maintains the shape of the cell. Magnification, X1,765.

Your study of this chapter will be complete when you can

1. state 2 tenets of the cell theory;
2. list several similarities and differences between prokaryotic cells and eukaryotic cells;
3. give several differences between bright-field light microscopy and transmission electron microscopy. Name several other types of microscopes that are available today;
4. explain, on the basis of cell volume to cell-surface relationships, why cells are so very small;
5. describe the structure of a prokaryotic cell, and give a function for each part mentioned;
6. describe the structure of the nucleus of a eukaryotic cell, and give a function for each part mentioned;
7. name the structures that form the endomembrane system, and tell how they are related to one another;
8. explain the relationship between chloroplasts and mitochondria; and describe the structure and function of each;
9. list 4 evidences for the endosymbiotic theory, which explains the origin of mitochondria and chloroplasts;
10. name the components of the cytoskeleton; describe the structure and the functions of each component;
11. contrast the structures of prokaryotic cells, eukaryotic animal cells, and eukaryotic plant cells.

**A**ll the organisms we see about us are made up of cells (fig. 5.1). The atoms and the molecules we studied previously were not alive, but the cell is. A **cell** is the smallest unit of living matter.

> All organisms are made up of cells, and a cell is the smallest unit of living matter.

A few cells, like a hen's egg or a frog's egg, are large enough to be seen by the naked eye, but most are not. This is the reason the study of cells did not begin until the invention of the first microscopes in the seventeenth century. It is not known exactly who was the first to see cells, but Anton van Leeuwenhoek of Holland is famous for observing tiny, one-celled living things that no one had seen before. Leeuwenhoek sent his findings to an organization of scientists called the Royal Society in London. Robert Hooke, an Englishman, confirmed Leeuwenhoek's observations and was the first to use the term *cell*. The tiny chambers he observed in the honeycomb structure of cork reminded him of the rooms, or cells, in a monastery. Naturally, then, he referred to the boundaries of these chambers as walls.

Although these early microscopists had seen cells, it was more than a hundred years—in the 1830s—before Matthias Schleiden published his theory that all plants are composed of cells and Theodor Schwann published a similar proposal concerning animals. These Germans based their ideas not only on their own work but on the work of all who had studied tissues under microscopes. Soon after, another German scientist, Rudolf Virchow, used a microscope to study the life of cells. He came to the conclusion that "every cell comes from a preexisting cell." By the middle of the nineteenth century, biologists clearly recognized that all organisms are composed of self-reproducing elements called cells.

> Cells are capable of self-reproduction. Cells come only from preexisting cells.

The previous 2 highlighted statements are often called the **cell theory,** but sometimes they are called the *cell doctrine*. Those who use the latter terminology want to make it perfectly clear that there are extensive data to support the cell theory and that it is universally accepted by biologists.

## Cell Variety

Since the early days of cell study, it has become apparent that a cell carries out the activities we associate with living things, such as those needed for *growth* and *reproduction*. Further, a cell's organized parts have specific functions that permit it to carry on these activities. Improved microscopy has vastly extended our ability to view the internal organization of a cell, and today's standard biochemical techniques allow us to determine the function of a cell's parts.

**Figure 5.1**
All organisms, whether plants or animals, are composed of cells. This is not readily apparent because a microscope is usually needed to see the cells. *a.* Corn plant. *b.* Light micrograph of leaf showing many individual cells. *c.* Rabbit. *d.* Light micrograph of the intestinal lining showing that it too is composed of cells. The dark-staining bodies are nuclei.

a.

b.

c.

d.

# Cell Biology: Technology and Techniques

**T**he study of cells is dependent upon the use of microscopes to detect cellular components and biochemical techniques to decipher their functions.

## Microscopes

In the *bright-field light microscope*, light rays passing through a specimen are brought to a focus by a set of glass lenses, and the resulting image is then viewed by the human eye. In the *transmission electron microscope*, electrons passing through a specimen are brought to a focus by a set of magnetic lenses, and the resulting image is projected onto a fluorescent screen or photographic film. We will use these statements to examine the differences between the 2 types of microscopes.

### The Means of Illumination

Almost everyone knows that the magnifying capability of an electron microscope is greater than that of a light microscope. A light microscope can magnify objects a few thousand times, but an electron microscope can magnify them hundreds of thousands of times. The difference lies in the means of illumination. The paths of light rays and electrons moving through space are wavelike, but the wavelength of electrons is much shorter than the wavelength of light. This difference in wavelength not only accounts for the electron microscope's greater magnifying capability, it also accounts for its greater resolving power. The greater the resolving power, the greater the detail eventually seen. *Resolution* is the *minimum* distance between 2 objects before they are seen as one larger object. If oil is placed between the sample and the objective lens of the light microscope, the resolving power is increased, and if ultraviolet light is used instead of visible light, it is also increased. But typically, a light microscope can resolve down to 0.2 μm, while the electron microscope can resolve down to 0.0001 μm. If the resolving power of the average human eye is set at one, then

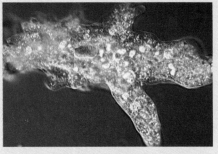
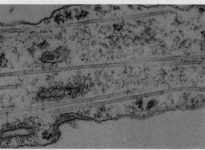
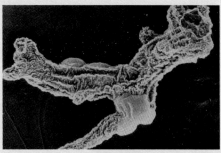

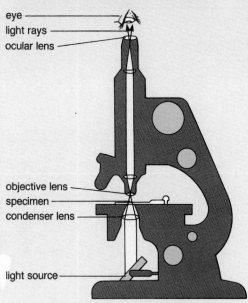

eye
light rays
ocular lens

objective lens
specimen
condenser lens

light source

a. Compound light microscope

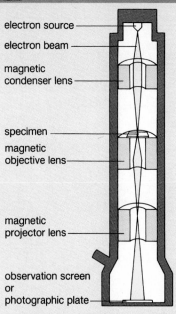

electron source
electron beam

magnetic
condenser lens

specimen
magnetic
objective lens

magnetic
projector lens

observation screen
or
photographic plate

b. Transmission electron microscope

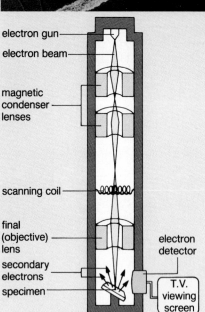

electron gun
electron beam

magnetic
condenser
lenses

scanning coil

final
(objective)
lens

secondary
electrons

specimen

electron
detector

T.V.
viewing
screen

c. Scanning electron microscope

## Figure 5.A

*a.* Light micrograph of *Amoeba proteus*, a single-celled organism. A light microscope uses light to view a specimen. The specimen can be living if it is thin enough to allow light to pass through. A light microscope does not magnify or distinguish as much detail as the electron microscope. *b.* Transmission electron micrograph (TEM) of a portion of a pseudopodium, an extension by which *Amoeba proteus* moves. Cytoskeleton elements are visible. The transmission electron microscope uses electrons to "view" the specimen. The specimen is nonliving and must be thin enough to allow electrons to pass through. The magnification and the amount of detail seen are far greater than with the light microscope. *c.* Scanning electron micrograph (SEM) of *Amoeba proteus*. The scanning microscope scans the surface of the specimen with an electron beam. The secondary electrons given off are collected and the result is a 3-dimensional image of the specimen. Magnification, ✕225.

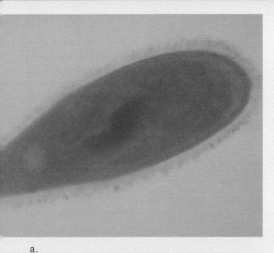

a.

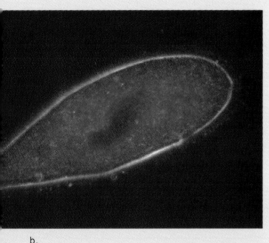

b.

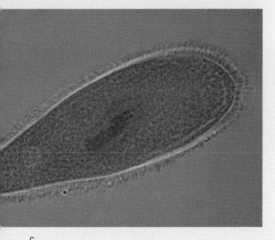

c.

## Figure 5.B
Various light microscopes can be used to view a specimen under different conditions. Magnification, X240. *a.* Bright-field micrograph of a paramecium. *b.* Dark-field micrograph of a paramecium. *c.* Phase-contrast micrograph of a paramecium.

that of the typical light microscope is about 500 and that of the electron microscope is 100,000 (fig. 5.A).

### The Lenses and the Means of Viewing the Object
Light rays can be bent (refracted) and brought to a focus as they pass through glass lenses, but electrons do not pass through glass. Electrons have a charge that allows them to be brought to a focus by magnetic lenses. The human eye utilizes light to see an object but cannot utilize electrons for the same purpose. Therefore, electrons leaving the specimen in the electron microscope are directed toward a screen or a photographic plate that is sensitive to their presence. Humans can then view the image on the screen or photograph.

### The Specimen
Organisms that are quite small and thin can be viewed alive with a light microscope but not with an electron microscope. Electrons randomly scatter as they pass through air; therefore, it is necessary to create a vacuum inside an electron microscope. Even with the light microscope, specimens are usually prepared for observation. Cells are killed, fixed so that they do not decompose, and embedded into a matrix. The matrix strengthens the specimen so that it can be thinly sliced. These sections are stained with colored dyes (light microscopy) or with electron-dense metals (electron microscopy).

This lengthy treatment is necessary for a couple of reasons: (1) light rays and electrons pass through thin materials only, and (2) the dyes create differences in density (e.g., light and dark) that help distinguish detail. Some colored dyes bind preferentially to certain structures, and this aids in their identification. The specimen appears colored because white light is composed of many colors of light and the pigments of the dye absorb certain colors and not others. For example, a green object absorbs all the colors except green light, which then remains for us to see.

One of the primary disadvantages of viewing dead, treated cells is that you can never be sure that the living cell is similar or that an artifact (structure not present when the cell is alive) has not been introduced.

### Other Microscopes
Notice that we made specific reference to the bright-field light microscope and to the transmission electron microscope. This is

obviously because there are other types of light microscopes and electron microscopes, each with their own particular advantage (fig. 5.B).

As mentioned, the light microscope has an advantage over the electron microscope because the untreated specimen can be viewed alive; however, special microscopes other than bright-field are needed to allow visualization of detail. *Phase-contrast* and *interference-contrast microscopy* capitalize on the fact that when light passes through a living cell, the phase of each light wave is altered according to the relative density of the cellular structure:

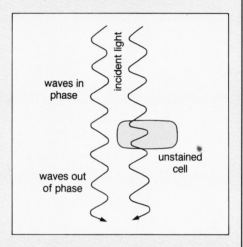

*Dark-field illumination*, as the name suggests, illuminates the object against a dark background. The light has to be directed from the side; only scattered light enters the microscope lens, and as a result, the cell appears vividly lit on a black field. A similar arrangement uses ultraviolet light to detect fluorescence in a specimen.

*The scanning electron microscope* (SEM) permits the development of 3-dimensional images. A narrow beam of electrons is scanned over the surface of the specimen, which is coated with a thin metal layer. The metal gives off secondary electrons that are collected by a detector to produce an image on a television screen. The 1986 Nobel Prize in physics was won by Gerd Binnig and Heinrich Rohrer for the invention of the *scanning tunneling microscope* (STM). This microscope uses a tiny probe that is brought to within a nanometer of the specimen's surface. A voltage applied between the probe's tip and the specimen creates a flow of electrons that tunnels through the nanometer gap between them. As the probe scans across the

***Continued on next page***

sample, the tip is moved up and down by the experimenter to keep the tunneling current constant. A computer keeps track of the up-and-down movement of the probe and produces a 3-dimensional image.

The STM does not work particularly well for biological molecules and cells. In the past few years, however, researchers have developed other types of scanning probe microscopes that can be used to view cells successfully. For example, the *atomic force microscope* (AFM) places the probe's tip right up against the surface of the sample. The amount of force required to keep the probe's tip in this position as it moves up and down also allows a computer to produce a 3-dimensional image. The microscope works best with individual biomolecules; to obtain good pictures of cells, investigators plan to build a low-temperature atomic force microscope for samples that will be firmed up by freezing them in liquid nitrogen. Other microscopes derived from the STM are in development. They include one designed to monitor ion flow into and out of living cells and the near-field optical microscope, which should allow experimenters to view viruses and chromosomes in vivo.

## Cell Fractionation

*Cell fractionation* is a means to separate cell components so that their individual biochemical composition and function can be

determined (fig. 5.C). First, it is necessary to break open a large number of similar type cells. The cells are usually placed in a *homogenizer,* a tube that contains a close-fitting pestle; when the pestle is rotated, the cells are broken. Second, the "freed" contents of the cells are subjected to a spinning action known as *centrifugation.* At a low

speed, large particles, like cell nuclei, settle out and are found in the sediment. Smaller particles are still in the fluid (supernatant), which can be poured into a fresh tube and subjected to centrifugation at a higher speed until the smallest particles have been separated out. The various cell fractions can then be biochemically analyzed.

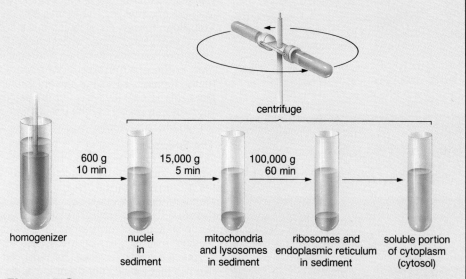

centrifuge

homogenizer | 600 g 10 min → | nuclei in sediment | 15,000 g 5 min → | mitochondria and lysosomes in sediment | 100,000 g 60 min → | ribosomes and endoplasmic reticulum in sediment | soluble portion of cytoplasm (cytosol)

**Figure 5.C**

Cell fractionation. Cells are broken open mechanically by the action of the pestle against the sides of the tube. Contents are centrifuged at ever-increasing speed to separate out first larger and then smaller components of the cell. (g = gravity)

---

Although there are different types of cells, they all have a barrier called the **plasma membrane,** which separates the contents of the cell from the environment outside the cell. The structure and the function of the plasma membrane are discussed at greater length in chapter 6. For now, we will mention that the plasma membrane regulates the passage of molecules into and out of the cell. The cell takes in nutrient molecules, which are used as building blocks and as an energy source so necessary to growth and repair. The cell also gives off waste molecules, which are end products of metabolism.

The plasma membrane is thin and weak, and alone it cannot give much strength to a cell. Some cells, such as those of plants and microorganisms (organisms that cannot be seen without a microscope), are strengthened by the addition of a **cell wall,** which protects the plasma membrane.

---

All cells are surrounded by a plasma membrane that separates them from their environment. Some cells also have a cell wall.

---

Cells are divided into 2 major groups, chiefly on the basis of structure. **Eukaryotic cells** have a true nucleus (*eu*—true; *karyon*—nucleus), a membrane-bounded compartment that houses

DNA within chromosomes, which are complex threadlike structures. The rest of the cell is also divided by membranes into compartments that perform various functions. Some of these are **organelles,** small membranous bodies whose structure suits their function.

The other major type of cell is called a **prokaryotic cell** because the cell lacks a true nucleus (*pro*—before; *karyon*—nucleus). The DNA is in a single chromosome located within a distinguishable region called the **nucleoid.** An internal membrane does not divide the cell into compartments, and there are few organelles.

The term **cytoplasm** refers to the cell contents found between the nucleus (or nucleoid) and the plasma membrane. The various structures within the cytoplasm are bathed by a semifluid medium known as **cytosol.** The cytosol in eukaryotic cells is now known to have an organized lattice of protein filaments called the cytoskeleton. The cytoskeleton is discussed on page 76.

---

There are 2 major types of cells: eukaryotic cells, which have a membrane-bounded nucleus, and prokaryotic cells, which do not have such a nucleus.

---

## Figure 5.2

Biological measurements. **a.** It takes a microscope to see cells because they are so very small. You can see cells with a light microscope, but not with much detail. It takes an electron microscope to make out most of the organelles in the cells. Note that a prokaryote (bacterium) is smaller than the nucleus of a human liver cell. **b.** The units of measurement common to cell biology and their equivalents.

## Cell Sizes

Cells are quite small. Most prokaryotic cells are from 1 μm–10 μm in diameter. Eukaryotic cells are much larger—10 μm–100 μm—but still too small to be seen by the naked eye (fig. 5.2). In fact, you could place as many as 10,000 prokaryotic cells and as many as 1,000 eukaryotic cells on the length of this 1 cm line —.

An explanation for why cells are so small and why the body of a plant or an animal is multicellular is found in a consideration of surface area-to-volume relationships. Suppose a cell is 1 mm in each dimension (height, width, depth); its surface area will be

$$6 (1 \times 1) = 6 \text{ mm}^2$$

The volume of the cell will be

$$1 \times 1 \times 1 = 1 \text{ mm}^3$$

If the linear dimensions of the cell are doubled, its surface area will be 24 mm² and its volume will be 8 mm³. Notice that as a result of the cell's doubling in size, the volume increased 8 times, but the surface area increased only 4 times.

Nutrients enter a cell and wastes exit a cell at the plasma membrane. A large cell requires more nutrients and produces more wastes than a small cell. Therefore, a large cell that is actively metabolizing cannot make do with proportionately less surface area than a small cell. Yet as previously mentioned, as a cell increases in size, surface area decreases dramatically in proportion to volume:

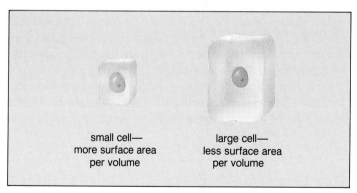

small cell—
more surface area
per volume

large cell—
less surface area
per volume

We would expect, then, that there would be a limit to how large an actively metabolizing cell can become. For example, consider that egg cells are among the largest known cells. A chick's egg is several centimeters in diameter, but the egg is not actively metabolizing and contains a large amount of storage material called yolk.

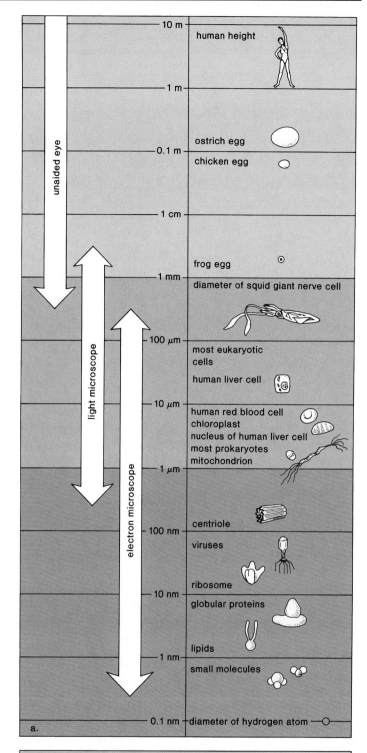

a.

| Measurements Used in Cell Biology | | | |
|---|---|---|---|
| Unit | Symbol | Scale | Seen by |
| centimeter | cm | 0.01 m | unaided eye |
| millimeter | mm | 0.1 cm | unaided eye |
| micrometer | μm | 0.001 mm | light microscope |
| nanometer | nm | 0.001 μm | electron microscope |

b.

**Figure 5.3**

Prokaryotic (i.e. bacterial) anatomy. A single chromosome is located within a region called the nucleoid. Ribosomes are the only cytoplasmic organelles. Prokaryotes usually have a cell wall and may also have a protective capsule. Magnification, X46,000.

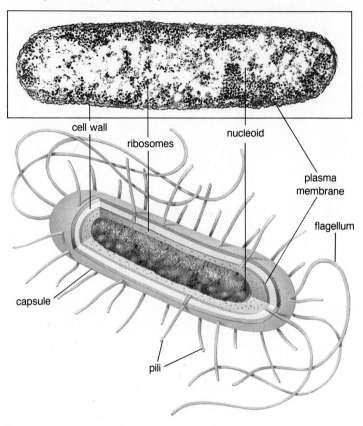

Once the egg is fertilized and metabolic activity begins, the egg divides repeatedly. Cell division provides the surface area needed for adequate exchange. Furthermore, cells that specialize in absorption have modifications to increase greatly the surface area per volume of the cell. The columnar cells along the surface of the intestinal wall have villi (villus, sing.) to increase their surface area. Modifications are necessary to increase surface area per volume.

> A cell needs a surface area (plasma membrane) that can adequately exchange materials with the environment. Surface area to volume relationships require that cells stay small.

There are other factors limiting the size of cells. The nucleus is the control center of the cell, and it can control only a certain amount of cytoplasm. Also, not only must there be an exchange of materials between the cell and the external environment, but materials must be made available to various parts of the cytoplasm. A eukaryotic cell is 10 times larger than a prokaryotic cell, and we should expect to see modifications for the distribution of molecules inside the cell. The compartmentalization of eukaryotic cells enhances their ability to make molecules available where they are needed.

## Prokaryotic and Eukaryotic Cell Differences

**Bacteria** are the only organisms that are prokaryotic cells. The bacteria are a very diverse group, and some members are photosynthetic. Among these are the **cyanobacteria,** which were formerly called blue-green algae.

Figure 5.3 illustrates the main features of prokaryotic anatomy. There is an exterior *cell wall* containing peptidoglycan, a large complex molecule consisting of polysaccharide polymers cross-linked by short chains of amino acids. In some bacteria, the cell wall is further surrounded by a gelatinous sheath, or *capsule*. Motile bacteria usually have long, very thin appendages called *flagella* that are composed of subunits of the protein called flagellin. The flagella, which rotate like propellers, rapidly move the bacterium in a fluid medium. Bacteria also have *pili*, which are short appendages that help bacteria attach themselves to an appropriate surface.

In prokaryotes, most genes are found within a single chromosome (loop of DNA) located within the nucleoid, but they may also have small accessory rings of DNA called *plasmids*. The cytoplasm lacks membranous organelles, but the plasma membrane may have folds and convolutions that extend into the cell. In addition, the photosynthetic cyanobacteria have light-sensitive pigments, usually within the membranes of flattened disks called *thylakoids*. Prokaryotic cytoplasm contains numerous granules known as *ribosomes*, which function in protein synthesis. These are smaller than the ribosomes of eukaryotic cells, and they are not attached to membranes.

| Bacteria are prokaryotic cells with these constant features. | |
| --- | --- |
| Outer boundary: | cell wall |
| | plasma membrane |
| Cytoplasm: | ribosomes |
| | thylakoids (cyanobacteria) |
| Nucleoid: | chromosome (DNA only) |

All cells aside from bacteria are eukaryotic. Eukaryotic organisms include algae, protozoa, and fungi, in addition to plants and animals. In this chapter, we will concentrate on the structure and the function of a typical animal cell (fig. 5.4) and a typical plant cell (fig. 5.5). Such drawings are composites based on data gathered from microscopic studies. In actuality, there is a great deal of variation among plant cells and animal cells.

The various organelles of eukaryotic cells have a structure that suits their function(s) (table 5.1). The cytoplasm is compartmentalized by these organelles and by other membranous structures. *Compartmentalization* keeps the cell organized and its various functions separate one from the other.

## Eukaryotic Cell Organelles

The **nucleus** is a prominent structure in the eukaryotic cell. With staining, it is usually large enough to be observed through the light microscope, although electron micrographs show much more detail. The nucleus is of primary importance because it is the

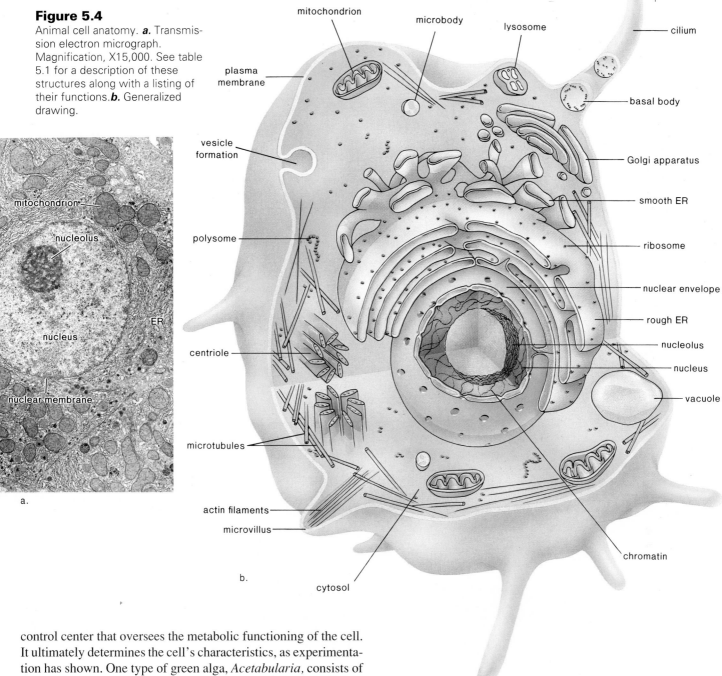

**Figure 5.4**
Animal cell anatomy. **_a._** Transmission electron micrograph. Magnification, X15,000. See table 5.1 for a description of these structures along with a listing of their functions.**_b._** Generalized drawing.

mitochondrion

microbody

lysosome

cilium

plasma membrane

basal body

vesicle formation

Golgi apparatus

smooth ER

polysome

ribosome

nuclear envelope

centriole

rough ER

nucleolus

nucleus

vacuole

microtubules

actin filaments

microvillus

chromatin

cytosol

mitochondrion

nucleolus

nucleus

ER

nuclear membrane

a.

b.

control center that oversees the metabolic functioning of the cell. It ultimately determines the cell's characteristics, as experimentation has shown. One type of green alga, _Acetabularia,_ consists of a single cell with a base, a stalk, and a cap (fig. 5.6). If the stalk and the cap are removed from the base (fig. 5.6_b_), they die, but the nucleus-containing base develops into a new organism. The importance of the nucleus is further exemplified when the base is combined with the stalk of a different species. The regenerated cap resembles the species cap of the nucleus, not the species cap of the stalk cytoplasm. Similarly, the nucleus of a frog's egg can be removed and the nucleus of another species can be introduced. The tadpole and eventually the frog that develops is like that of the introduced nucleus.

The nucleus is the control center of the cell.

It is clear that the nucleus is the control center of the cell because it contains the hereditary material DNA (deoxyribonucleic acid). As mentioned in the previous chapter, DNA carries the instructions for the proper sequencing of amino acids in proteins. The proteins of a cell determine its structure and the functions it can perform. The nucleus directs what proteins will be present and in that way determines the structure and coordinates the activities of the cell.

When you look at the nucleus, even in an electron micrograph, you cannot see DNA molecules but you can see chromatin.

## Figure 5.5

Plant cell anatomy. **a.** Generalized drawing. **b.** Transmission electron micrograph. Magnification, X20,000. See table 5.1 for a description of the structures along with a listing of their functions.

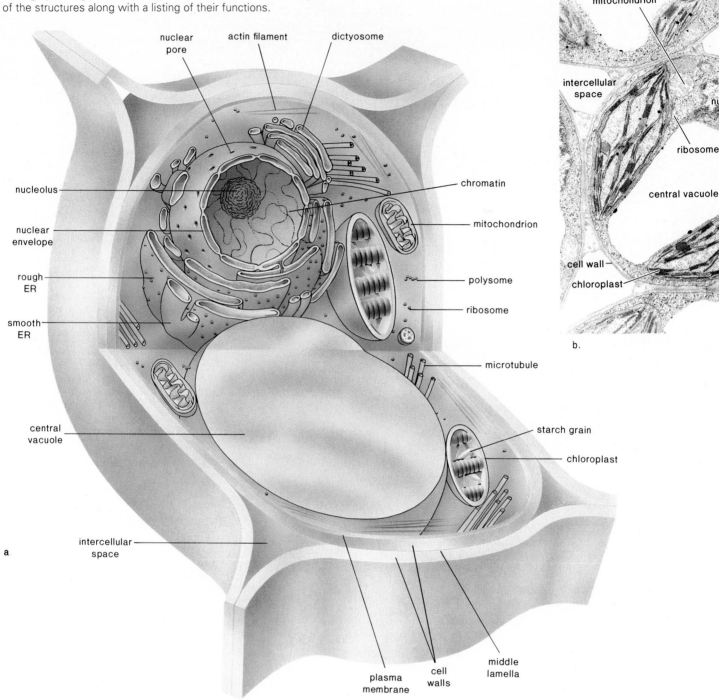

a

**nuclear pore**    **actin filament**    **dictyosome**

**nucleolus**

**nuclear envelope**

**rough ER**

**smooth ER**

**central vacuole**

**intercellular space**

**chromatin**

**mitochondrion**

**polysome**

**ribosome**

**microtubule**

**starch grain**

**chloroplast**

**plasma membrane**    **cell walls**    **middle lamella**

b.

**mitochondrion**

**intercellular space**

**nucleus**

**ribosome**

**central vacuole**

**cell wall**

**chloroplast**

---

**Chromatin** looks grainy, but actually it is a threadlike material that undergoes coiling into rodlike structures called **chromosomes** just before the cell divides. Chemical analysis shows that chromatin, and therefore chromosomes, contains DNA, much protein, and some RNA. Chromatin is immersed in a semifluid matrix called the *nucleoplasm*. Nucleoplasm has a different pH from the cytosol, and this suggests that it has a different composition.

Most likely, too, when you look at an electron micrograph of a nucleus, you will see one or 2 regions that look darker than the rest of the chromatin. These are **nucleoli**, where another type of RNA, called ribosomal RNA (rRNA), is produced. (Ribosomes are small bodies in the cytoplasm that contain ribosomal RNA and proteins.)

**Table 5.1**

Eukaryotic Structures in Animal Cells and Plant Cells

| Name | Composition | Function |
|------|-------------|----------|
| Cell wall* | Contains cellulose fibrils | Support and protection |
| Plasma membrane | Bilayer of phospholipid with embedded proteins | Selective passage of molecules into and out of cell |
| Nucleus | Nuclear envelope surrounding the nucleoplasm, chromatin, and nucleoli | Control of growth and reproduction |
| Nucleolus | Concentrated area of chromatin, RNA, and proteins | Ribosomal formation |
| Ribosomes (polysomes) | Protein and RNA in 2 subunits | Protein synthesis |
| Endoplasmic reticulum | Membranous flattened channels and tubular canals | Synthesis and/or modification of proteins and other substances, and transport by vesicle formation |
| Rough ER | Studded with ribosomes | Protein synthesis |
| Smooth ER | Having no ribosomes | Various; lipid synthesis in some cells |
| Golgi apparatus[d] (dictyosome)* | Stack of membranous saccules | Processing and packaging of molecules, (e.g., glycoproteins) |
| Vacuoles* and vesicles | Membranous sacs | Storage of substances |
| Lysosomes[d] | Membranous vesicles containing digestive enzymes | Intracellular digestion |
| Microbodies | Membranous vesicles containing specific enzymes | Various metabolic tasks |
| Mitochondria | Inner membrane (cristae) within outer membrane | Cellular respiration |
| Chloroplasts | Inner membrane (grana) within 2 outer membranes | Photosynthesis |
| Cytoskeleton | Microtubules, and actinfilaments | Shape of cell and movement of its parts |
| Cilia and flagella | 9 + 2 pattern of microtubules | Movement of cell |
| Centrioles[d] | 9 + 0 pattern of microtubules | Form basal bodies, which give rise to microtubules |

*Plant cells
[d]Animal cells

**Figure 5.6**

The *Acetabularia* are single-celled algae up to 5 cm long which lend themselves to regeneration experiments. *a.* Photo of *Acetabularia mediterranea* commonly called mermaid's wineglass. *b.* Experiments to demonstrate that the nucleus is the control center of the cell. (1) In this experiment, the stalk and the cap which do not have a nucleus, die, but the base, with a nucleus, regenerates. (2) In this experiment, the regenerated cap resembles that of the species of the nucleus, not that of the species of the cytoplasm.

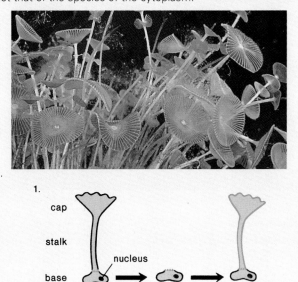

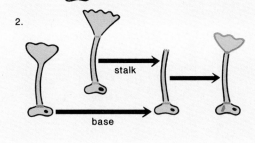

The nucleus is separated from the cytoplasm by a double membrane known as the **nuclear envelope** (fig. 5.7). You might think that the presence of this envelope would prevent materials from passing to and from the nucleus, but electron micrographs show nuclear *pores* in the nuclear envelope. These pores are of sufficient size (100 nm) to permit the passage of ribosomal subunits. High-power electron micrographs show that the pores have a complex structure (fig. 5.7c). Therefore, they should not be regarded as simple openings. They have associated proteins that can regulate the passage of materials to and from the nucleus.

The structural features of the nucleus include

Chromatin (chromosomes): DNA and proteins
Nucleolus: chromatin and ribosomal subunits
Nuclear envelope: double membrane with pores

## Figure 5.7

Anatomy of the nucleus. *a.* The fluid portion of the nucleus is the nucleoplasm, where chromatin is found. The nucleolus is a special region of chromatin where rRNA is produced and joined with proteins from the cytoplasm to form the subunits of ribosomes. The nucleus is bounded by a double membrane which contains pores. *b.* Freeze-fracture (chap. 6) preparation of the nuclear envelope of a rat pancreas cell showing the existence of nuclear pores. Magnification, X145,000. *c.* Each nuclear pore contains a pore complex consisting of 8 granules and possible projections that block the opening. These complexes are believed to regulate the passage of materials into and out of the nucleus. Two proposed versions of the pore complex are shown.

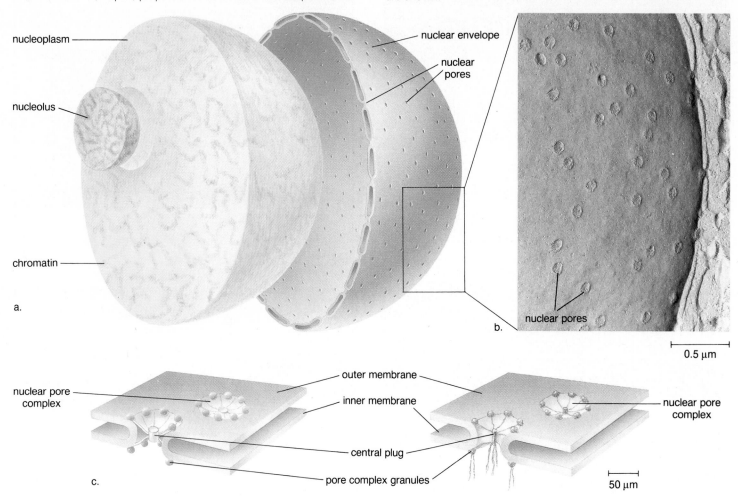

## Ribosomes

**Ribosomes** are small particles composed of RNA and proteins. Unlike many of the organelles discussed in this chapter, ribosomes are found in both prokaryotes and eukaryotes. The ribosomes found in eukaryotic cells are about one-third larger than those found in prokaryotic cells. Both types of ribosomes are composed of 2 subunits, a large subunit and a small subunit. Each subunit has its own mix of proteins and rRNA. Ribosomes perform a very important function because they help carry out protein synthesis.

Several ribosomes synthesizing the same protein are called a *polysome*. Polysomes can lie free within the cytosol, and they can be attached to the endoplasmic reticulum, a membranous system of saccules and channels in the cytoplasm. Cells that produce proteins for secretion, such as those that produce digestive enzymes for the small intestine, contain a few million ribosomes and an extensive endomembrane system.

Ribosomes are small organelles that are involved in protein synthesis. A polysome contains a group of ribosomes, each one involved in producing a copy of the same protein.

## Endomembrane System

The endomembrane system contains several membranous structures. Many types of enzymatic reactions occur all the time in cells, and the endomembrane system keeps each reaction restricted to a particular region. For example, the endomembrane system keeps the reactions that occur in the cytosol separate from those that occur in the endoplasmic reticulum and Golgi apparatus.

## Figure 5.8

Rough ER. **a.** Electron micrograph of a mouse hepatocyte (liver) cell shows a cross section of many flattened vesicles with ribosomes attached to the side that abuts the cytosol. Magnification, ✕75,000. **b.** The 3-dimensions of the organelle. **c.** Model of a single ribosome illustrates that each one is actually composed of 2 subunits. **d.** Method by which the endoplasmic reticulum acts as a transport system.

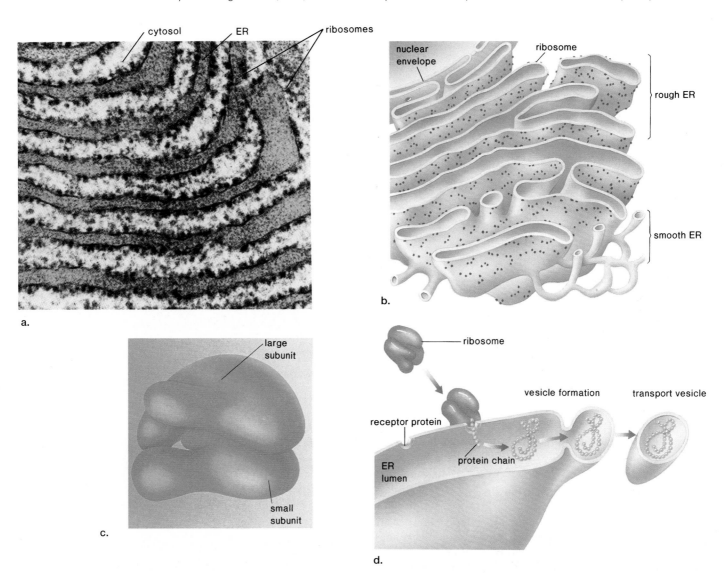

a.

b.

c.

d.

### Endoplasmic Reticulum

The **endoplasmic reticulum** (*endoplasmic*—inside the cytoplasm; *reticulum*—network) is continuous with the nuclear envelope. Its complicated system of membranous channels and saccules may be studded with ribosomes, in which case it is called **rough ER,** or it may lack ribosomes, in which case it is called **smooth ER** (fig. 5.8).

Rough ER is involved in protein synthesis, modification of newly formed proteins, production of new membrane, and transport of these proteins and membrane to other locations within the cell. Proteins produced by polysomes attached to rough ER start entering the ER lumen even while protein synthesis is occurring (fig. 5.8*d*). Once inside the lumen they can be modified; for

example, they can be shortened or have a sugar chain attached to them. These proteins can be incorporated into ER membrane or they can move along until they reach smooth ER.

Smooth ER can have various functions depending on the particular cell. Sometimes it specializes in the production of steroid hormones. For example, it is abundant in cells of the testes and the adrenal cortex, both of which produce steroid hormones. In the liver it helps detoxify drugs; in muscle cells it acts as a storage site for calcium ions that are released when contraction occurs.

Regardless of any specialized function, smooth ER also forms *transport vesicles* (little vacuoles) in which large molecules are moved within the cell (fig. 5.8*d*). When transport vesicles fuse with another part of the endomembrane system, any contents of the vesicle are discharged and the receiving membrane is enlarged in this manner:

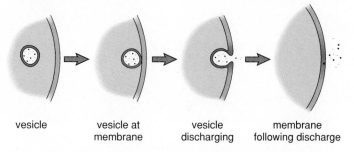

vesicle       vesicle at       vesicle       membrane
          membrane    discharging   following discharge

Often transport vesicles are on their way to the Golgi apparatus.

Rough ER specializes in protein synthesis. Smooth ER has a function suited to a particular type cell; sometimes it specializes in lipid synthesis. The ER is a transport system. Molecules moving through the system are eventually enclosed in vesicles that usually move to the Golgi apparatus.

### Golgi Apparatus

In most cells the **Golgi apparatus** consists of a stack of 3 to 20 slightly curved saccules (flattened vacuoles) (fig. 5.9). In fact a stack of pancakes can be compared to the appearance of the Golgi apparatus except that there are vesicles at the ends of its saccules. The Golgi apparatus receives and processes molecules that are to be transported about the cell. It is involved in the packaging, storage, and distribution of these molecules about or from the cell.

In animal cells, the Golgi apparatus has a "forming face" nearest the endoplasmic reticulum and a "maturing face" pointing toward the plasma membrane. This terminology is consistent with the observation that vesicles are received from smooth ER at the forming face of the Golgi apparatus and vesicles leave the apparatus at the maturing face. After a vesicle is received at the forming face, its molecular contents are discharged into the first of the saccules. Thereafter, the molecules move through the Golgi in vesicles pinched off the saccules. In the meantime, the molecules undergo all sorts of modifications. A sugar chain or a phosphate or a sulfate group can be added, for example. Finally, the molecule is packaged in a vesicle that will carry it either to the plasma membrane for secretion or into a vesicle such as a lysosome (fig. 5.10).

The Golgi apparatus processes, packages, and distributes molecules about or from the cell. Therefore, it is involved in the secretion of molecules.

Researchers are trying to determine what causes molecules to be either secreted or kept in the cell. Just as letters are addressed, perhaps molecules carry some type of identification that determines their final destination in the cell. We do know that proteins that enter the rough ER lumen have a lead sequence of amino acids that causes the polysome to attach to the ER in the first place. Perhaps there are other types of amino acid sequences that act as signals to the Golgi apparatus. Or perhaps there are receptors inside the Golgi, and once a molecule attaches to a receptor its fate is determined.

**Figure 5.9**
Golgi apparatus in an animal cell. The Golgi receives protein-containing vesicles from the smooth ER at its forming face. These proteins are processed while they pass from saccule to saccule of the central region. Finally, they are repackaged in vesicles that leave the maturing face. Some of these vesicles fuse with the plasma membrane allowing secretion to occur, and others are lysosomes, which contain digestive enzymes. **a.** Electron micrograph from liver cells of a mammal. Magnification, X62,000. **b.** Artist representation of Golgi apparatus.

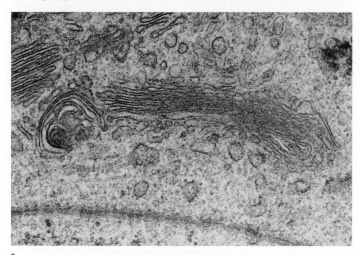

a.

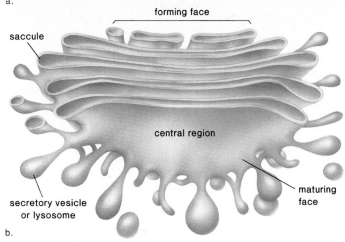

forming face

saccule

central region

secretory vesicle
or lysosome

maturing
face

b.

In a plant cell, the Golgi apparatus has a different name and orientation, and it performs other functions. It is called a **dictyosome** and is found in special regions of the cell where its services seem to be needed. For example, the number of dictyosomes increases during cell division because they form vesicles containing polysaccharides needed for a cell plate. The cell plate develops into a cell wall, separating the 2 new cells. Most likely, dictyosomes also provide vesicles that fuse with developing plasma membrane.

In plant cells, dictyosomes provide the material to form new cell walls and new membrane following cell division.

## Figure 5.10

Functions of the endomembrane system. Proteins made at the rough ER move through the lumen of the canals and tubules until they are transported in vesicles from the smooth ER to the Golgi apparatus. The Golgi apparatus produces vesicles that move to the plasma membrane and lysosomes, which contain digestive enzymes. Macromolecules, which enter a cell by vesicle formation, are broken down when the vesicle fuses with a lysosome to give a secondary lysosome.

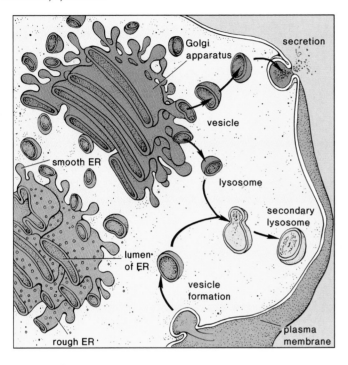

## Figure 5.11

Lysosomes in a kidney cell of a mammal. The small dark body is a primary lysosome, which has not yet fused with an incoming vesicle. The larger bodies are secondary lysosomes, which have fused with such vesicles. The dark appearance of the lysosomes results from staining for a particular enzyme, acid phosphatase, whose presence is the test for this organelle. Magnification, X92,000.

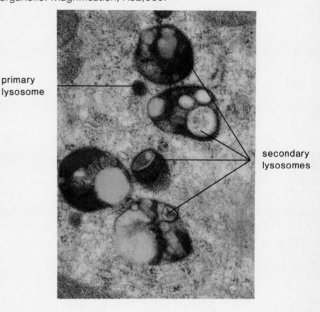

Lysosomes are membrane-bounded vesicles that contain specific enzymes. Lysosomes are produced by Golgi apparatuses, and their hydrolytic enzymes digest macromolecules from various sources.

### Lysosomes

**Lysosomes** are membrane-bounded vesicles produced by the Golgi apparatus that contain hydrolytic digestive enzymes (fig. 5.11).

Sometimes macromolecules are brought into a cell by vesicle formation at the plasma membrane (fig. 5.10). A lysosome can fuse with such a vesicle and can digest its contents into simpler subunits that then enter the cytoplasm. Some white blood cells defend the body by engulfing bacteria that are then enclosed within vesicles. When lysosomes fuse with these vesicles, the bacteria are digested. It should come as no surprise then that even parts of a cell are digested by its own lysosomes (called autodigestion). Normal cell rejuvenation most likely takes place in this manner, but autodigestion is also important during development. For example, when a tadpole becomes a frog, lysosomes are utilized to digest away the cells of the tail. The fingers of a human embryo are at first webbed, but the fingers are freed from one another following lysosomal action.

Occasionally, a child is born with a metabolic disorder involving a missing or inactive lysosomal enzyme. In these cases, the lysosomes fill to capacity with a macromolecule that cannot be broken down. The cells become so full of these lysosomes that the child dies. Someday soon it may be possible to supply these children with the missing enzyme.

### Microbodies

**Microbodies** are similar to lysosomes because they are vesicles bounded by a single membrane and because they contain specific enzymes (fig. 5.12). In this way, they also function to compartmentalize the cell. Although microbodies are believed to be quite varied, 2 types are noteworthy: peroxisomes and glyoxysomes.

Peroxisomes are microbodies that contain enzymes for transferring hydrogen atoms to oxygen, forming hydrogen peroxide ($H_2O_2$), a toxic molecule that is immediately broken down to water by the enzyme catalase. Peroxisomes are abundant in cells that are metabolizing lipids and in liver cells that are metabolizing alcohol. They are believed to help detoxify the alcohol.

Glyoxysomes have been observed in leaves that are carrying on photosynthesis. In this case, they contain enzymes that can metabolize some of the molecules involved in the photosynthetic process. They are also seen in germinating seeds, where they are believed to convert oils into sugars used as nutrients by the growing plant.

It is possible that the endoplasmic reticulum produces microbodies, but we know that their enzymes are produced by free ribosomes within the cytosol. Afterwards, the enzymes enter the microbodies directly from the cytosol.

### Vacuoles

A **vacuole** is a large membranous sac; a vesicle is smaller. Animal cells have vacuoles, but they are much more prominent in plant cells. Typically, plant cells have one or 2 large vacuoles so filled with a watery fluid that they give added support to the cell. Most of the central area of the cell is occupied by vacuole, and the remaining cell contents are pushed to the sides (fig. 5.5).

Vacuoles are most often storage areas, but they can have specialized functions. Plant vacuoles contain not only water, sugars, and salts but also pigments and toxic substances. The pigments are responsible for many of the red, blue, or purple colors of flowers and some leaves. The toxic substances help protect a plant from herbivorous animals. (As long as the substance is contained within the vacuole, it will not harm the plant.) The vacuoles present in protozoans are quite specialized, and they include contractile vacuoles and digestive vacuoles (fig. 24.2).

---

The organelles of the endomembrane system are as follows:

Endoplasmic reticulum: synthesis and/or modification and transport of proteins and other substances
Rough ER: protein synthesis
Smooth ER: lipid metabolism, in particular
Golgi apparatus: processing and packaging of protein molecules
Lysosomes: intracellular digestion
Microbodies: various metabolic tasks
Vacuoles: storage areas

Figure 5.10 shows how certain of these organelles are related to each other.

---

# Energy-Related Organelles

Life is possible only because of a constant input of energy to maintain the structure of the cell. Chloroplasts and mitochondria are the 2 eukaryotic membranous organelles that specialize in converting energy into a form that can be used by the cell.

Photosynthesis, which occurs in **chloroplasts,** is the process by which solar energy is converted to chemical-bond energy within carbohydrates. Photosynthesis can be represented by this equation:

$$\text{light energy} + \text{carbon dioxide} + \text{water} \longrightarrow \text{carbohydrate} + \text{oxygen}$$

Here the word *energy* stands for solar energy, the ultimate source of energy for cellular organization. Only plants, algae, and certain bacteria are capable of carrying on photosynthesis. (Bacteria, however, being prokaryotic, do not have chloroplasts.)

Cellular respiration, which occurs in **mitochondria,** is the process by which the chemical energy of carbohydrates is converted to that of ATP (adenosine triphosphate), the common

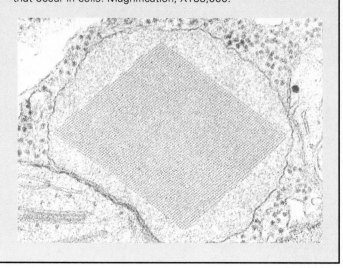

**Figure 5.12**
Peroxisome in a tobacco leaf. The crystalline, squarelike core of this microbody cross section is believed to consist of the enzyme catalase, which breaks down hydrogen peroxide to water. Microbodies carry out various metabolic reactions and are another means of compartmentalizing enzymatic reactions that occur in cells. Magnification, X155,000.

carrier of chemical energy in cells. Cellular respiration can be represented by this equation:

$$\text{carbohydrate} + \text{oxygen} \longrightarrow \text{carbon dioxide} + \text{water} + \text{energy}$$

Here the word *energy* stands for ATP molecules. When a cell needs energy, ATP supplies it. ATP energy is used for synthetic reactions, active transport, and all energy-requiring processes in cells. All organisms carry on cellular respiration, and all organisms except bacteria have mitochondria. Chloroplasts use the sun's energy to produce carbohydrates, and carbohydrate-derived products are broken down in mitochondria to build up ATP molecules.

---

Chloroplasts and mitochondria supply the eukaryotic cell with usable energy. Energy is needed for various cellular activities, all of them helpful for maintaining the structure of cells.

---

## Chloroplasts

Chloroplasts are about 4 μm–6 μm in diameter and 1 μm–5 μm in length; they belong to a group of plant organelles known as plastids. Among the plastids are also the *amyloplasts,* which store starch, and the *chromoplasts,* which contain red and orange pigments.

A chloroplast is bounded by a double membrane, and inside there is even more membrane organized into flattened sacs called **thylakoids.** The thylakoids are piled up like stacks of coins, and each stack is called a **granum.** There are membranous connections between the grana called *lamellae* (fig. 5.13). The fluid-filled space about the grana is called the **stroma.**

### Figure 5.13

Chloroplast structure. *a.* Electron micrograph. Magnification, X40,000. *b.* Generalized drawing in which the outer and inner membranes have been cut away to reveal the grana.

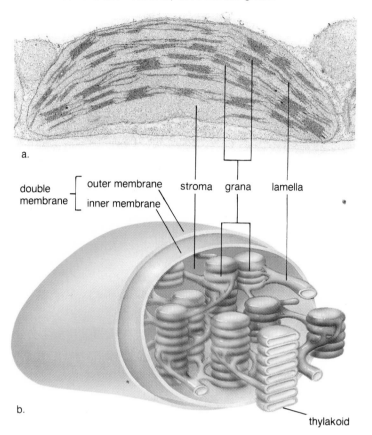

a.

double membrane [ outer membrane / inner membrane ]   stroma   grana   lamella

b.

thylakoid

### Figure 5.14

Mitochondrion structure. *a.* Electron micrograph. Magnification, X70,000. *b.* Generalized drawing in which the outer membrane and portions of the inner membrane have been cut away to reveal the cristae.

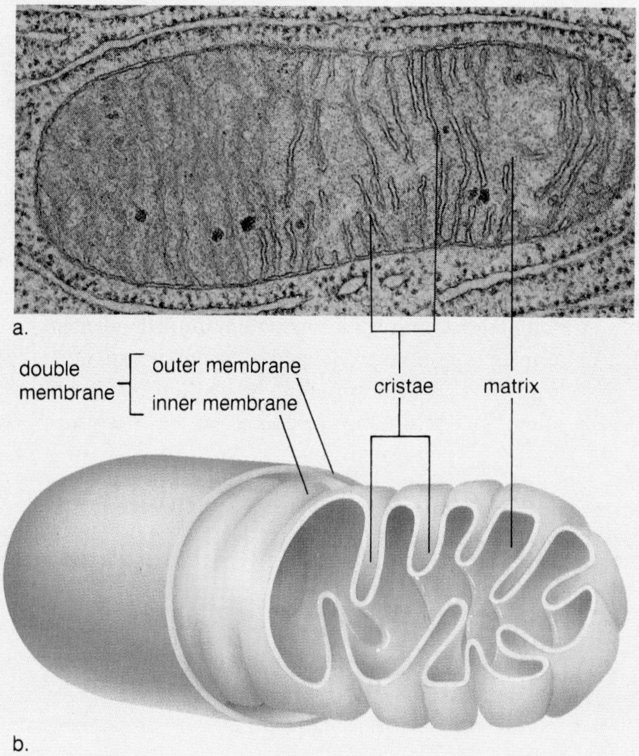

a.

double membrane [ outer membrane / inner membrane ]   cristae   matrix

b.

Photosynthesis requires pigments that capture solar energy and enzymes that synthesize carbohydrate. The green pigments, called chlorophylls, are the major pigments found within the thylakoid membrane of grana. Enzymes are found within the stroma. This basically tells you how chloroplasts are organized to carry out photosynthesis. As mentioned previously, the cyanobacteria, which are prokaryotic, usually possess cytoplasmic thylakoids to which photosynthetic pigments are bound.

Chloroplasts have their own DNA and ribosomes and can produce certain proteins. They also reproduce themselves by division. In other words, a chloroplast is similar in many respects to a photosynthetic prokaryotic organism, an observation to which we will return.

## Mitochondria

Most mitochondria are between 0.5 μm and 1.0 μm in diameter and 7 μm in length, although the size and the shape can vary.

Mitochondria, like chloroplasts, are bounded by a double membrane. The inner of these 2 membranes is folded to form little shelves, called **cristae,** which project into the **matrix,** an inner space filled with a gel-like fluid (fig. 5.14). Analysis reveals that the matrix contains enzymes that break down carbohydrate-derived products, while ATP production occurs at the cristae.

Like chloroplasts, mitochondria also contain their own DNA and ribosomes and can produce a few of their own proteins. They also reproduce themselves by division.

> Chloroplasts and mitochondria are membranous organelles whose structure is suited to compartmentalizing the processes that occur within them.

## Origin of Energy-Related Organelles in Eukaryotic Cells

It is believed that the first eukaryotic cells did not have energy-related organelles; instead, they were acquired through the process of endosymbiosis. The **endosymbiotic theory** states that chloroplasts and mitochondria were originally prokaryotes that came to reside inside a eukaryotic cell, establishing a symbiotic relationship known as mutualism (p. 761). In other words, today's mitochondria may be derived from an aerobic (oxygen-using) bacterium that was taken up by a eukaryotic cell in exchange for a ready supply of nutrients. Chloroplasts may be derived from a photosynthetic bacterium that permitted its host to carry on photosynthesis (fig. 5.15).

The evidence for the endosymbiotic theory is as follows:

1. Mitochondria and chloroplasts are similar to bacteria in size and in structure.
2. Both organelles are bounded by a double membrane—the outer membrane may be derived from the engulfing vesicle, and the inner one may be derived from the plasma membrane of the original prokaryote.
3. Mitochondria and chloroplasts contain a limited amount of genetic material and are capable of self-reproduction. Their DNA is a circular loop like that of bacteria.
4. Although most of the proteins within mitochondria and chloroplasts are now produced by the eukaryotic host, they do have their own ribosomes and they do produce some proteins. Their ribosomes resemble those of bacteria.

> The endosymbiotic theory states that mitochondria and chloroplasts are derived from bacteria that took up residence in a eukaryotic cell(s).

## Cytoskeleton: Cell Shape and Movement

The **cytoskeleton** is a network of protein elements that extends throughout the cytosol in eukaryotic cells (fig. 5.16). Prior to the 1970s, it was believed that the cytosol was an unorganized mixture of biomolecules. High-voltage electron microscopes, which can penetrate thicker specimens, showed, however, that the cytosol was instead highly organized. It contains 3 types of protein elements: microtubules, actin filaments, and intermediate filaments.

The name *cytoskeleton* is convenient in that it allows us to compare the cytoskeleton to the bones and muscles of an animal. The bones and muscles give an animal structure and allow it to move. Similarly the cytoskeleton of a cell is responsible for its shape and movement. Various cells such as red blood cells, muscle cells, and nerve cells have a distinctive shape that allows easy recognition. The cytoskeleton maintains the shape of a cell, but at the same time it also allows movement, as when, for example, vesicles move from the Golgi apparatus to the plasma membrane or when chloroplasts circulate in large plant cells. The cytoskeletal elements seem to act as "highways" along which the organelles (and perhaps molecules) are propelled in a certain direction only. In addition, we know that cells have appendages, such as cilia, that can move or even at times make the entire cell move. Some cells can move by creeping in a forward direction.

Muscles are attached to bones, and it is possible that enzymes are attached to the cytoskeleton in a way that permits them to act in a sequential manner. The term *cytoskeleton* is inappropriate, however, if it makes us think of something rigid and nonchanging. On the contrary, the cytoskeleton appears to undergo

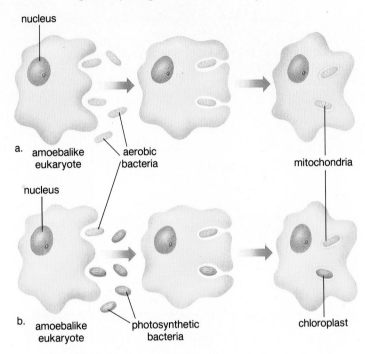

continuous and even rapid changes that allow the cell to change its shape as is necessary when cells creep forward or when embryonic cells become specialized during development.

In summary the cytoskeleton has these functions:

Maintenance of cell shape (e.g., red blood cells, nerve cells, and muscle cells have different shapes)
Change in cell shape (e.g., embryonic cells become specialized during development)
Movement of cell parts (e.g., organelles move about; cilia beat)
Movement of the cell (e.g., some cells creep; some are propelled by cilia or flagella)

> The cytoskeleton contains 3 types of elements: microtubules, actin filaments, and intermediate filaments, which are responsible for cell shape and movement.

## Microtubules

**Microtubules** are small cylinders about 25 nm in diameter and from 200 nm-25 $\mu$m in length. Their length can change because microtubules can assemble (polymerize) and disassemble (depolymerize). Microtubules are made of a globular protein called tubulin. When assembly occurs, 2 tubulin molecules come together as dimers and then the dimers form the hollow cylinder that is the microtubule (fig. 5.16). Microtubules have 13 rows of tubulin dimers. In many cells the regulation of microtubule assembly is under the control of a *microtubule organizing*

## Figure 5.16

The cytoskeleton. *a.* Overview showing the general placement of the cytoskeletal elements. *b.* Intermediate filaments maintain the shape of the cell and give support to various structures. *c.* Microtubules radiate out from the microtubule organizing center and may serve as tracts for organelle movement. *d.* Actin filaments are present as bundles or a mesh network. They are often involved in cellular movements of all sorts.

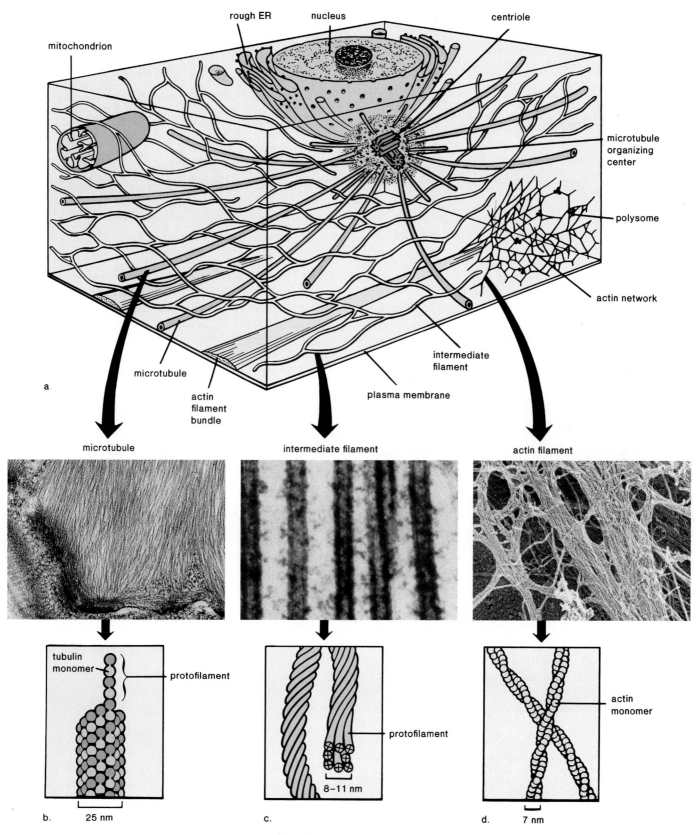

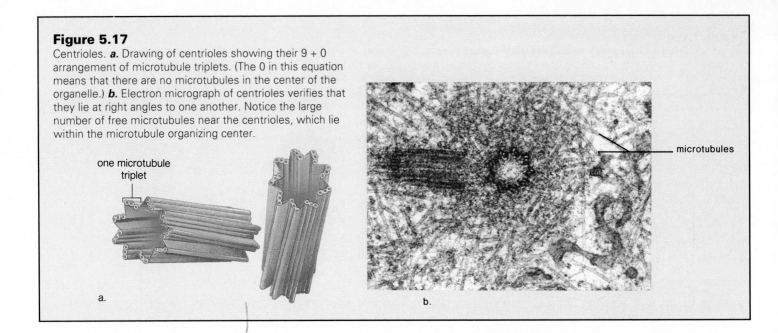

**Figure 5.17**
Centrioles. **a.** Drawing of centrioles showing their 9 + 0 arrangement of microtubule triplets. (The 0 in this equation means that there are no microtubules in the center of the organelle.) **b.** Electron micrograph of centrioles verifies that they lie at right angles to one another. Notice the large number of free microtubules near the centrioles, which lie within the microtubule organizing center.

one microtubule triplet

microtubules

a.

b.

*center,* which lies near the nucleus. The microtubules radiate out from the center, reinforcing the shape of the cell and acting as tracts along which organelles can move. The movement of vesicles from the Golgi apparatus to the plasma membrane is believed to be associated with microtubules. It is well known that during cell division, microtubules are involved in the movement of chromosomes (p. 160). During cell division, and indeed at all times, microtubules are forming and reforming as assembly and disassembly occur.

In animal cells, but not plant cells, the microtubule organizing center contains 2 centrioles lying at right angles to one another. **Centrioles** are short cylinders with a 9 + 0 pattern of microtubule triplets, that is, a ring having 9 sets of triplets with none in the middle (fig. 5.17). Before an animal cell divides, the centrioles replicate and the members of each pair are also at right angles to one another. During cell division the centriole pairs separate so that each new cell receives one pair.

*Cilia* and *flagella* are hairlike projections that can move either in an undulating fashion, like a whip, or stiffly, like an oar. Cells that have these organelles are capable of movement. For example, single-celled paramecia move by means of cilia; sperm cells move by means of flagella. The cells that line our upper respiratory tract are ciliated. The cilia sweep debris trapped within mucus back up into the throat, where it can be swallowed. This action helps keep the lungs clean.

Cilia are much shorter than flagella, but they have a similar construction and this construction is quite different from that of prokaryotic flagella. Eukaryotic cilia and flagella are membrane-bounded cylinders enclosing a matrix area. In the matrix are 9 microtubule doublets arranged in a circle around 2 central microtubules. This is called the 9 + 2 pattern of microtubules. Cilia and flagella move when the microtubule doublets slide past one another (fig. 5.18).

Each cilium and flagellum has a basal body lying in the cytoplasm at its base. Basal bodies have the same circular arrangement of microtubule triplets as centrioles and are believed to be derived from them. It is possible that basal bodies organize the microtubules within cilia and flagella, but we know that cilia and flagella grow by the addition of tubulin dimers to their tips.

Centrioles have a 9 + 0 pattern of microtubules and give rise to basal bodies that organize the 9 + 2 pattern of microtubules in cilia and flagella.

## Actin Filaments

Actin filaments (formerly called microfilaments) are long, extremely thin fibers (about 7 nm in diameter) that occur in bundles or meshlike networks. The actin filament contains 2 chains of globular actin molecules twisted about one another in a helical manner (fig. 5.16). Actin filaments, like microtubules, can assemble and disassemble.

Actin filaments are well known for their role in muscle contraction, when they interact with other filaments composed of myosin. Actin filaments, however, assist movement in nearly all eukaryotic cells. For example, actin filaments are seen in the microvilli that project from intestinal cells. Their presence most likely accounts for the ability of microvilli alternately to shorten and extend into the intestine. In plant cells, they form the tracts that cause the cytoplasm and thus the chloroplasts to circulate or stream in a particular direction.

Working in conjunction with myosin molecules, actin filaments form a constricting ring that becomes ever smaller when an animal cell divides into 2 cells. Also, it has been discovered that the presence of a network of fibers lying beneath the plasma membrane probably accounts for the formation of pseudopodia, extensions that allow certain cells to move in an amoeboid fashion (fig. 5.19).

## Figure 5.18

**a.** A sperm is propelled by a flagellum that contains microtubules (**b**). Magnification, ×500. **c.** A basal body with a 9 + 0 pattern of microtubule triplets is at the base of a flagellum. (Notice that there are no central microtubules within a ring of 9 triplets.) **d.** A flagellum has a 9 + 2 pattern of microtubule doublets. (Notice there are 2 central microtubule singlets with a ring of 9 doublets.) Compare the cross section of the basal body to the flagellum and note that in place of the third microtubule, the outer doublets have dynein side arms. Dynein is an enzyme that splits ATP. **e.** Movement of flagellum. In the presence of ATP, the dynein side arms reach out and connect to their neighboring doublet and push against it. If it were not for the 2 central microtubule singlets, each doublet would simply slide past its neighbor, but because of the radial spokes connecting them to the central microtubules, a bending occurs instead.

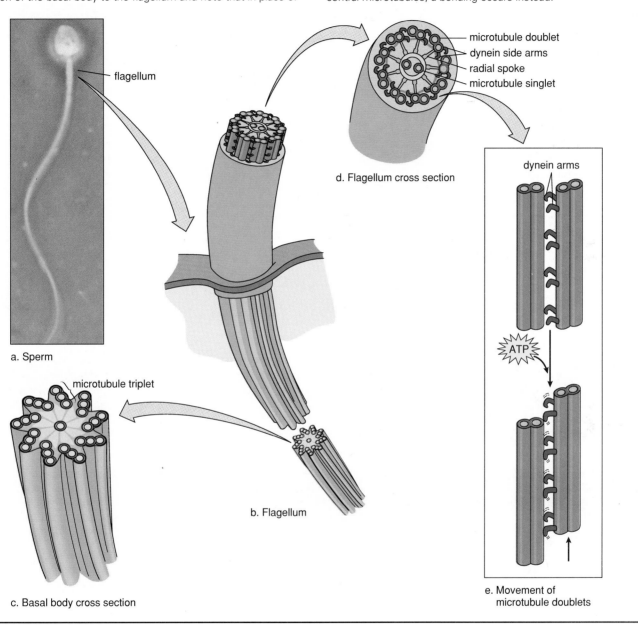

- flagellum
- microtubule doublet
- dynein side arms
- radial spoke
- microtubule singlet

d. Flagellum cross section

dynein arms

ATP

a. Sperm

b. Flagellum

microtubule triplet

c. Basal body cross section

e. Movement of microtubule doublets

## Intermediate Filaments

Intermediate filaments (8 nm–11 nm in diameter) are intermediate in size between microtubules and microfilaments. They are ropelike polymers of fibrous proteins, but the specific type varies according to the tissue. In the skin, the filaments are made of the protein keratin, and they give great mechanical strength to cells. In all cells there are intermediate filaments that support the nuclear envelope.

Unlike the other 2 types of cytoskeletal elements, intermediate filaments do not assemble and disassemble. Therefore, they seem to be particularly important in maintaining the shape of cells and may have the function of supporting the other elements of the cytoskeleton.

## Figure 5.19

*a.* Pseudopodium formation in a cell that moves in an amoeboid manner. There is a meshlike network of actin filaments beneath the plasma membrane. In one region the network changes from a gel state to a sol state as the actin filaments disassemble. When water from the rest of the cell moves into the region due to increased osmotic pressure, a pseudopodium forms. *b.* Reassembly of the actin filaments causes a return of the gel state and gives structure to the newly formed pseudopodium. The actin binding protein holds the rigid actin filaments together so that network results.

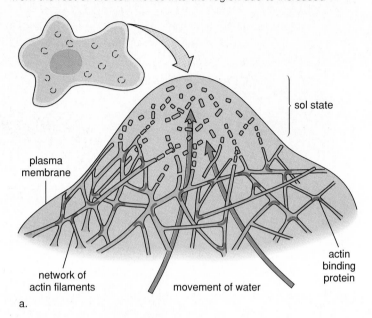

sol state

plasma membrane

network of actin filaments

movement of water

actin binding protein

a.

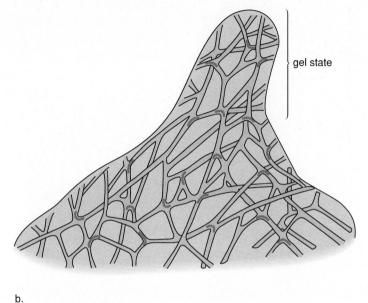

gel state

b.

## Summary

1. All organisms are composed of cells, the smallest units of living matter. Cells are capable of self-reproduction, and existing cells come only from preexisting cells.

2. There are 2 major groups of cells: prokaryotic and eukaryotic. Both types have a plasma membrane and cytoplasm. Eukaryotic cells also have a nucleus and various organelles. Prokaryotic cells have a nucleoid that is not bounded by a nuclear envelope. They also lack most of the other organelles that compartmentalize eukaryotic cells. Table 5.2 summarizes the differences between prokaryotic and eukaryotic cells.

3. Cells are very small and are measured in small units of the metric system. The plasma membrane regulates exchange of materials between the cell and the external environment. Cells must remain small in order to have an adequate amount of plasma membrane per cell volume.

4. The nucleus of eukaryotic cells, represented by animal and plant cells, is bounded by a nuclear envelope containing pores. These pores serve as passageways between the cytoplasm and the nucleoplasm. Within the nucleus, the chromatin undergoes coiling into chromosomes at the time of cell division. The nucleolus is a special region of the chromatin where ribosomal RNA (rRNA) is produced and where proteins from the cytoplasm gather to form ribosomal subunits. These subunits are joined in the cytoplasm.

5. Ribosomes are organelles that function in protein synthesis. They can be bound to endoplasmic reticulum or exist free within the cytosol. Several ribosomes are involved in synthesizing the same type protein. The complex is called a polysome.

6. The endomembrane system includes the endoplasmic reticulum (both rough and smooth), the Golgi apparatus (dictyosome in plant cells), the lysosomes, and other types of vesicles and vacuoles. The endomembrane system serves to compartmentalize the cell and keep the various biochemical reactions separate from one another. Proteins that are made in the cytoplasm quite often enter the rough ER, where they may be modified before proceeding to the lumen of the smooth ER. The smooth ER has various metabolic functions depending on the cell, but it also forms vesicles that carry proteins and other substances to different locations, particularly to the Golgi apparatus. The Golgi apparatus processes proteins and repackages them into lysosomes, which carry out intracellular digestion, or into vesicles that fuse with the plasma membrane. Following fusion, secretion occurs. Also part of the endomembrane system are microbodies that have special enzymatic functions, and the large single plant cell vacuole, which not only stores substances but lends support to the plant cell when fluid-filled.

7. Cells require a constant input of energy to maintain their structure. Chloroplasts capture the energy of the

The Cell

## Table 5.2
Comparison of Prokaryotic Cells and Eukaryotic Cells

| | Prokaryotic Cells | Eukaryotic Cells | |
| --- | --- | --- | --- |
| | | Animal | Plant |
| Size | Smaller (1-10 μm in diameter) | Larger (10-100 μm in diameter) | |
| Plasma membrane | yes | yes | yes |
| Cell wall | usually (peptidoglycan) | no | yes (cellulose) |
| Nuclear envelope | no | yes | yes |
| Nucleolus | no | yes | yes |
| DNA | yes (single loop) | yes (chromosomes) | yes (chromosomes) |
| Mitochondria | no | yes | yes |
| Chloroplasts | no | no | yes |
| Endoplasmic reticulum | no | yes | yes |
| Ribosomes | yes (smaller) | yes | yes |
| Vacuoles | no | yes (small) | yes (usually large, single vacuole) |
| Golgi apparatus | no | yes | yes |
| Lysosomes | no | always | often |
| Microbodies | no | usually | usually |
| Cytoskeleton | no | yes | yes |
| Centrioles | no | yes | no (in higher plants) |
| 9 + 2 cilia or flagella | no | often | no (in flowering plants) yes (in ferns, cycads, and bryophytes) |

sun and carry on photosynthesis, which produces carbohydrate. Carbohydrate products are broken down in mitochondria as ATP is produced. This is an oxygen-requiring process called cellular respiration. It is believed that mitochondria and chloroplasts are derived from prokaryotes that took up residence in eukaryotic cells.

8. The cytoskeleton contains microtubules, intermediate filaments, and microfilaments. These are involved in maintaining cell shape and in the movement of cellular components. Microtubules are made up of 13 rows of tubulin dimers. They are found free within the cytosol but also are present in centrioles, cilia, and flagella. Centrioles give rise to basal bodies, which lie at the base of cilia and flagella. Actin filaments are 2 long chains of actin arranged in a helical manner. Intermediate filaments contain protein fibers and are intermediate in size between microtubules and microfilaments.

### Writing Across the Curriculum

In order to practice writing skills, students should write out the answers to any or all of the study questions and the critical thinking questions. The study questions are sequenced in the same order as the text. Suggested answers to the critical thinking questions are in appendix D.

## Study Questions

1. What are the 2 basic tenets of the cell theory?
2. What are the contrasting advantages of light microscopy and electron microscopy?
3. What similar features do prokaryotic cells and eukaryotic cells have? What is their major difference?
4. Contrast the size of prokaryotic and eukaryotic cells, and tell why cells are so small.
5. Roughly sketch a prokaryotic cell, label its parts, and state a function for each of these.
6. Describe an experiment that supports the concept of the nucleus as the control center of the cell.
7. Distinguish between the nucleolus, ribosomal RNA, and ribosomes.
8. Describe the structure and the function of the nuclear envelope and the nuclear pores.
9. Trace the path of a protein from rough ER to the plasma membrane.
10. Give the overall equations for photosynthesis and cellular respiration, contrast the 2, and tell how they are related.
11. Draw a diagram that explains the endosymbiotic theory.
12. What are the 3 components of the cytoskeleton? What are their structures and functions?

## Objective Questions

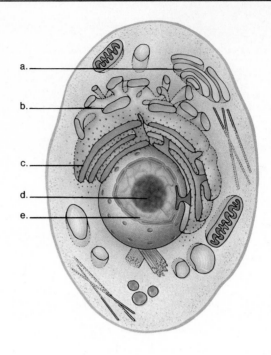

a. _____

b. _____

c. _____

d. _____

e. _____

1. The small size of cells is best correlated with
   a. the fact they are self-reproducing.
   b. their prokaryotic versus eukaryotic nature.
   c. an adequate surface area for exchange of materials.
   d. All of these.
2. Which of these is not a true comparison of the light microscope and the transmission electron microscope?

   Light————Electron

   a. Uses light to "view" object—uses electrons to "view" object.
   b. Uses glass lenses for focusing—uses magnetic lenses for focusing.
   c. Specimen must be killed and stained————specimen may be alive and nonstained..
   d. Magnification is not as great—magnification is greater.
3. Which of these best distinguishes a prokaryotic cell from a eukaryotic cell?
   a. Prokaryotic cells have a cell wall, but eukaryotic cells never do.
   b. Prokaryotic cells are much larger than eukaryotic cells.
   c. Prokaryotic cells have flagella, but eukaryotic cells do not.
   d. Prokaryotic cells do not have a membrane-bounded nucleus, but eukaryotic cells do have such a nucleus.
4. Which of these is not found in the nucleus?
   a. functioning ribosomes
   b. chromatin that condenses to chromosomes
   c. nucleolus that produces ribosomal RNA (rRNA)
   d. nucleoplasm instead of cytoplasm
5. Vesicles from the smooth ER most likely are on their way to the
   a. rough ER.
   b. lysosomes.
   c. Golgi apparatus.
   d. plant cell vacuole only.

6. Lysosomes function in
   a. protein synthesis.
   b. processing and packaging.
   c. intracellular digestion.
   d. lipid synthesis.
7. Mitochondria
   a. carry on cellular respiration.
   b. break down ATP to release energy for cells.
   c. contain grana and cristae.
   d. All of these.
8. The endosymbiotic theory explains
   a. the origin of the nucleus.
   b. where the endomembrane system came from.
   c. why prokaryotic cells are different from eukaryotic cells.
   d. the origin of mitochondria and chloroplasts.
9. Which organelle releases oxygen?
   a. ribosomes
   b. Golgi apparatus
   c. mitochondria
   d. chloroplasts

10. Cilia and flagella contain
    a. microtubules.
    b. microfilaments.
    c. tubulin dimers.
    d. Both a and c.
11. Label these parts of the cell that are involved in protein synthesis and modification. Give a function for each structure.
12. Study the example given. Then for each other organelle listed, state another that is structurally and functionally related. Tell why you paired these 2 organelles.
    a. The nucleus can be paired with *nucleoli* because nucleoli are found in the nucleus. Nucleoli occur where chromatin is producing rRNA.
    b. mitochondria
    c. centrioles
    d. endoplasmic reticulum

## Concepts and Critical Thinking

**1.** *All organisms are made up of cells.*

What data would you use to convince someone that all organisms are composed of cells? What data would you have to find to nullify the cell theory?

**2.** *Cells are compartmentalized and highly organized.*

Substantiate this concept by referring to table 5.1.

**3.** *Life begins at the cellular level of organization.*

Which organelles contribute to the ability of the cell to maintain its structure and grow? What are the functions of these organelles?

## Selected Key Terms

**cell** (sel) 61
**plasma membrane** (plaz'ma mem'brān) 64
**cell wall** (sel wawl) 64
**eukaryotic cell** (u"kar-e-ot'ik sel) 64
**organelle** (or"gan-el') 64
**prokaryotic cell** (pro"kar-e-ot'ik sel) 64
**nucleoid** (nu'kle-oid) 64
**cytoplasm** (si'to-plazm") 64
**cytosol** (si'to-sol) 64
**bacterium** (bak-te're-um) 66

**nucleus** (nu'kle-us) 66
**chromatin** (kro'mah-tin) 68
**chromosome** (kro'mo-sōm) 68
**nucleolus** (nu-kle'o-lus) 68
**nuclear envelope** (nu'kle-ar en've-lōp) 69
**ribosome** (ri'bo-sōm) 70
**endoplasmic reticulum** (en"do-plas'mik rē-tik'u-lum) 71
**Golgi apparatus** (gol'je ap"ah-ra'tus) 72
**lysosome** (li'so-sōm) 73
**microbody** (mi"kro-bod'e) 73

**chloroplast** (klo'ro-plast) 74
**mitochondrion** (mi"to-kon'dre-an) 74
**cytoskeleton** (si"to-skel'ē-ton) 76
**microtubule** (mi'kro-tu'būl) 76
**centriole** (sen'tri-ōl) 78

# 6

# Membrane Structure and Function

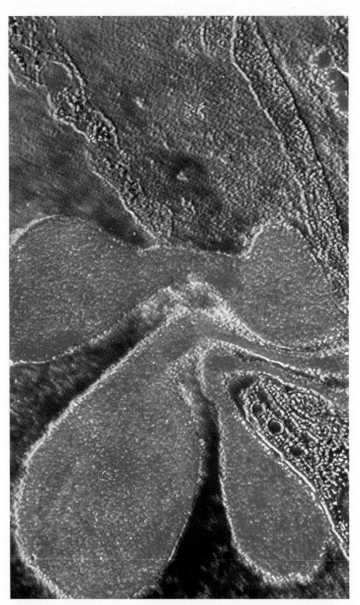

The plasma membrane is flexible. What greater proof than this photo of red blood cells squeezing through the walls of a small blood capillary?

Your study of this chapter will be complete when you can

1. contrast the "sandwich" model of membrane structure with the "fluid-mosaic" model, and cite the evidence supporting these models;
2. describe the arrangement of the lipid component, and give a function for each type of lipid in the membrane;
3. describe the arrangement of the protein component, and give several functions for these proteins;
4. define diffusion and osmosis, and explain their relevance to cell biology;
5. describe the appearance of a plant cell and an animal cell in isotonic, hypotonic, and hypertonic solutions;
6. name 2 types of transport by carrier proteins, and give an example for each;
7. contrast endocytosis and exocytosis. Name 3 types of endocytosis and differentiate among them;
8. describe the composition of the plant cell wall;
9. list and describe 3 types of junctions that occur between animal cells. Name and describe one type of junction between plant cells.

**A**ll cells have a plasma membrane that serves as an interface between the interior, which is alive, and the exterior, which is nonliving. An intact plasma membrane is absolutely essential to a cell, and if by chance the plasma membrane is disrupted, the cell loses its contents and dies. The plasma membrane was absolutely essential to the evolution of the first cell or cells. Only when a membrane is present can there be a cell at all.

The interior and exterior environments of a living cell are largely fluids, even if the cell is within a multicellular organism. Quite often the composition of intracellular fluid (i.e., cytosol) is not quite the same as that of the extracellular fluids. The plasma membrane functions to keep the intracellular fluid relatively constant. It regulates the entrance and exit of molecules into and out of the cell, and in this way the intracellular fluid remains compatible with the continued existence of the cell.

## Studies of Membrane Structure

At the turn of the century, investigators noted that lipid-soluble molecules entered cells more rapidly than water-soluble molecules. This prompted them to suggest that lipids are a component of the plasma membrane. Later, chemical analysis disclosed that the plasma membrane contains phospholipids (fig. 6.1). In 1925, E. Gorter and G. Grendel measured the amount of phospholipid extracted from red blood cells and determined that there is just enough to form a bilayer around the cells.[1] They further suggested that the nonpolar (hydrophobic) tails are directed inward and the polar (hydrophilic) heads were directed outward:

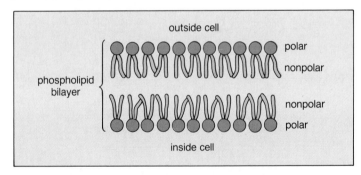

The presence of lipids cannot account for all the properties of the plasma membrane. For example, to account for the permeability of the membrane to certain nonlipid substances, J. Danielli and H. Davson suggested in the 1940s that globular proteins are also a part of the membrane. They proposed the "sandwich" model of membrane structure in which the phospholipid bilayer is a filling

[1]Later investigators found that Gorter and Grendel reported too low an amount of lipid and underestimated the surface area of red blood cells; however, since the 2 errors canceled each other, they still came to the correct conclusion.

### Figure 6.1

The plasma membrane contains phospholipid molecules. **a.** Each phospholipid molecule is composed of glycerol bonded to 2 fatty acid chains and a chain that contains a phosphate group and a nitrogen group. **b.** The molecule arranges itself in such a way that the phosphate and nitrogen-containing chain forms a polar (hydrophilic) head, and the fatty acid chains form nonpolar (hydrophobic) tails. The presence of a double bond causes a "kink" in a fatty acid tail.

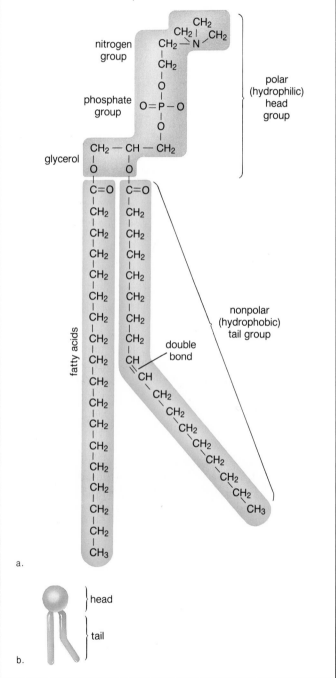

between 2 layers of proteins arranged to give channels through which polar substances may pass:

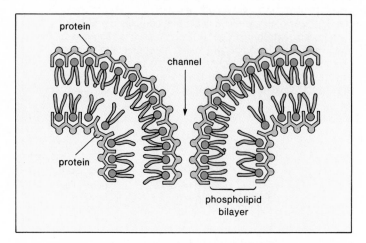

By the late 1950s, electron microscopy had advanced to allow viewing of the plasma membrane and other membranes in the cell. Since the membrane has a sandwichlike appearance (fig. 6.2), J. D. Robertson assumed that the outer dark layer (stained with heavy metals) contained protein plus the hydrophilic heads of the phospholipids. The interior was simply the hydrophobic tails of these molecules. Robertson went on to suggest that all membranes in various cells have basically the same composition. This proposal was called the "unit membrane" model.

This model of membrane structure was accepted for at least 10 years, even though investigators began to doubt its accuracy. For example, not all membranes have the same appearance in electron micrographs, and they certainly do not have the same function. The inner membrane of a mitochondrion is coated with rows of particles and functions in cellular respiration; it therefore has a far different appearance and function from the plasma membrane. Finally, in 1972, S. Singer and G. Nicolson introduced the **fluid-mosaic model** of membrane structure, which proposes that the membrane is a bilayer of phospholipids with the consistency of olive oil, in which protein molecules are either partially or wholly embedded. The proteins are scattered throughout the membrane, forming a mosaic pattern (fig. 6.3). The fluid-mosaic model of membrane structure is supported especially by electron micrographs of freeze-fractured membranes.

---

The fluid-mosaic model of membrane structure is widely accepted at this time.

---

Our perception of the plasma membrane has changed over the years. There have been a series of models, each one developed to suit the evidence available at that time. This illustrates that scientific knowledge is always subject to change, and modifications are made whenever new data are presented. A model is useful because it pulls together the available data and suggests new avenues of research.

## Figure 6.2

Membrane structure. **a.** The red blood cell plasma membrane typically has a 3-layered appearance in electron micrographs; there are 2 outer dark layers and a central light layer. Magnification, X415,000. **b.** In 1960, Robertson's unit membrane model was widely accepted. This model proposed that the outer dark layers in electron micrographs were made up of protein and polar heads of phospholipid molecules, while the inner light layer was composed of the nonpolar tails. The Robertson model was challenged, particularly by the Singer and Nicolson fluid-mosaic model, which puts protein molecules within the lipid bilayer. **c.** A technique, called freeze-fracture, allows an investigator to view the interior of the membrane. Cells are rapidly frozen in liquid nitrogen and then fractured with a knife. The fracture often splits the membrane in 2, just in the middle of the lipid bilayer. Platinum and carbon are applied to the fractured surface to produce a faithful replica that is observed by electron microscopy. **d.** The electron micrograph is not as smooth as it would be if the unit membrane model was correct. Instead, the micrograph shows the presence of particles as is expected by the fluid-mosaic model.

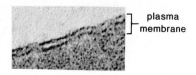

a. Electron micrograph of red blood cell plasma membrane

plasma membrane

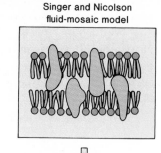

b. Robertson unit membrane

Singer and Nicolson fluid-mosaic model

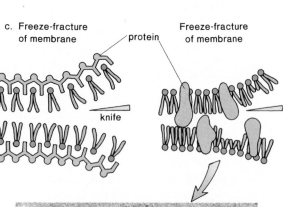

c. Freeze-fracture of membrane

protein

Freeze-fracture of membrane

knife

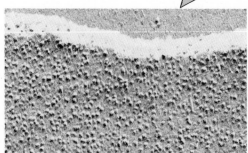

d. Electron micrograph of freeze-fractured membrane shows presence of particles

**Figure 6.3**

Fluid-mosaic model of the plasma membrane. The plasma membrane is composed of a phospholipid bilayer with embedded proteins. The hydrophilic heads of the phospholipids are at the surfaces of the membrane, and the hydrophobic tails make up the interior of the membrane. Note the asymmetry of the membrane; for example, carbohydrate chains project externally and cytoskeleton filaments attach to proteins on the cytoplasmic side of the plasma membrane.

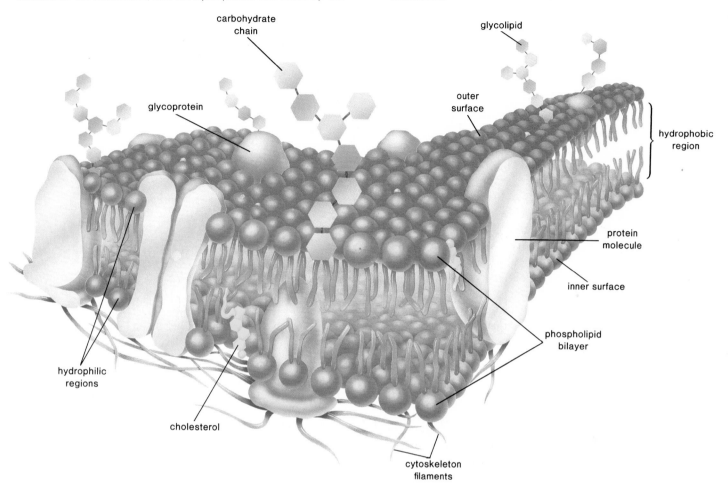

## Fluid-Mosaic Model of the Plasma Membrane

The fluid-mosaic model of membrane structure has 2 components, lipids and proteins (fig. 6.3). The lipids form the matrix of the membrane, and the proteins carry out all of its functions.

### Lipid Component

Most of the lipids in the plasma membrane are **phospholipids** (fig. 6.1). In a laboratory container, phospholipids spontaneously form a bilayer in which the hydrophilic (polar) heads face water on both sides and the hydrophobic tails avoid contact with water. The hydrophobic interior portions of the phospholipid bilayer are never exposed to water because it spontaneously forms a sphere—becoming a barrier just like the plasma membrane:

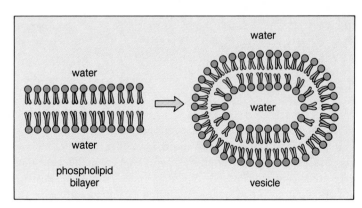

At body temperature, the phospholipid bilayer of the plasma membrane has the consistency of olive oil. The greater the concentration of unsaturated fatty acid residues, the more liquid the bilayer. In each monolayer, the hydrocarbon tails wiggle, and the entire phospholipid molecule can move sideways, even exchanging places with its neighbor. (Phospholipid molecules rarely flip-flop from one layer to the other, because this would require the hydrophilic head to move through the hydrophobic center of the membrane.) The fluidity of a phospholipid bilayer means that cells are pliable. Imagine if they were not—the nerve cells in your neck would crack whenever you nodded your head!

The plasma membrane is asymmetrical. For example, the types of phospholipids in each half of the lipid bilayer can be different. Phospholipids can vary especially by the kind of nitrogen group they have, and certain kinds are more likely to be found in the outer half of the membrane than in the inner half. *Glycolipids* are another type of lipid found in the plasma membrane. They are like phospholipids except that the hydrophilic head is made up of a variety of sugars joined to form a straight or branching carbohydrate chain. As figure 6.3 shows, glycolipids contribute to the asymmetry of the plasma membrane because the carbohydrate chain is always directed externally, an observation that offers a clue to their function. It is possible that glycolipids function as part of a receptor or help mark the cell as belonging to a particular individual. They might also regulate the action of plasma membrane proteins involved in the growth of the cell, and if so they might have a role in the occurrence of cancer.

In animal cells, **cholesterol** is a major membrane lipid, equal in amount to phospholipids. Cholesterol has both a hydrophilic and a hydrophobic end and arranges itself in this manner:

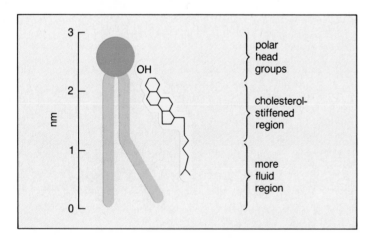

Cholesterol makes the membrane more impermeable to most biological molecules. Small molecules, like amino acids and sugars, do not pass through the membrane easily because they are polar, and the center of the membrane is nonpolar. Passage of molecules through the membrane is dependent upon the protein component.

The phospholipid bilayer portion of the plasma membrane forms a hydrophobic impermeable barrier that prevents the movement of polar (most biological) molecules through the plasma membrane.

## Protein Component

The proteins associated with the plasma membrane are either attached to its inner surface or embedded in the phospholipid bilayer. Proteins often have hydrophobic and hydrophilic regions; hydrophobic regions are within the membrane, while the hydrophilic regions project from both surfaces of the bilayer:

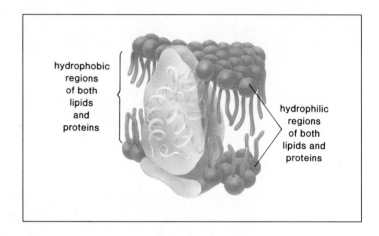

Although proteins are oriented so that a particular hydrophilic region always projects from either the outer or inner surface, they can move laterally in the fluid lipid bilayer. This has been demonstrated by the experiment explained in figure 6.4.

The protein composition of the plasma membrane varies among cells—red blood cell plasma membrane can have 20%-40% lipid and 60%-80% protein, for example. Also, the proteins on the outer and inner surfaces of the membrane can differ, and only those on the inner surface are attached to cytoskeletal filaments. This shows that the plasma membrane is asymmetrical, as does the placement of glycoproteins. *Glycoproteins,* like glycolipids, have an attached carbohydrate chain that only projects externally from the membrane.

Glycoproteins make cell-to-cell recognition possible and serve as the "fingerprints" of the cell. They enable the immune system to recognize its own organs and, unfortunately, are the cause of an immune system response to transplanted organs. As we will discuss in more detail, certain plasma membrane proteins are involved in the passage of molecules through the membrane. Some of these have a *channel* through which a substance simply can move across the membrane; others are *carriers* that combine with a substance and help it to move across the membrane. Still other proteins are *receptors;* each type of receptor has a specific shape that allows a specific molecule to bind to it. The binding of a molecule, such as a hormone, can influence the metabolism of the cell. Viruses often must attach to

## Figure 6.4

Experiment to demonstrate that plasma membrane proteins move about. Mouse cells and human cells are exposed to their respective fluorescent antibodies. Proper laboratory treatment causes the cells to fuse. Immediately after fusion, microscopic observation shows that the mouse and human glycoproteins are still separated from one another. Forty minutes after fusion, the glycoproteins are completely intermixed. This shows that the proteins are able to move within the plasma membrane.

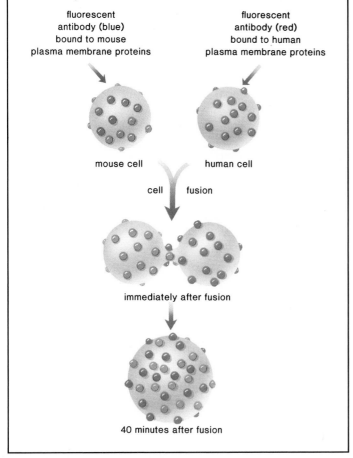

fluorescent antibody (blue) bound to mouse plasma membrane proteins

fluorescent antibody (red) bound to human plasma membrane proteins

mouse cell

human cell

cell fusion

immediately after fusion

40 minutes after fusion

receptors before they enter a cell. Some proteins have an *enzymatic function* and carry out metabolic reactions. Figure 6.5 depicts the various functions of membrane proteins.

# Movement of Molecules into and out of Cells

If the plasma membrane were completely permeable, all substances would move across it with no difficulty. Since it can be observed that all substances cannot move across the plasma membrane, it is correct to say that the plasma membrane is semipermeable. It can further be observed, however, that proteins are often involved in allowing some selected substances to move across the membrane. In addition, vesicle formation brings only certain macromolecules into cells. Because the plasma membrane is able to regulate which substances enter and exit the cell, it is said to be **selectively permeable.**

> The plasma membrane is selectively permeable—it has special mechanisms to regulate the passage of most molecules into and out of the cell.

Certain small molecules are simply able to diffuse across the plasma membrane.

## Diffusion

Ions and molecules in air and in water constantly move about, seemingly at random. It can be demonstrated, however, that there is a net movement of molecules from a region of greater concentration to a region of lesser concentration. For example, when a few crystals of dye are placed in water, the dye molecules move in various directions, but their net movement is in a direction away from the concentrated source (fig. 6.6). The net movement of the water molecules is in the opposite direction. Eventually, the dye is evenly dissolved in the water, resulting in a colored solution. The dye and water molecules continue to move about, but there is no net movement in any one direction.

**Diffusion** is the movement of molecules from the area of greater concentration to the area of lesser concentration. Diffusion occurs spontaneously, and no energy is required to bring it about; however, diffusion does require a difference in concentrations.

A few substances freely diffuse across plasma membranes. For example, the respiratory gases, oxygen and carbon dioxide, diffuse into and out of cells. After you have taken in a breath of air, there is a greater concentration of oxygen in the alveoli of your lungs than there is in the blood capillaries surrounding the lungs (fig. 6.7). Oxygen diffuses down this *concentration gradient* from the lungs into the blood. Water is another molecule that simply diffuses across cell membranes. This has important biological consequences, which we will examine next.

> Only a few types of small molecules freely diffuse through the plasma membrane from the area of greater concentration to the area of lesser concentration.

### Osmosis

The diffusion of water across a selectively permeable membrane has been given a special name: it is called **osmosis.** To illustrate osmosis, a thistle tube covered at one end by a selectively permeable membrane is placed in a beaker of distilled water. Inside the tube is a solution containing both a solute (e.g., sugar molecules) and a solvent (water) (fig. 6.8). Obviously, the lesser solute and greater water concentration is outside the tube, so there will be a net movement of water from outside the tube to inside through the membrane. The solute is unable to pass out of the tube because the membrane is impermeable to it; therefore, the level of the solution within the tube rises (fig. 6.8c). The eventual height of the solution

## Figure 6.5

Membrane protein diversity. These are some of the functions performed by proteins found in the plasma membrane.

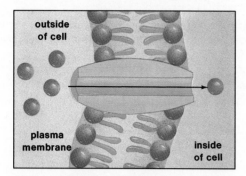

### Channel Protein

A protein that allows a particular molecule or ion to cross the plasma membrane freely as it enters or exits the cell. Recently, it has been shown that cystic fibrosis, an inherited disorder, is caused by a faulty chloride ($Cl^-$) channel. When the channel is not functioning normally, a thick mucus collects in airways and in pancreatic and liver ducts.

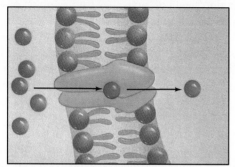

### Carrier Protein

A protein that selectively interacts with a specific molecule or ion so that it can cross the plasma membrane to enter or exit the cell. The carrier protein that transports sodium ($Na^+$) ions and potassium ($K^+$) ions across the plasma membrane requires ATP energy. The inability of some persons to use up energy for sodium-potassium transport has been suggested as the cause of their obesity.

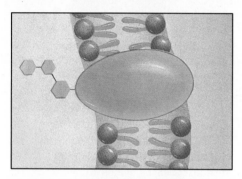

### Cell Recognition Protein

A glycoprotein that identifies the cell. For example, the MHC (major histocompatibility complex) glycoproteins are different for each person, so organ transplants are difficult to achieve. Cells with foreign MHC glycoproteins are attacked by blood cells responsible for immunity.

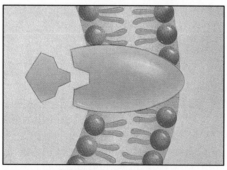

### Receptor Protein

A protein that is shaped in such a way that a specific molecule can bind to it. Recently, it has been shown that Pygmies are short not because they do not produce enough growth hormone, but because their plasma membrane growth hormone receptors are faulty and cannot interact with growth hormone.

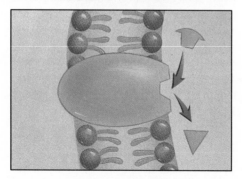

### Enzymatic Protein

A protein that catalyzes a specific reaction. For example, there is a plasma membrane protein called adenylate cyclase that is involved in ATP metabolism. Polluted water can contain cholera bacteria, which release a toxin that interferes with the proper functioning of adenylate cyclase. So many sodium ions and so much water leave intestinal cells that the individual dies from severe diarrhea.

**Figure 6.6**

Process of diffusion is spontaneous and no energy is required to bring it about. **a.** When dye crystals are placed in water they are concentrated in one area. **b.** The dye dissolves in the water, and there is a net movement of dye molecules away from the area of concentration. There is a net movement of water molecules in the opposite direction. **c.** Eventually, the water and dye molecules are equally distributed throughout the container.

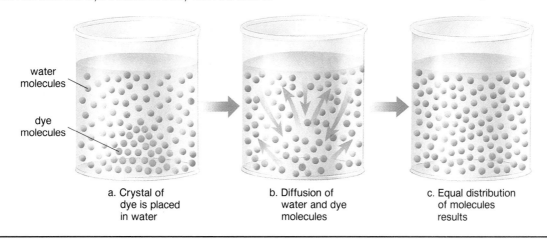

water molecules

dye molecules

a. Crystal of dye is placed in water

b. Diffusion of water and dye molecules

c. Equal distribution of molecules results

**Figure 6.7**

Oxygen (dots) diffuses into the capillaries because there is a greater concentration of oxygen in the alveoli (air sacs) of the lungs than in the capillaries.

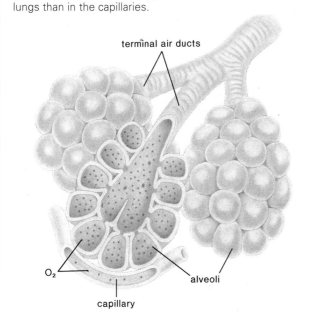

terminal air ducts

$O_2$

alveoli

capillary

2. A difference in solute and water concentrations exists on the 2 sides of the membrane.

3. The membrane is impermeable (does not permit passage) to the solute particles.

4. The membrane is permeable (permits passage) to the water, which moves from the area of greater water (lesser solute) concentration to the area of lesser water (greater solute) concentration.

5. An osmotic pressure is present; the amount of liquid increases on the side of the membrane with the greater solute concentration.

These considerations will be important as we discuss osmosis in relation to cells placed in different solutions. The plasma membrane is not completely impermeable to solutes such as sugars and salts, but the difference in permeability between water and these solutes is so great that cells do have to cope with the osmotic movement of water.

> Osmosis is the diffusion of water across a selectively permeable membrane. The presence of osmotic pressure is evident when there is an increased amount of water on the side of the membrane that has the greater solute concentration.

### Tonicity

Cells can be placed in solutions that are isotonic, hypotonic, or hypertonic (table 6.1 and fig. 6.9). In an **isotonic solution,** a cell neither gains nor loses water because the concentration of solute and water is the same on both sides of the plasma membrane (*iso*—same as). Most animal cells normally live under isotonic conditions. For example, the tissue fluid that surrounds your cells is normally isotonic to them.

in the tube indicates the degree of **osmotic pressure** caused by the flow of water from the area of greater water concentration to the area of lesser water concentration.

Notice the following elements in this illustration of osmosis:

1. A selectively permeable membrane separates a solution from pure water.

## Figure 6.8

Osmosis demonstration. *a.* A thistle tube, covered at the broad end by a selectively permeable membrane, contains a sugar solution. The beaker contains only solvent (water). *b.* The solute (green circles) is unable to pass through the membrane, but the water passes through in both directions. There is a net movement of water toward the inside of the thistle tube, where the solute concentration is greater and the water concentration is lesser. *c.* In the end, the level of the solution rises in the thistle tube until a pressure equivalent to osmotic pressure builds.

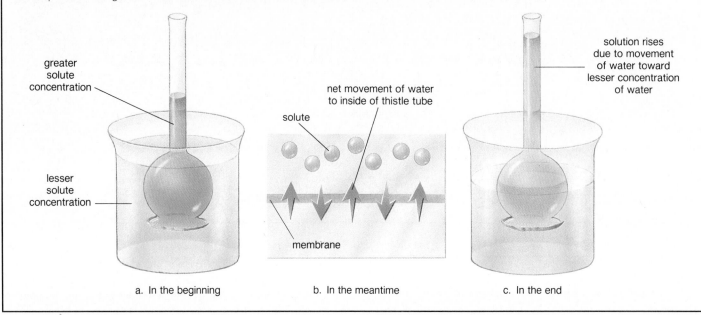

greater solute concentration

lesser solute concentration

net movement of water to inside of thistle tube

solute

membrane

solution rises due to movement of water toward lesser concentration of water

a. In the beginning     b. In the meantime     c. In the end

In a **hypotonic solution,** a cell tends to gain water because the concentration of solute is lesser (*hypo*—less than) and the concentration of water is greater than that of the cell. Plants and algae that live in freshwater ponds live under hypotonic conditions. The cytoplasm and central vacuoles gain water, and the plasma membrane pushes against the rigid cell wall. The resulting pressure, called **turgor pressure,** helps give internal support to the cell. Even plants that live on land are dependent upon turgor pressure to keep them from wilting. Freshwater protozoans are not protected from osmotic swelling by a rigid cell wall, but they have a contractile vacuole that rapidly pumps out excess water (fig. 6.10).

In a **hypertonic solution,** a cell tends to lose water because the concentration of solute is greater (*hyper*—more than) and the concentration of water is less than that of the cell. When plant cells are placed in a hypertonic solution, it is possible to see that the cytoplasm has lost water and has undergone plasmolysis because the plasma membrane pulls away from the cell wall. Some organisms, such as marine fishes, live in a hypertonic environment. The gills of these fishes extrude salt from the blood, and the blood remains isotonic to the cells.

Cells can die when placed in a solution of unfavorable tonicity. For example, red blood cells will lyse (burst) when placed in a very hypotonic solution and will undergo crenation (dry up) when placed in a very hypertonic solution (fig. 6.9*e* and *f*). Organisms that live in environments with unfavorable tonicities have evolved mechanisms to keep the tonicity of cellular cytoplasm within a normal range.

### Table 6.1
Effect of Osmosis on a Cell

| Tonicity of Solution | Concentrations | | Net Movement of Water | Effect on Cell |
|---|---|---|---|---|
| | *Solute* | *Water* | | |
| Isotonic | Same as cell | Same as cell | None | None |
| Hypotonic | Less than cell | More than cell | Cell gains water | Swells, turgor pressure |
| Hypertonic | More than cell | Less than cell | Cell loses water | Shrinks, plasmolysis |

When a cell is placed in an isotonic solution, it neither gains nor loses water. When a cell is placed in a hypotonic solution (lesser solute concentration than isotonic) the cell gains water. When a cell is placed in a hypertonic solution (greater solute concentration than isotonic), the cell loses water and the cytoplasm shrinks.

## Transport by Carriers

The presence of the plasma membrane impedes the passage of all but a few substances. Yet biologically useful molecules do enter and exit the cell at a rapid rate because there is a transport system

## Figure 6.9

Osmosis in plant and animal cells. The arrows indicate the net movement of water. In an isotonic solution, a cell neither gains nor loses water; in a hypotonic solution, a cell gains water; and in a hypertonic solution, a cell loses water.

**Plant Cells**

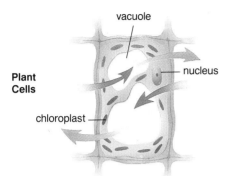

vacuole

nucleus

chloroplast

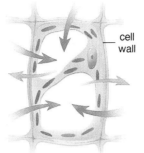

cell wall

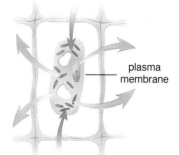

plasma membrane

a. Under isotonic conditions, there is no net movement of water.

b. In a hypotonic environment, vacuoles fill with water, turgor pressure develops, and chloroplasts are seen next to cell wall.

c. In a hypertonic environment, vacuoles lose water, cytoplasm shrinks (called plasmolysis), and chloroplasts are seen in center of cell.

**Animal Cells**

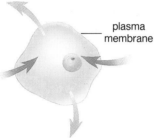

plasma membrane

d. Under isotonic conditions, there is no net movement of water.

e. In a hypotonic environment, water enters cell, which may burst (lyse) due to osmotic pressure.

f. In a hypertonic environment, water leaves the cell, which shrivels (crenation occurs).

## Figure 6.10

Many organisms have evolved mechanisms to cope with the tonicity of the surrounding medium. A paramecium **a.** lives in fresh water, which is hypotonic to it. It has contractile vacuoles that expel water from the cell. Magnification, X107. **b.** Water from the cytoplasm flows into tubules, filling vacuole. **c.** The vacuole collapses, appearing to "contract," as water is released through a temporary opening in the plasma membrane.

contractile vacuole

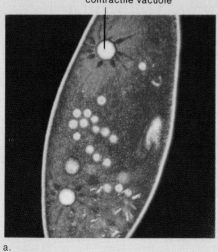

a.

b.

c.

Membrane Structure and Function

## Figure 6.11

Facilitated transport. A carrier protein speeds the rate at which the solute crosses the plasma membrane in the direction of decreasing concentration. Note that the protein carrier undergoes a change in shape (called a conformational change) as it moves a solute across the membrane.

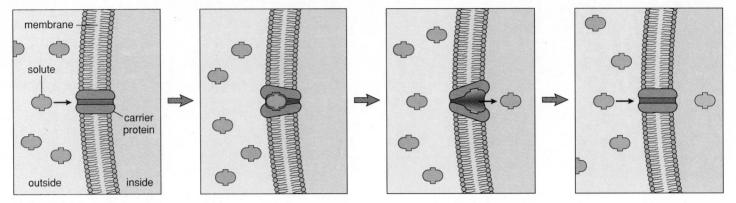

in the plasma membrane that delivers them in a timely manner. Some of the proteins that span the plasma membrane act as **carriers** to transport these molecules. Carrier proteins are specific, and each can combine with only a certain type of molecule, which is then transported across the plasma membrane. It is not completely understood how carrier proteins function, but after they combine with a substance, they are believed to undergo a *conformational change,* a change in shape that causes the substance to be moved across the membrane.

> Some of the proteins in the plasma membrane are carriers that transport biologically useful molecules into and out of the cell.

### Facilitated Transport

A model for **facilitated transport** suggests that after a carrier has assisted the movement of a molecule to one side of the membrane, it is free to assist the passage of another similar molecule (fig. 6.11). This form of transport is involved in carrying various sugars, amino acids, and nucleotides into or out of a cell according to concentration gradients. Because the molecule is still traveling down its concentration gradient, no significant energy input is required.

### Active Transport

Due to **active transport,** molecules or ions accumulate either inside or outside the cell, even when this is the region of greater concentration. For example, the plasma membranes of plant root cells are able to extract inorganic ions from the soil due to active transport. We already mentioned that in marine fishes the cells in the gills are able to extrude salt so that the blood remains isotonic to body cells. In our body, the thyroid gland accumulates iodine and the kidneys actively reabsorb sodium ions ($Na^+$) so that water follows by osmosis. If this did not happen, we would be in constant danger of dehydration because of water loss by way of the kidneys. As described in this chapter's reading, the chambered nautilus uses a similar mechanism to maintain buoyancy.

Both a protein carrier and an expenditure of energy are needed for active transport. The necessary energy is produced by the breakdown of ATP (adenosine triphosphate). Therefore, it is not surprising that cells involved primarily in active transport, such as kidney cells, have a large number of mitochondria near the membrane where active transport is occurring. Proteins involved in active transport are often called *pumps* because just as a water pump uses energy to move water against the force of gravity, proteins use energy to move a substance against its concentration gradient.

One type of pump that is active in all cells, but especially associated with nerve and muscle cells, moves sodium ions ($Na^+$) to the outside of the cell and potassium ions ($K^+$) to the inside of the cell. These 2 events are presumed to be linked, and the carrier protein is called a **sodium-potassium pump.** ATP energy released by an ATPase enzyme (an enzyme that breaks down ATP) is believed to be necessary to bring about a change in shape of the carrier, allowing it to combine alternately with $Na^+$ and $K^+$ (fig. 6.12).

Sodium ions and potassium ions are charged particles; the action of the pump causes the plasma membrane to be positively charged on the outside and negatively charged on the inside. Therefore, the membrane has an electrical gradient; this, along with any concentration gradient, influences the movement of charged particles. Charged particles tend to move according to their *electrochemical gradient.*

The sodium gradient, established by the sodium-potassium pump, provides the energy for an animal cell to take up other types of molecules, such as sugars and amino acids. For example, there are carriers in the plasma membrane of an animal cell that cotransport both sodium ions and an amino acid. Because there are so many more sodium ions outside the cell than inside the cell, the carrier automatically transports sodium ions, and consequently the amino acid as well, even if there is

# Chambered Nautilus

The chambered nautilus (fig. 6.A), a mollusk that takes its name from its multichambered shell, still occurs today in deep tropical oceans, but it was much more prevalent some 65 million years ago. Along with closely related forms, it dominated the oceans. The advantage these animals must have had was the ability to achieve buoyancy. Thus, they were no longer confined to the ocean floor. Recently, scientists discovered that as the nautilus grows, it empties liquid from each newly formed chamber of its spiral shell. This makes it light enough to float in water.

Research has been conducted into the puzzle of how the nautilus empties its chambers. A complex tubular organ, the siphuncle, contains blood vessels spirals in the same manner as the shell (b and c). Water from each newly formed chamber makes its way to these blood vessels, but by what means? The blood pressure within these vessels should cause water to *leave* the blood vessels, not enter them!

Electron micrographs have provided a possible explanation. They show that the wall of the siphuncle (d) has a folded tissue lining containing many small grooves (e). If the cells making up this tissue use active transport to carry solutes from the chamber liquid to the grooves, it would establish a concentration gradient causing water to follow passively by osmosis. This process is called *local osmosis* because it depends on the buildup of solutes by active transport in an isolated area, such as the grooves. The cells about these grooves do indeed contain many mitochondria that can supply the energy needed for active transport of the solute into the grooves (f) so that water will follow passively by means of osmosis. After being withdrawn, the water moves from the grooves of the tubular organ to collecting channels that funnel it to the blood vessels.

Although the nautilus can empty its chambers, it never attempts to fill them. This observation is consistent with the hypothesis that local osmosis accounts for how the chambered nautilus empties the chambers of its shell so that they contain only gas and not liquid.

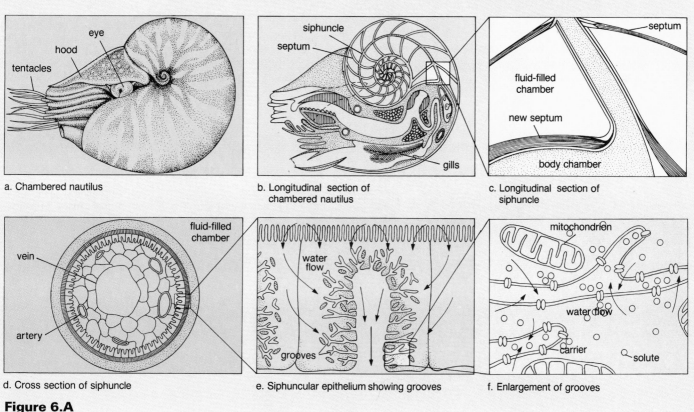

a. Chambered nautilus

b. Longitudinal section of chambered nautilus

c. Longitudinal section of siphuncle

d. Cross section of siphuncle

e. Siphuncular epithelium showing grooves

f. Enlargement of grooves

**Figure 6.A**
Aspects of chambered nautilus anatomy and physiology.

already a greater concentration of amino acid inside the cell than outside. The sodium-potassium pump establishes a gradient that is utilized for the uptake of nutrient molecules; therefore, we can understand why it apparently requires one-third of all our energy expenditure!

During facilitated transport (no energy required), small molecules follow their concentration gradients. During active transport (energy required), small molecules go against their concentration gradient.

## Figure 6.12

The sodium-potassium pump. The same carrier protein transports sodium ions (Na⁺) to the outside of the cell and potassium ions (K⁺) to the inside of the cell because it undergoes an ATP dependent conformational change. The result is both a concentration gradient and an electrical gradient for these ions across the plasma membrane. Three sodium ions are carried outward for every 2 potassium ions carried inward; therefore, the inside of the cell is negatively charged compared to the outside.

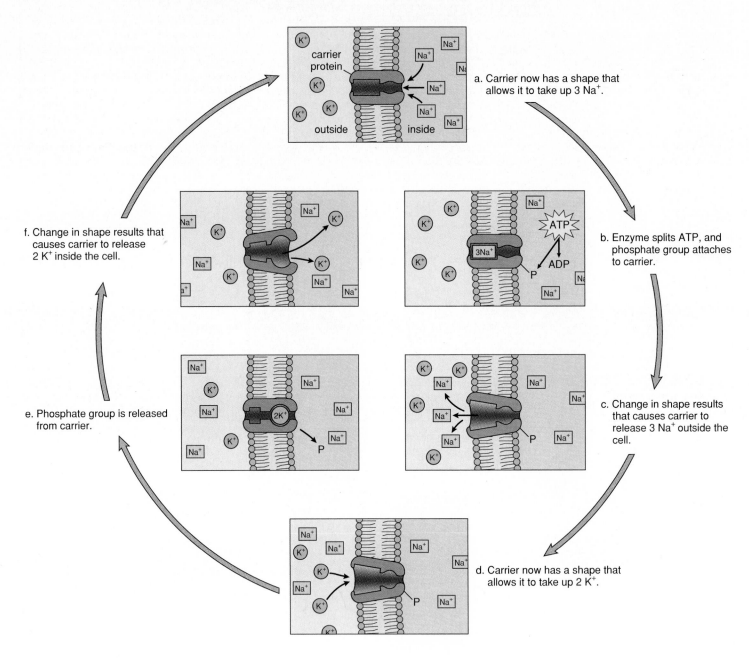

a. Carrier now has a shape that allows it to take up 3 Na⁺.

b. Enzyme splits ATP, and phosphate group attaches to carrier.

c. Change in shape results that causes carrier to release 3 Na⁺ outside the cell.

d. Carrier now has a shape that allows it to take up 2 K⁺.

e. Phosphate group is released from carrier.

f. Change in shape results that causes carrier to release 2 K⁺ inside the cell.

## Endocytosis and Exocytosis

Some substances are too large to be transported by protein carriers; instead they are taken into the cell by **endocytosis:**

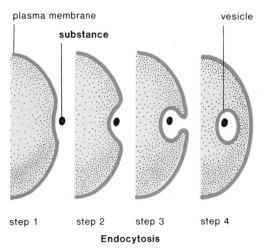

**Endocytosis**

When the material taken in is quite large, the process is called **phagocytosis** (cell eating), and the result is formation of a vacuole. Phagocytosis is common in amoeboid-type cells such as the cells known as macrophages, which take in bacteria and worn-out red blood cells (fig. 6.13).

When macromolecules are taken in by endocytosis, the process is called **pinocytosis** (cell drinking), and the result is formation of a vesicle. Both phagocytic vacuoles and pinocytic vesicles can fuse with lysosomes, whose enzymes digest their contents.

The opposite of endocytosis is **exocytosis.** During exocytosis, a vesicle fuses with the plasma membrane, discharging its contents:

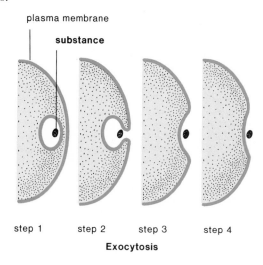

**Exocytosis**

### Figure 6.13
Macrophages are large cells that act as scavengers. These "big eaters" take in (phagocytize) all sorts of debris, including worn out red blood cells, as shown here.

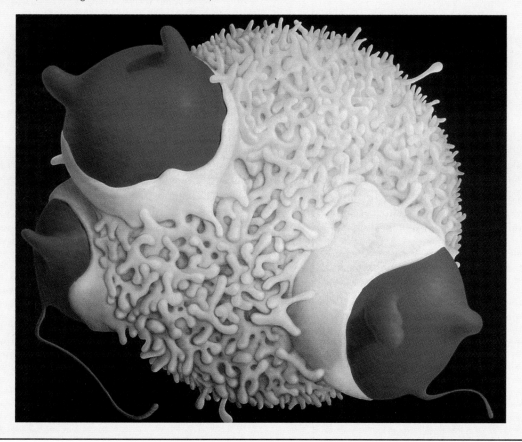

Membrane Structure and Function

**Figure 6.14**

Receptor-mediated endocytic cycle takes up membrane when endocytosis occurs and returns it when exocytosis occurs. ***a.*** (1) The receptors in the coated pits combine only with a specific substance, called a ligand. (2) The vesicle that forms is at first coated with the structural protein clathrin but soon the vesicle loses its coat. (3) Ligands leave the vesicle. (4) When exocytosis occurs membrane is returned to the plasma membrane. ***b.*** Electron micrographs of a coated pit in the process of forming a vesicle.

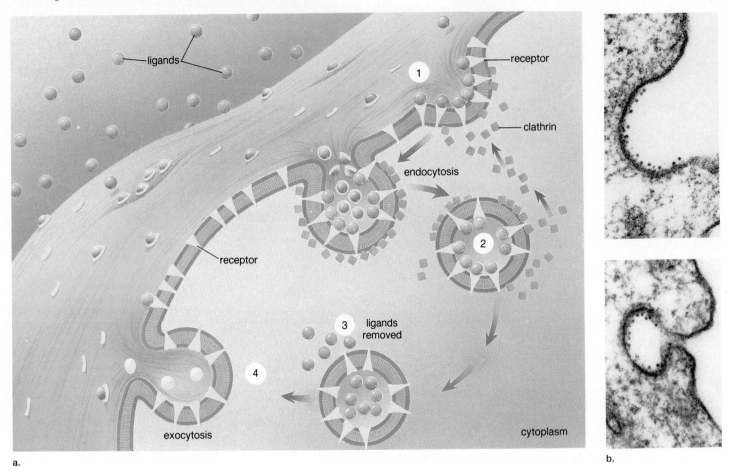

a.

b.

As noted in chapter 5, vesicles formed at the Golgi apparatus secrete cell products in this manner. Exocytosis adds plasma membrane to a cell, and endocytosis takes it away. It can be reasoned that the amount of plasma membrane lost by endocytosis must be replaced by exocytosis unless the cell is growing. It is possible in that case that exocytosis would occur simply for the purpose of adding on more plasma membrane.

### Receptor-Mediated Endocytic Cycle

Receptor-mediated endocytosis is a form of pinocytosis that is very specific because it involves the use of plasma membrane receptors. A macromolecule that binds to a receptor is called a ligand (fig. 6.14). The binding of ligands to specific receptor sites causes the receptors to gather at one location before endocytosis occurs. This location is called a coated pit because there is a layer of fibrous protein, called clathrin, on the cytoplasmic side. Clathrin is a protein designed to form lattices around membranous vesicles. When a vesicle forms, it also is coated, but soon it loses its coat. At this point the ligands can directly enter the cell or else end up in lysosomes, which digest them to smaller molecules, which enter the cell. In any case, the receptors return to the plasma membrane and exocytosis occurs.

The importance of receptor-mediated endocytosis is exemplified by the occurrence of a genetic disease. Normally cells take up cholesterol, which is carried in the blood by a lipoprotein called low density lipoprotein (LDL). When cells need more cholesterol for membrane production they produce receptors for LDL. After LDL molecules bind to receptors, receptor-mediated endocytosis occurs. Later, the receptors are returned to the plasma membrane and lysosomes disengage cholesterol from LDL. This whole process goes awry in individuals who lack a gene or inherit a faulty gene for the LDL receptor. Because cholesterol is unable to enter their cells, it builds up and forms plaque on blood vessel walls leading to cardiovascular disease and heart attacks. Children with this genetic disorder have been known to have heart attacks even as early as 6 years old.

## Table 6.2
### Passage of Molecules into and out of Cells

| Name | Direction | Requirements | Examples |
|------|-----------|--------------|----------|
| *Diffusion* | Toward lesser concentration | — | Lipid-soluble molecules Water Gases |
| *Transport* | | | |
| Facilitated | Toward lesser concentration | Carrier | Sugars and amino acids |
| Active | Toward greater concentration | Carrier plus energy | Sugars, amino acids, and ions |
| *Endocytosis* | | | |
| Phago-cytosis | Toward inside | Vesicle formation | Cells and subcellular material |
| Pinocytosis | Toward inside | Vesicle formation | Macromolecules |
| Receptor-mediated | Toward inside | Vesicle formation | Macromolecules |
| *Exocytosis* | Toward outside | Vesicle fuses with plasma membrane | Macromolecules |

Endocytosis and exocytosis are other ways that substances can enter and exit cells. Table 6.2 summarizes the various ways that molecules pass into and out of cells.

# Modifications of the Cell Surface

The plasma membrane is the outer living boundary of the cell, but many cells have an additional component that is formed outside the membrane. For example, animal cells have an extracellular matrix and plant cells have a cell wall.

## Extracellular Matrix of Animal Cells

An extracellular matrix is a meshwork of fibrous proteins and polysaccharides that is laid down by cells and is found in intercellular spaces. The carbohydrate chains of glycolipids and glycoproteins are a part of the extracellular matrix, but there are also various other types of macromolecules that are secreted by but are not part of the cell itself. Some of these molecules (e.g., unattached glycoproteins) form a gel in which other molecules (e.g., collagen) form fibers. Animal cells like to adhere, that is, stick to, a surface, and the extracellular matrix provides a suitable surface for adherence. The extracellular matrix not only keeps the cells stationary, it also helps them keep their usual shape so necessary to performing their usual functions.

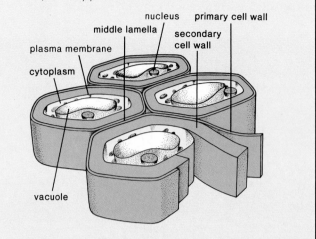

### Figure 6.15
Plant cell wall. All plant cells have a primary cell wall and some have a secondary cell wall. The cell wall, which lends support to the cell, is freely permeable.

## Cell Walls of Plant Cells

In addition to a plasma membrane, plant cells are surrounded by a porous cell wall that varies in thickness, depending on the function of the cell (fig. 6.15). All plant cells have a primary cell wall, whose best-known component is cellulose polymers united into thread-like microfibrils that form fibrils (fig. 4.8). The cellulose fibrils form a framework whose spaces are filled by noncellulose molecules. Among these are found pectic substances, which allow the wall to stretch when the cell is growing, and hemicelluloses, which harden the wall when the cell is mature. A layer of pectic substances, called the middle lamella, also occurs between plant cells. Some cells in woody plants have a secondary wall that forms inside the primary cell wall. The secondary wall has a greater quantity of cellulose fibrils than the primary wall, and these layers are laid down at right angles to one another. Lignin, a substance that adds strength, is another common ingredient of secondary cell walls in woody plants.

The plasma membrane in animals is supported by an extracellular matrix, and in plants it is supported by a cell wall. Both of these lend strength to the cell and do not interfere with the functions of the plasma membrane.

## Junctions between Cells

The plasma membrane of cells sometimes interacts with other cells. The junctions that occur between cells are an example of such cellular interaction.

### Animal Cells

Three types of junctions are seen between animal cells: spot desmosome, tight junction, and gap junction (fig. 6.16). In a *spot desmosome,* internal cytoplasmic plaques, firmly attached to the

## Figure 6.16

Cell-to-cell junctions between animal cells. ***a.*** In a spot desmosome, filaments pass from one cell to the other at a region where plaques of dense material are reinforced by cytoskeleton filaments. ***b.*** In a tight junction, adjacent proteins are bonded directly together, preventing any passage of materials in the space between the cells. ***c.*** In gap junctions, donut-shaped proteins from each cell join to form tiny channels. These allow small molecules to pass from cell to cell. These junctions span the "gap" between cells.

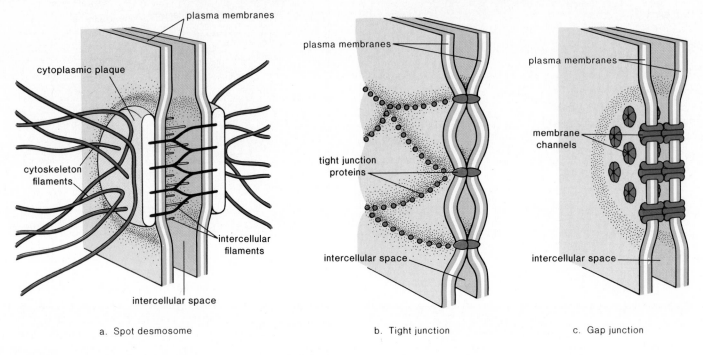

a. Spot desmosome

b. Tight junction

c. Gap junction

## Figure 6.17

Cell-to-cell junction between plant cells. At the plasmodesmata, the plasma membrane of adjacent cells passes through the cell wall to form channels, which allow substances to pass from one cell to the other. ***a.*** Electron micrograph. ***b.*** Drawing.

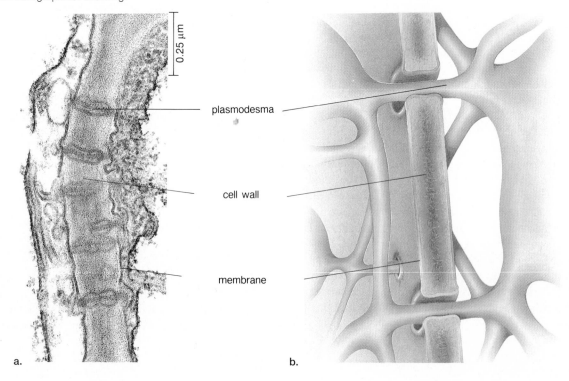

a.

b.

cytoskeleton within each cell, are joined by intercellular filaments. Plasma membranes of adjacent cells is even more closely joined in a *tight junction,* where plasma membrane proteins actually attach to each other, producing a zipperlike fastening. These 2 junctions lend strength and support to the cells. In addition, there is another type of junction that allows cells to communicate. A *gap junction* is formed when 2 identical plasma membrane channels join. The channel of each cell is lined by 6 plasma membrane proteins. A gap junction lends strength to the cells, but it also allows small molecules and ions to pass between them.

All 3 of these junctions are observed between cells that line the lumen (cavity) of the small intestine. These cells pump the products of digestion out of the intestinal lumen and then deposit them in the blood. The presence of tight junctions, in particular, means that the digestion products must enter the cells. Tight junctions also restrict the lateral movement of proteins in the plasma membrane, allowing the surface facing the lumen of the intestine to maintain a different mix of proteins than does the

surface adjacent to the underlying blood vessel. These surfaces carry on different functions depending on the proteins they contain.

---

In animal cells the plasma membrane is joined by desmosomes and tight junctions. Gap junctions allow small molecules to pass from one cell to the other.

---

### Plant Cells

In plant tissues, the cytoplasm of neighboring cells can be connected by numerous narrow channels that pass through the cell wall. These channels are bounded by plasma membrane and contain cytoplasmic strands, called **plasmodesmata,** plus a tightly constricted tubular portion of endoplasmic reticulum. Plasmodesmata allow direct exchange of materials between neighboring plant tissue cells (fig. 6.17).

---

In plant cells, plasmodesmata are strands of cytoplasm that allow small molecules to pass from one cell to the other.

---

## Summary

1. The fluid-mosaic model of membrane structure developed by Singer and Nicolson was preceded by several other models. Electron micrographs of freeze-fractured membranes support the fluid-mosaic model and not Robertson's unit membrane concept based on the Danielli and Davson sandwich model.

2. There are 2 components of the membrane, lipids and proteins. In the lipid bilayer, phospholipids are arranged with their hydrophilic (polar) heads at the surfaces and their hydrophobic tails in the interior. The lipid bilayer has the consistency of oil but acts as a barrier to the entrance and exit of most biological molecules. Glycolipids and glycoproteins are involved in marking the cell as belonging to a particular individual.

3. The hydrophobic portion of a protein lies in the lipid bilayer of the plasma membrane, and the hydrophilic portion occurs at the surfaces. Proteins act as receptors, carry on enzymatic reactions, join cells together, form channels, or act as carriers to move substances across the membrane.

4. Some molecules (lipid-soluble compounds, water, gases) simply diffuse across the membrane from the area of greater concentration to the area of lesser concentration. No energy is required for diffusion to occur.

5. The diffusion of water across a selectively permeable membrane is called osmosis. Water moves across the membrane into the area of lesser water (greater solute) content. When cells are in an isotonic solution, they neither gain nor lose water; when they are in a hypotonic solution, they gain water; and when they are in a hypertonic solution, they lose water.

6. Other molecules are transported across the membrane by carrier proteins that span the membrane.

7. During facilitated transport, a protein carrier assists the movement of a molecule down its concentration gradient. No energy is required.

8. During active transport, a protein carrier acts as a pump that causes a substance to move against its concentration gradient. The sodium-potassium pump carries $Na^+$ to the outside of the cell and $K^+$ to the inside of the cell. Energy in the form of ATP molecules is required for active transport to occur.

9. Larger substances can enter and exit from a membrane by endocytosis and exocytosis. Endocytosis includes both phagocytosis and pinocytosis. Exocytosis involves secretion.

10. The receptor-mediated endocytic cycle makes use of receptor molecules in the plasma membrane.

Once specific substances (e.g., nutrients) bind to their receptors, the entire molecule is drawn into a coated pit, which becomes a coated vesicle. After losing the coat, and after the movement of the nutrients into the cytoplasm, the receptor-containing vesicle fuses with the plasma membrane.

11. Plant cells have a freely permeable cell wall whose main component is cellulose.

12. Junctions between animal cells include spot desmosomes and tight junctions, which help to hold cells together, and gap junctions, which allow passage of small molecules between cells. Plant cells are joined by small channels that span the cell wall and contain plasmodesmata, strands of cytoplasm, which allow materials to pass from one cell to another.

---

### Writing Across the Curriculum

In order to practice writing skills, students should write out the answers to any or all of the study questions and the critical thinking questions. The study questions are sequenced in the same order as the text. Suggested answers to the critical thinking questions are in appendix D.

---

## Study Questions

1. Describe the fluid-mosaic model of membrane structure as well as the models that preceded it. Cite the evidence that either disproves or supports these models.
2. Tell how the phospholipids are arranged in the plasma membrane. What other lipids are present in the membrane, and what functions do they have?
3. Describe how proteins are arranged in the plasma membrane. What are their functions? Describe an experiment that indicates that proteins can move laterally in the membrane.
4. What is diffusion, and what substances can diffuse through a selectively permeable membrane?
5. Describe an experiment that measures osmotic pressure.
6. Tell what happens to an animal cell and a plant cell when placed in isotonic, hypotonic, and hypertonic solutions.
7. Why do substances have to be assisted through the plasma membrane? Contrast movement by facilitated transport with movement by active transport.

## Objective Questions

1. Electron micrographs following freeze-fracture of the plasma membrane indicate that
   a. the membrane is a phospholipid bilayer.
   b. some proteins span the membrane.
   c. protein is found only on the surfaces of the membrane.
   d. glycolipids and glycoproteins are antigenic.
2. A phospholipid molecule has a head and 2 tails. The tails are found
   a. at the surfaces of the membrane.
   b. in the interior of the membrane.
   c. both at the surfaces and the interior of the membrane.
   d. spanning the membrane.
3. Energy is required for
   a. active transport.
   b. diffusion.
   c. facilitated transport.
   d. All of these.
4. When a cell is placed in a hypotonic solution,
   a. solute exits the cell to equalize the concentration on both sides of the membrane.
   b. water exits the cell toward the area of lower concentration.
   c. water enters the cell toward the area of higher solute concentration.
   d. solute exits and water enters the cell.
5. A protozoan's contractile vacuole is more likely to be active
   a. in a hypertonic environment.
   b. in an isotonic environment.
   c. in a hypotonic environment.
   d. when endocytosis is occurring.
6. Active transport
   a. requires a protein carrier.
   b. moves a molecule against its concentration gradient.
   c. requires a supply of energy.
   d. All of these.
7. The sodium-potassium pump
   a. helps establish an electrochemical gradient across the membrane.
   b. concentrates sodium on the outside of the membrane.
   c. utilizes a protein carrier and energy.
   d. All of these.
8. Receptor-mediated endocytosis
   a. is no different from phagocytosis.
   b. brings specific nutrients into the cell.
   c. helps to concentrate proteins in vesicles.
   d. All of these.
9. Plant cells
   a. always have a secondary cell wall, and the primary one may disappear.
   b. have channels between cells that allow strands of cytoplasm to pass from cell to cell.
   c. develop turgor pressure when water enters the nucleus.
   d. do not have cell-to-cell junctions like animal cells.
10. Label this diagram of the plasma membrane. Why are the phospholipid "tails" hydrophobic?

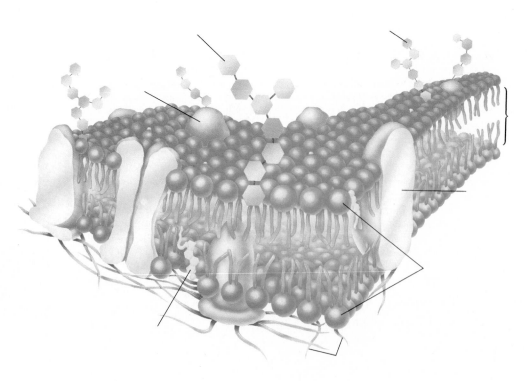

**11.** Write hypotonic solution or hypertonic solution beside each cell. Justify your conclusions.

a. _____

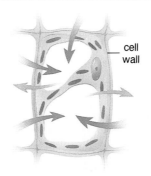

cell wall

b. _____

## Concepts and Critical Thinking

**1.** *All cells have a membrane that separates their contents from the extracellular environment and serves to maintain their integrity.*

In a structural sense, how does the plasma membrane maintain the integrity of the cell? In a functional sense, how does the plasma membrane maintain the integrity of the cell?

**2.** *To remain alive, cells must have an extracellular environment that is compatible with their continued existence.*

How does the phenomenon of osmosis demonstrate this concept?

**3.** *Cells of multicellular organisms function as a unit.*

For cells to function as a unit, communication is necessary. How does the plasma membrane help distantly located cells communicate with each other? How does the plasma membrane help adjacently located cells communicate with each other?

## Selected Key Terms

**fluid-mosaic model** (floo'id mo-za'ik mod'el) 86
**phospholipid** (fos"fo-lip'id) 87
**cholesterol** (ko-les'ter-ol) 88
**selectively permeable** (sē-lek'tiv-le per'me-ah-b'l) 89
**diffusion** (di-fu'zhun) 89
**osmosis** (oz-mo'sis) 89
**osmotic pressure** (os-mot'ik presh'ur) 91
**isotonic solution** (i"so-ton'ik sol-u'shun) 91

**hypotonic solution** (hi"po-ton'ik sol-u'shun) 92
**turgor pressure** (tur'gor presh'ur) 92
**hypertonic solution** (hi"per-ton'ik sol-u'shun) 92
**carrier** (kar'ē-er) 94
**active transport** (ak'tiv trans'port) 94
**sodium-potassium pump** (so'de-um po-tas'e-um pump) 94

**endocytosis** (en"do-si-to'sis) 97
**phagocytosis** (fag"o-si-to'sis) 97
**pinocytosis** (pi"no-si-to'sis) 97
**exocytosis** (ex"o-si-to'sis) 97

# 7

## Cellular Energy

A lubber grasshopper feeds on leaves, acquiring the materials and energy it needs to maintain its organization. The organic molecules in cells are constantly being built up and broken down. Overall, there is a loss of useful energy, which must then be replaced from an outside source.

Your study of this chapter will be complete when you can

1. state 2 energy laws that have important consequences for living things;
2. explain, on the basis of these laws, why living things need an outside source of energy;
3. describe 2 types of metabolic reactions (pathways) that are found in all cells;
4. describe the structure and function of enzymes and the conditions that affect the yield of enzymatic reactions;
5. describe the regulation of enzymatic reactions (pathways) by the process of inhibition. Compare competitive inhibition, noncompetitive inhibition, and feedback inhibition;
6. describe the function of the coenzymes $NAD^+$ and FAD, and state their role in the production of ATP by the electron transport system in mitochondria. Contrast this electron transport system to the one in chloroplasts;
7. describe the function of the coenzyme $NADP^+$;
8. explain, with the aid of a diagram, how degradative pathways drive synthetic pathways, emphasizing the role of ATP and $NADP^+$. Tell why an outside source of matter and energy should be included in the diagram;
9. describe the structure and function of ATP and how it is produced by chemiosmotic phosphorylation.

We can readily see that a butterfly's wing (fig. 7.1) is highly organized, but it takes a knowledge of cells to realize that they are also organized. All cells contain the same type of carbon-based molecules, and in eukaryotic cells, some of these molecules make up the organelles. It is the organelles (see fig. 5.4 and fig. 5.5) that show us that a cell is just as organized in its own way as the butterfly's wing. Such organization cannot be maintained without a supply of energy. As an analogy, consider that your room quickly becomes a total mess unless you keep doing work to keep it straight (organized). Similarly, a cell must continually use energy to first form macromolecules from unit molecules and then to arrange the macromolecules into its structural components.

## Nature of Energy

**Energy** is defined as the capacity to do work, that is, to bring about a change. There are several different forms of energy. For example, you use your own mechanical energy to straighten up your room; light energy allows you to see what you are doing; electrical energy makes the motor turn in your vacuum cleaner; and heat energy maintains a comfortable room temperature.

Another form of energy, not as obvious as these, is chemical-bond energy, the energy that is stored in the bonds that hold atoms together. For example, the chemical-bond energy in food provides the energy for your muscles to do mechanical work. Just as chemical-bond energy in food can be transformed into the mechanical energy used by your muscles, other forms of energy

**Figure 7.1**
The complex organization of living things is dramatized by the pattern of color in a butterfly's wing. But the unseen cells that make up the wing are also highly organized, and the nectar the butterfly obtains from the flower is needed to maintain the organization of both the cells and the wing.

Celluar Energy

can be transformed into another. For example, electrical energy is transformed into light energy within a light bulb, and oil is transformed into electrical energy at a power plant.

A power plant does not create energy, it merely transforms it. Do transformations ever cause energy to be used up? We realize that matter is never created or destroyed. When we take materials and build a vacuum cleaner, we know that even if we throw the vacuum cleaner away, the materials are still not used up. On page 33, it was mentioned that chemical equations must always be balanced—the same number of atoms are present before and after a reaction—because mass cannot be used up. But what about energy? Can it ever be used up?

## First Energy Law

As you have probably guessed, not only can energy never be created, it also can never be used up. The same amount of energy has always been present in the universe and will always be present. The first energy law states:

> Energy can be transformed from one form into another, but it cannot be created or destroyed.

We have what appears to be a paradox, however, because it often seems as if energy has been "lost" when transformations occur. For example, cars need to be refueled and animals must continue eating. The answer to the paradox is that neither automobiles nor organisms are isolated. It is said that they are a *system* that has *surroundings*. The surroundings supply the gasoline (or the food), and when the gasoline is used to do work, it results in heat that is accepted by the surroundings:

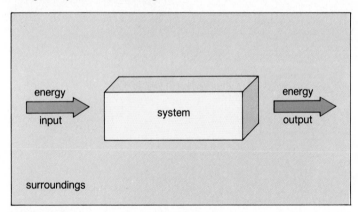

The first energy law can be interpreted to mean that if we put X amount of energy into a system (automobile, organism), then eventually X amount of energy will leave the system even if work has been done. Energy usually leaves in the form of heat.

## Second Energy Law

Although energy was not lost in our examples, something seems to have been lost. To determine what this is, consider that gasoline can be burned to power a car, but the heat given off is no longer able to perform this task. The amount of *useful* energy, then, has changed. Similarly, once food energy is used to power

muscles, this energy is no longer available because it has been converted to heat. The second energy law states:

> When one form of energy is transformed into another form, some useful energy is always lost as heat; therefore, energy cannot be recycled.

The second energy law implies that only processes that decrease the amount of useful energy occur naturally, or spontaneously. The word *spontaneous* does not mean that the process has to happen all at once, it simply means that it happens naturally over time. For example, your room tends to become messy, not neat; water tends to flow downhill, not up; and after death, organisms decay and eventually disintegrate. In other words, the amount of disorder (i.e., entropy) is always increasing in the universe. If this is the case, how do we account for an orderly room or the presence of highly organized structures, such as living cells and organisms? Obviously, these systems must have a continual input of energy, and this continual input of energy eventually increases the entropy of the universe.

While we have been drawing an analogy between nonliving systems (a car, your room) and a living system (cell or organism), there is a remarkable difference between the 2 types of systems. Although living things are continually taking in useful energy and putting out heat just like the nonliving system, the work performed in between these 2 events maintains their organization and even allows them to grow. A living organism represents stored energy in the form of chemical compounds:

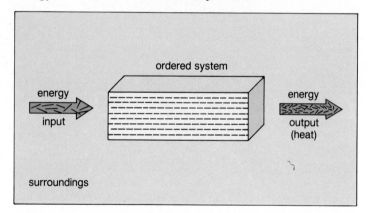

## Implication for Living Things

Life does not violate the energy laws discussed earlier, since each living thing merely represents a temporary storage area for useful energy. The useful energy stored in one organism can be used by another to maintain its own organization. Because of this, energy flows through a community of organisms (fig. 7.2). As transformations of energy occur, useful energy is lost to the environment in the form of heat, until finally useful energy is completely used up. Since energy cannot recycle, there is a need for an ultimate source of energy. This source, which continually supplies all living things with energy, is the sun. The entire universe is tending toward disorder, but in the meantime, solar energy is sustaining all living things. Photosynthesizing organisms like plants and algae are able

## Figure 7.2

The loss of useful energy in a community of organisms. About 2% of the solar energy reaching the earth is taken up by plants. This is the energy that allows them to make their own food. Herbivores obtain their food by eating plants, and carnivores obtain food by eating other animals. Whenever the energy content of food is used by organisms it is eventually converted to heat. With death and decay all the energy temporarily stored in organisms returns as heat to the atmosphere.

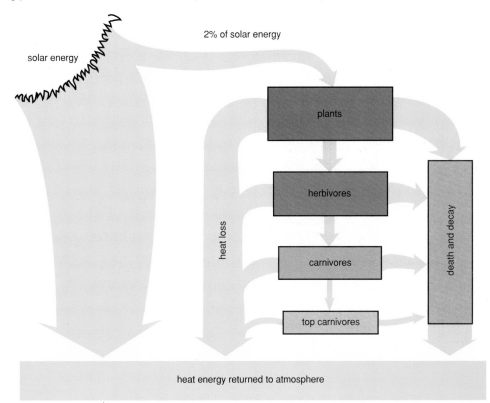

to capture less than 2% of the solar energy that reaches the earth. They use this energy to make their own organic food, which then becomes the food for all other types of living things also. Without photosynthesizing organisms, life as we know it could not exist.

> Living things continually lose useful energy to the environment, but a new supply comes to them in the form of organic food produced by photosynthesizing organisms.

### Energy-Balance Sheet

Only 42% of solar energy reaches the earth's surface; the rest is absorbed by or reflected into the atmosphere and becomes heat.

Of this usable portion, only about 2% is eventually utilized by plants; the rest becomes heat.

Of this, only 0.1%-1.6% is ever incorporated into plant material; the rest becomes heat.*

Of this, only 20% is eaten by herbivores; a large proportion of the remainder becomes heat.*

Of this, only 30% is ever eaten by carnivores; a large proportion becomes heat.*

**Conclusion:** Most of the available energy is never utilized by living things.

*Plant and animal remains became the fossil fuels that we burn today to provide energy. Eventually this energy becomes heat.

## Cellular Metabolism

The food an organism eats becomes the nutrient molecules of its cells. Within cells, these molecules take part in a vast array of chemical reactions, collectively termed the **metabolism** of the cell. So far, we have stressed that nutrient molecules can be used as a source of energy. For this to occur, they have to be broken down, and in fact, they have to be oxidized. Recall that oxidation is the removal of an electron either alone or in the form of a hydrogen atom. An oxidation reaction is a *degradative* process that releases energy. In general, we can represent it like this:

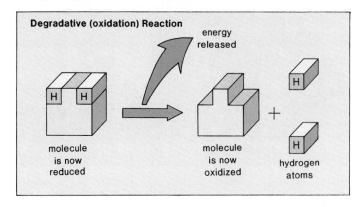

Degradative (oxidation) Reaction

energy released

molecule is now reduced

molecule is now oxidized

hydrogen atoms

We have already indicated that a cell needs energy to maintain its structure and to grow. In other words, some of the nutrient molecules present in cells are used to form the structure of the cell. For example, amino acids are the unit molecules for proteins, and fatty acids are part of the lipids found in membranes. There are also *synthetic* reactions in which reduction, the gain of electrons, occurs. Reduction can occur when hydrogen atoms are added to a molecule:

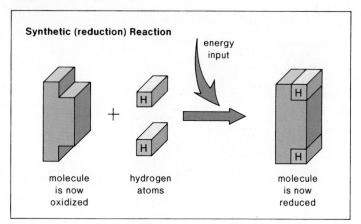

**Synthetic (reduction) Reaction**

energy input

molecule is now oxidized + hydrogen atoms → molecule is now reduced

Do you suppose it would be possible to take the energy released by degradative reactions and use it for synthetic reactions? Further, would it be possible to use the hydrogen atoms produced in certain degradative reactions for certain synthetic reactions? The answer to these questions is *yes*, and we are now going to explore how these processes occur inside cells.

## Pathways

At any one time there are thousands of reactions occurring in cells. Compartmentalization, which we discussed in detail in chapter 5, helps to segregate different types of reactions but more segregation is needed. Within the different organelles, reactions are organized into *metabolic pathways*. Some of these pathways are largely degradative ones that release energy; for example, carbohydrates are broken down in mitochondria (p. 74). Other pathways are largely synthetic ones that require energy input; for example, lipids are synthesized in smooth ER.

In general, metabolic pathways can be represented by this diagram:

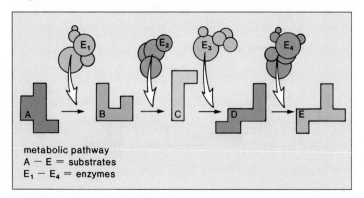

metabolic pathway
A – E = substrates
$E_1$ – $E_4$ = enzymes

## Table 7.1
Enzymes Named for Their Substrate

| Substrate | Enzyme |
|---|---|
| Lipid | Lipase |
| Urea | Urease |
| Maltose | Maltase |
| Ribonucleic acid | Ribonuclease |
| Lactose | Lactase |

In this pathway, A-E are products of the previous reaction and reactants of the next reaction, while $E_1$–$E_4$ identify different enzymes. A reactant in an enzymatic reaction is a **substrate** for that enzyme. A is the beginning substrate, and E is the end product of the pathway.

This simplified representation of a metabolic pathway does not indicate that the intermediate molecules (B-D) can also be starting points for other pathways. Degradative pathways can even interconnect with synthetic pathways, and in this way degradative pathways sometimes provide not only the energy but also the molecules needed for synthesis of cell components.

Some pathways (reactions) in cells are degradative ones that release energy and some are synthetic ones that require energy input.

## Enzymes

As indicated previously, every reaction within a metabolic pathway requires an enzyme. **Enzymes** are organic catalysts, usually globular protein molecules, that speed up chemical reactions without being permanently changed by them. Some RNA molecules, particularly in the nucleus, can be enzymes, and these are called *ribozymes*. Every enzyme is very specific in its action and can speed up only one particular reaction or one type of reaction. Table 7.1 indicates that the name of an enzyme is often formed by adding *ase* to the name of its substrate. Some enzymes are named for the action they perform: for example, a dehydrogenase is an enzyme that removes hydrogen atoms from its substrate.

*No reaction can occur in a cell unless its own enzyme is present and active.* For example, if enzyme 2 (i.e., $E_2$) in the preceding diagram is missing or not functioning, the pathway shuts down at B. Since enzymes are so necessary in cells, their mechanism of action has been studied extensively.

### Energy of Activation
Molecules frequently do not react with one another unless they are activated in some way. In the laboratory, activation is very often achieved by heating the reaction flask to increase the number of effective collisions between molecules. The energy that must be

## Figure 7.3

Enzymes speed up the rate of chemical reactions because they lower the required energy of activation. ***a.*** Energy of activation ($E_a$) that is required for a reaction to occur when an enzyme is not available. ***b.*** Required energy of activation ($E_a$) is much lower when an enzyme is available.

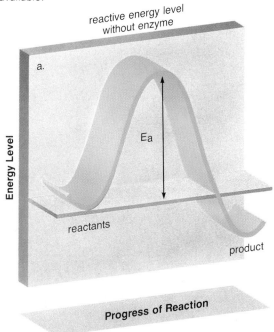

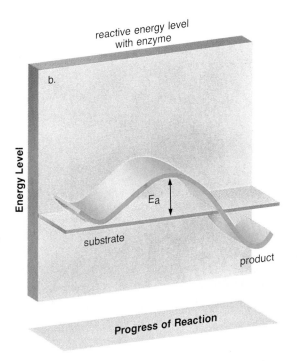

supplied to cause molecules to react with one another is called the *energy of activation* ($E_a$). Figure 7.3*a* and *b* compares $E_a$ when an enzyme is not present to when an enzyme is present, illustrating that enzymes lower the energy of activation. For example, the hydrolysis of casein (the protein found in milk) requires an energy of activation of 20,600 Kcal/mole[1] in the absence of an enzyme, but only 12,600 Kcal/mole is required if the appropriate enzyme is present. Enzymes lower the energy of activation by forming an enzyme-substrate complex.

### Enzyme-Substrate Complex

The following equation, which is pictorially shown in figure 7.4, is often used to indicate that an enzyme forms a complex with its substrate:

$$E + S \rightarrow ES \rightarrow E + P$$

enzyme  substrate  enzyme-substrate  product
complex

In most instances only one small part of the enzyme, called the **active site,** complexes with the substrate(s) (fig. 7.4). It is here that the substrate and enzyme fit together, seemingly

like a key fits a lock; however, it is now known that the active site undergoes a slight change in shape in order to accommodate the substrate(s) more perfectly. This is called the *induced fit model* because as binding occurs, the enzyme is induced (undergoes a slight alteration) to achieve optimum fit. The change in shape of the active site facilitates the reaction that now occurs. After the reaction has been completed, the product(s) is released, and the active site returns to its original state. Only a small amount of enzyme is actually needed in a cell because enzymes are used repeatedly.

Some enzymes do more than simply complex with their substrate(s); they actually participate in the reaction. Trypsin digests protein by breaking peptide bonds (fig. 7.5). The active site of trypsin contains 3 amino acids with R groups that actually interact with members of the peptide bond first to break the bond and then to introduce the components of water. This illustrates that the formation of the enzyme-substrate complex is very important in speeding up the reaction.

Sometimes it is possible for a particular reactant(s) to produce more than one type of product(s). The presence or absence of an enzyme determines which reaction takes place. For example, if substance A can react to form either product B

[1]The Kcal (Kcalorie) is a common way to measure heat, and a mole is the molecular weight of a substance expressed in grams.

## Figure 7.4

*a.* The enzyme and substrates before the reaction begins. *b.* Formation of the enzyme-substrate complex holds the substrates together in such a way that they can react. *c.* Following the reaction the enzyme returns to its prior configuration.

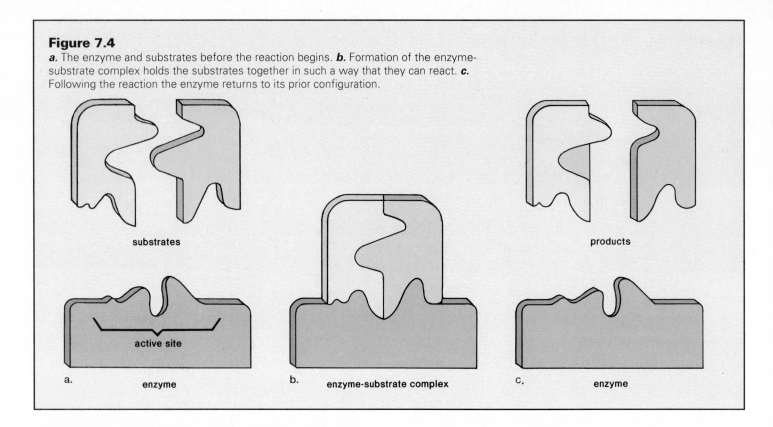

substrates

products

active site

a.  enzyme

b.  enzyme-substrate complex

c.  enzyme

## Figure 7.5

Computer-generated model of the functional backbone of the enzyme trypsin. The 3 R groups shown in green are in the active site. They interact with members of the peptide bond in order to hydrolyze it.

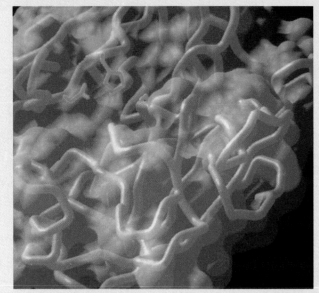

or product C, then the enzyme that is present and active—$E_1$ or $E_2$—determines which product is produced:

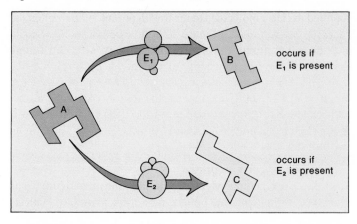

occurs if $E_1$ is present

occurs if $E_2$ is present

Enzymes are proteins that speed up chemical reactions by lowering the energy of activation. They do this by forming an enzyme-substrate complex.

## Conditions Affecting Enzymatic Reactions

Enzymatic reactions proceed quite rapidly. For example, the breakdown of hydrogen peroxide into water and oxygen can occur 600,000 times a second when the enzyme catalase (an enzyme found in peroxisomes, p. 73) is present. How quickly an enzyme works, however, is affected by certain conditions.

## Figure 7.6

Enzyme activity is affected by temperature and pH. **a.** At first, as with most chemical reactions, the rate of an enzymatic reaction doubles with every 10°C rise in temperature. In this graph, which is typical of a mammalian enzyme, the rate is maximum at about 40°C and then it decreases until it stops altogether, indicating that the enzyme is now denatured. **b.** Pepsin, an enzyme found in the stomach, acts best at a pH of about 2, while trypsin, an enzyme found in the small intestine, prefers a pH of about 8. Other pHs do not properly maintain the shape that enables enzymes to bind with their substrate.

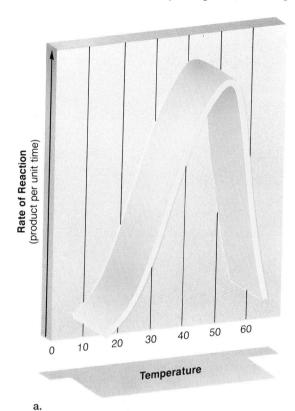

a.

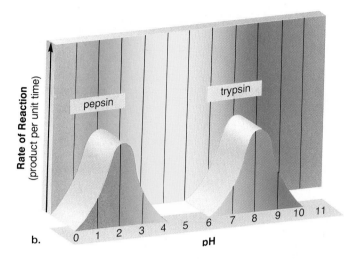

### Substrate Concentration

Generally, an enzyme's activity increases with greater substrate concentration because there are more collisions between substrate molecules and the enzyme. More substrate molecules fill active sites for a larger proportion of time, and so more product results per unit time. But when the concentration of substrate is so great that the enzyme's active sites are almost continuously filled with substrate, the enzyme's rate of activity cannot increase. Maximum velocity has been reached.

### Temperature and pH

A higher temperature generally results in an increase in enzyme activity. As the temperature rises, the movement of enzyme molecules and substrate molecules increases, and there are more effective collisions between them. If the temperature rises beyond a certain point, however, enzyme activity eventually levels out and then declines rapidly because the enzyme is **denatured** (fig. 7.6a). As shown in figure 4.17, an enzyme's shape changes during denaturation. Therefore, an enzyme can no longer bind substrate molecules efficiently.

A change in pH can also affect enzyme activity (fig. 7.6b). Each enzyme has an optimal pH that helps maintain its normal configuration. Recall that the tertiary structure of a protein is dependent on interactions, such as hydrogen bonding, between R groups (p. 53). A change in pH can alter the ionization of these side chains and disrupt the normal interactions, and denaturation eventually occurs. Again, the enzyme is then unable to combine efficiently with its substrate.

> Adequate substrate concentration and optimal temperature and pH all increase the speed of enzyme activity.

### Regulation of Enzyme Activity

While these factors affect enzyme activity, the cell also has built-in mechanisms to control directly both enzyme concentration and activity. First of all, cells are able to regulate whether an enzyme is present at all, but since this type of regulation involves control of protein synthesis, it is not discussed until chapter 17. Cells also have ways to control the level of activity of enzymes that have already been synthesized and are present in the cytoplasm.

### Inhibition

*Inhibition* is a common means by which cells regulate enzyme activity. In *competitive inhibition*, another molecule is so close in shape to the enzyme's substrate that it can compete with the true substrate for the enzyme's active site. This molecule inhibits the reaction because only the binding of the true substrate results in a product. In *noncompetitive inhibition*, a molecule binds to an enzyme but not at the active site. This other binding site is called the *allosteric site* (*allo*—other; *steric*—space or structure). In this instance also, the molecule is an inhibitor because when it binds there is a shift in the 3-dimensional structure of the enzyme that prevents the substrate from binding to the active site. In cells, inhibition is usually reversible; that is, the inhibitor is not permanently bound to the enzyme.

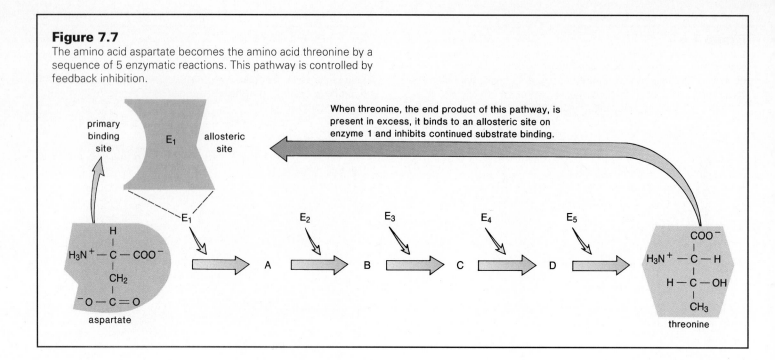

**Figure 7.7**

The amino acid aspartate becomes the amino acid threonine by a sequence of 5 enzymatic reactions. This pathway is controlled by feedback inhibition.

---

During competitive inhibition, an inhibitor binds to the active site; during noncompetitive inhibition, an inhibitor binds to an allosteric site.

---

Irreversible inhibition of enzymes also occurs, due to the presence of a poison. For example, penicillin causes the death of bacteria due to irreversible inhibition of an enzyme needed to form the bacterial cell wall. In humans, hydrogen cyanide irreversibly binds to a very important enzyme (cytochrome oxidase) present in all cells, and this accounts for its lethal effect on the human body.

### Feedback Inhibition

The activity of almost every enzyme in a cell is regulated by feedback inhibition. **Feedback inhibition** is an example of a common biological control mechanism called negative feedback. Just as excessively high temperature can cause a furnace to shut off, so a product produced by an enzyme can inhibit its activity. When the product is in abundance, it binds competitively with its enzyme's active site (fig. 7.4c); as the product is used up, inhibition is reduced and more product can be produced. In this way, the concentration of the product is always kept within a certain range.

Most enzymatic *pathways* are also regulated by feedback inhibition, but in these cases the end product of the pathway binds at an allosteric site on the *first* enzyme of the pathway (fig. 7.7). This binding shuts down the pathway, and no more product is produced.

---

In feedback inhibition, the inhibitor is either the product of an enzymatic reaction that binds to the active site or the end product of a metabolic pathway that binds to an allosteric site on the first enzyme of the pathway.

---

## Coenzymes

Many enzymes require a nonprotein *cofactor* to assist them in carrying out their function. Some cofactors are ions; for example, magnesium ($Mg^{++}$), potassium ($K^+$), and calcium ($Ca^{++}$) are often involved in enzymatic reactions. Some other cofactors, called **coenzymes,** are organic molecules that bind to enzymes and serve as carriers for chemical groups or electrons.

Usually coenzymes participate directly in the reaction. For example, the coenzyme tetrahydrofolate can carry a methyl group ($CH_3$). Sometimes it works in conjunction with the enzyme thymidylate synthase and donates a methyl group to form thymine, one of the bases found in DNA. In humans, this coenzyme is not present unless the diet includes the vitamin folic acid. **Vitamins** are relatively small organic molecules that are required in trace amounts in our diet and in the diet of other animals for synthesis of coenzymes that affect health and physical fitness.

The presence of other coenzymes in cells is similarly dependent on our intake of vitamins. For example,

| Vitamin | Coenzyme |
| --- | --- |
| Niacin | $NAD^+$ |
| $B_2$ (riboflavin) | FAD |
| $B_1$ (thiamine) | Thiamine pyrophosphate |
| Pantothenic acid | Coenzyme A (CoA) |
| $B_{12}$ | Cobamide coenzymes |

A deficiency of any one of these vitamins results in a lack of the coenzyme listed and therefore a lack of certain enzymatic actions. In humans, this eventually results in vitamin-deficiency symptoms. For example, niacin deficiency results in a skin disease called pellagra, and riboflavin deficiency results in cracks at the corners of the mouth.

## Figure 7.8

Secondary structure of an enzyme that degrades alcohol, called alcohol dehydrogenase. NAD⁺ is a coenzyme for this enzyme, and therefore the enzyme has a binding site not only for its substrate (alcohol) but also for NAD⁺. The NAD⁺ binding site is in green and yellow; the substrate-binding site is in blue. A bound NAD⁺, in purple, lies close to the substrate-biding site.

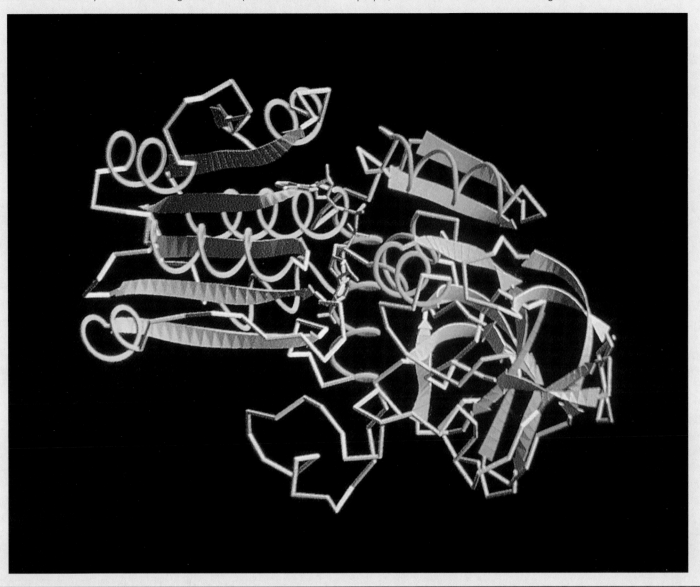

Coenzymes are nonprotein molecules that assist enzymes in performing their reactions. Coenzyme synthesis requires an intake of vitamins.

### NAD⁺ and FAD

**NAD⁺** (nicotinamide adenine dinucleotide) is a coenzyme that carries electrons and quite often works in conjunction with enzymes called dehydrogenases (fig. 7.8). An enzyme of this type removes 2 hydrogen atoms ($2e^- + 2H^+$) from its substrate; both electrons but only one hydrogen ion are passed to NAD⁺—NAD⁺ becomes NADH:

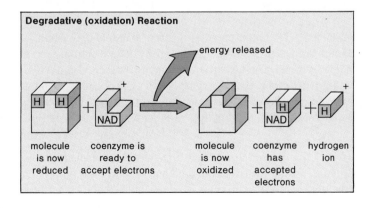

**Degradative (oxidation) Reaction**

energy released

molecule is now reduced + coenzyme is ready to accept electrons → molecule is now oxidized + coenzyme has accepted electrons + hydrogen ion

**Figure 7.9**

Electron transport system, in which electrons pass from carrier to carrier. With each oxidation reaction some energy is released and a portion of it is trapped for the purpose of producing ATP molecules.

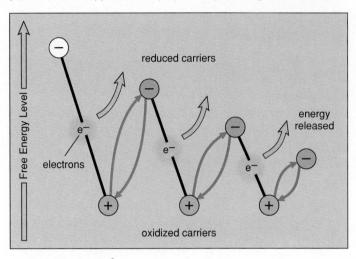

This is a degradative (oxidation) reaction and energy is released; but by comparing this equation to the similar one on page 107, you can see that much less energy has been lost because the red arrow is smaller. The energy has been transferred to NAD$^+$ in the form of high-energy electrons. Another coenzyme called **FAD** (flavin adenine dinucleotide) also carries high-energy electrons, but it accepts both hydrogen ions (to become FADH$_2$).

Both NAD$^+$ and FAD are involved in **cellular respiration,** a metabolic pathway by which carbohydrates are oxidized in a step-by-step manner to carbon dioxide and water. At various junctions along the way, these coenzymes accept electrons from certain substrates and carry them to an **electron transport system** consisting of membrane-bounded carriers that pass electrons from one carrier to another. High energy electrons are delivered to the system and low energy electrons leave it. Every time an electron is transferred, oxidation occurs and energy is released; this energy is ultimately used to produce ATP molecules (fig. 7.9).

> NAD$^+$ and FAD are electron carriers that take electrons to the electron transport system in mitochondria. This system uses the energy of oxidation to produce ATP molecules.

Chloroplasts also have an electron transport system for producing ATP. In this case the electrons that fuel the system are taken from water. Solar energy energizes these electrons, and as they pass from carrier to carrier, the energy is released and ATP is built up. In the end, the coenzyme NADP$^+$ (nicotinamide adenine dinucleotide phosphate) accepts the electrons and becomes reduced to NADPH.

### NADP$^+$

**NADP$^+$** (nicotinamide adenine dinucleotide phosphate) has a structure similar to NAD$^+$, but it contains a phosphate group that is lacking in NAD$^+$. This may signal enzymes that its function is slightly different. It does carry electrons and a hydrogen ion just like NAD$^+$, but its electrons are used to bring about reduction synthesis:

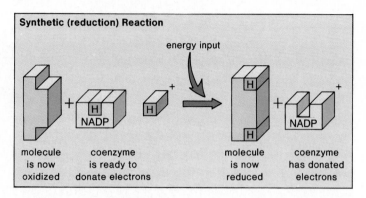

**Synthetic (reduction) Reaction**

energy input

molecule is now oxidized | coenzyme is ready to donate electrons | molecule is now reduced | coenzyme has donated electrons

During **photosynthesis** within chloroplasts, carbon dioxide is reduced to a carbohydrate. NADPH supplies the necessary electrons and ATP supplies the necessary energy to bring about this reduction.

NADPH also functions outside of chloroplasts. Within the cytosol, plant and animal cells use a special degradative pathway to produce NADPH. This NADPH is used during the synthesis of many necessary molecules in cells. Energy too is required, but for this purpose, ATP produced by mitochondria is utilized.

> NADP$^+$ is an electron carrier that participates along with ATP in synthetic reactions.

## Metabolism Revisited

We stated it is possible to harness the energy and hydrogens provided by degradative reactions for synthetic reactions. This is possible *only* because ATP carries energy and NADPH carries electrons between degradative pathways and synthetic pathways (fig. 7.10).

Degradative pathways do run synthetic pathways as long as the cell receives an outside source of energy and matter. The energy laws discussed earlier in this chapter tell us that life is only possible because organisms (cells) are able to *store* temporarily some of the energy that flows through them.

Of all the organisms, only photosynthesizers such as plants are able to make organic food by utilizing *solar* energy. The green arrows in figure 7.10 illustrate the process of photosynthesis. In chloroplasts, solar energy is used to produce the ATP and the NADPH that are used to reduce carbon dioxide to carbohydrates. These and the other macromolecules produced by plants supply plants and eventually all other living things with a source of organic food.

Plants and animals use organic food as both an energy source and a source of unit molecules (blue arrows). When used as an energy source, carbohydrates are degraded (oxidized) to carbon dioxide and water, which are excreted. Heat is given off in the process, and some of the energy is used to produce ATP. Degradation also produces a supply of NADPH. Again ATP is a carrier for energy and NADPH is a carrier for electrons utilized

## Figure 7.10

The degradative part of photosynthesis drives the synthetic part, and the degradative part of general metabolism drives the synthetic part. Degradation (oxidation) provides electrons and energy that are carried to synthetic reactions by NADPH and ATP, respectively.

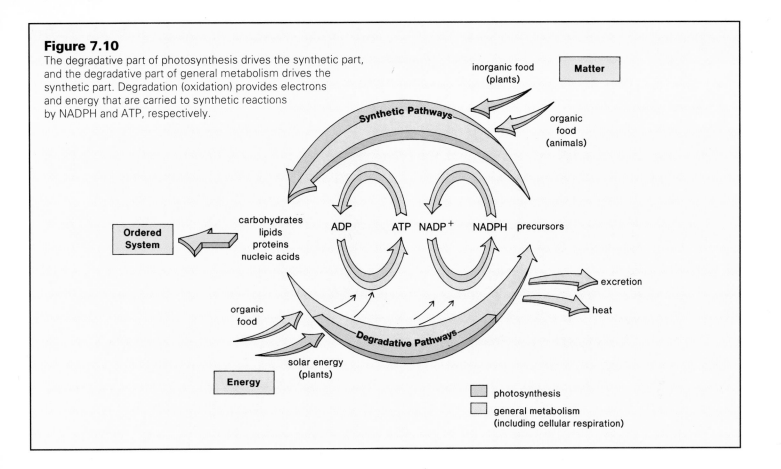

## Figure 7.11

ATP reaction. ATP, the common energy carrier in cells, is a high-energy molecule as represented by the wavy lines between the phosphate groups (P). When the last phosphate group is removed, the energy released is used by the cell for various purposes. ATP can be reformed if enough energy is supplied to rejoin ADP and (P).

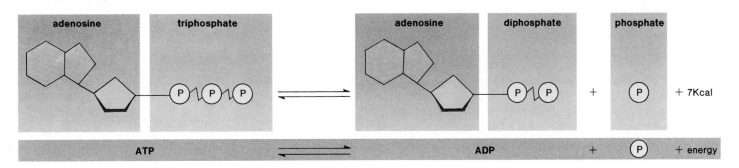

in synthetic (reduction) reactions. In this way, the flow of energy through the organism (cells) is utilized for growth and maintenance of the organism.

The role of enzymes in metabolism must not be forgotten. They make "warm chemistry" possible. Degradative reactions occur spontaneously (naturally) but only if sufficient energy of activation is supplied. Enzymes lower the energy of activation, making it possible for these reactions to occur very quickly at body temperature. Since the enzymes involved in synthesis are not the

same as those used for degradation, however, the cell needs the coenzyme NADP+ to carry electrons and ATP to carry energy from degradative to synthetic reactions.

## ATP

**ATP** (adenosine triphosphate) is the common energy currency of cells; when cells require energy, they "spend" ATP (fig. 7.11). You may think that this causes our bodies to produce a lot of ATP, and it

**Figure 7.12**

The coupling of ATP buildup and breakdown to other reactions. Energy-releasing reactions are shown as gears with arrows going down; energy-requiring reactions are shown as gears with arrows going up. Degradation of food is coupled to the production of ATP (*far left*); breakdown of ATP releases energy to drive energy-requiring reactions (*right*). Very often this involves the transfer of a phosphate group from ATP to an intermediate molecule in the reaction (*lower right*).

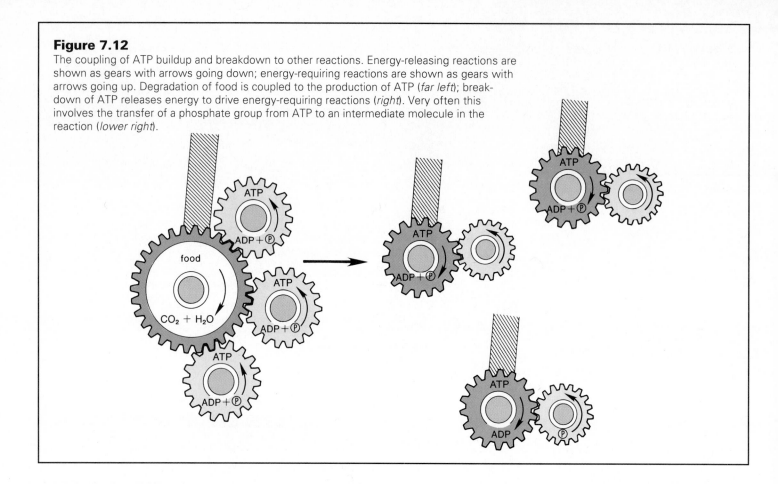

does; the average male needs to produce about 8 kg (over 17 lb) of ATP an hour! Obviously, it would be impossible to carry around even a few hours supply of ATP, and indeed the body has on hand only about 50 g (1.8 oz) at any time. The answer to the paradox is that ATP is constantly being recycled from **ADP** (adenosine diphosphate) and $\textcircled{P}$. The cell's entire supply of ATP is recycled about once each minute.

## Structure of ATP

ATP is a nucleotide composed of the base adenine and the sugar ribose (together called adenosine) and 3 phosphate groups. ATP is called a "high-energy" compound because its phosphate groups are easily removed. Under cellular conditions, the amount of energy released is about 7.3 Kcal per mole. The wavy line between the phosphate bonds indicates the high potential energy of these bonds.[2]

The use of ATP as a carrier of energy derived from degradation has some advantages:

1.  The energy of degradation can be captured in a step-by-step manner that prevents waste (see fig. 7.10).
2.  It provides a common energy currency that can be used in many different types of reactions.
3.  When ATP becomes ADP + $\textcircled{P}$ the amount of energy released is just about enough for the biological purposes mentioned in the following section and so little energy is wasted.

[2]The high-energy content of ATP does not reside simply in the phosphate bond; rather it comes from the complex interaction of the atoms within the molecule. The use of the wavy line is for convenience only.

(1) and (3) are only possible because most reactions (e.g., ATP breakdown and synthetic buildup) are *coupled;* that is, they take place at the same time and at the same place and usually utilize the same enzyme (fig. 7.12). During the coupling process, a phosphate group is often transferred to an intermediate compound. For example, during active transport, a phosphate group is transferred to a membrane-bounded protein (see fig. 6.12).

## Function of ATP

Recall that at various times we have mentioned at least 3 uses for ATP.

**Chemical work.** Supplies the energy needed to synthesize macromolecules that make up the cell.

**Transport work.** Supplies the energy needed to pump substances across the plasma membrane.

**Mechanical work.** Supplies the energy needed to cause muscles to contract, cilia and flagella to beat, chromosomes to move, and so forth.

All organisms, both prokaryotic and eukaryotic, use ATP. This illustrates the chemical unity of all living things.

> ATP is a carrier of energy in cells. It is the common energy currency because it supplies energy for many different types of reactions.

## Formation of ATP

Cells have 2 ways to make ATP. The first, called **substrate-level phosphorylation,** occurs in the cytosol. During this process, a high-energy substrate transfers a phosphate group to ADP, forming ATP:

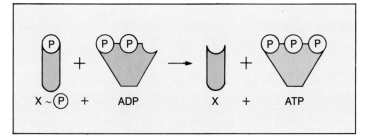

Substrate-level phosphorylation is not the primary way of forming ATP, however; most ATP is made within mitochondria and chloroplasts by way of the electron transport system mentioned earlier (fig. 7.9). In this case, the phosphorylation that occurs is called **chemiosmotic phosphorylation.**

## Chemiosmotic Phosphorylation

For many years it was known that ATP synthesis was somehow coupled to the electron transport system, but the exact mechanism could not be determined. Peter Mitchell, a British biochemist, received a Nobel Prize in 1978 for his *chemiosmotic theory* of ATP production in both mitochondria and chloroplasts.

In mitochondria and chloroplasts, the carriers of the electron transport system are located within a membrane. According to the chemiosmotic theory, hydrogen ($H^+$) ions tend to collect on one side of the membrane because they are pumped there by certain carriers. This establishes an electrochemical gradient across the membrane that can be used to provide energy for ATP production. Particles, called *ATP synthase complexes,* span the membrane. Each complex contains a channel protein that allows hydrogen ions to flow down their electrochemical gradient. The flow of hydrogen ions through the channel provides the energy for the ATP synthase enzyme to produce ATP from ADP + $\textcircled{P}$ (fig. 7.13). This process is called chemiosmotic phosphorylation because chemical and osmotic events have been joined to allow ATP synthesis.

The chemiosmotic synthesis of ATP was supported by a now-famous experiment performed by Andre Jagendorf of Cornell University utilizing chloroplasts (fig. 7.14). The experiment shows that ATP production is indeed tied only to a hydrogen ion gradient.

Most of the energy for ATP synthesis is derived from a hydrogen ion gradient established across a membrane by an electron transport system.

### Figure 7.13

The energy released by the passage of electrons down the electron transport system is used to pump hydrogen ions ($H^+$) across a membrane in mitochondria and chloroplasts. This establishes an electrochemical gradient that is utilized for ATP synthesis. When the hydrogen ions flow back across the membrane through a channel protein in the membrane that is also an ATP synthase, ATP is synthesized.

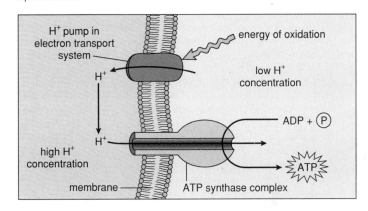

### Figure 7.14

An experiment performed by Andre Jagendorf in 1966 supports the chemiosmotic theory. An electrochemical gradient was established by first soaking isolated chloroplasts in an acid medium and then abruptly changing the pH to basic. ATP production occurred after ADP and $\textcircled{P}$ were added. This all occurred in the dark, proving that ATP synthesis is not directly linked to the chloroplast electron transport system, which works only in the light.

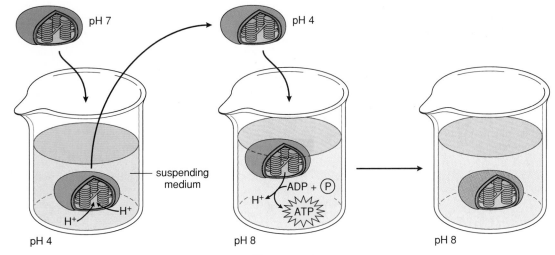

H$^+$ diffuses into chloroplasts so that they become pH 4.

ADP and $\textcircled{P}$ are added to beaker. As H$^+$ diffuses out of chloroplasts, ATP is produced.

Chloroplast and medium have same pH. H$^+$ electrochemical gradient no longer exists and ATP production ceases.

# Summary

1. There are 2 energy laws that are basic to understanding energy-use patterns in organisms (cells). The first states that energy cannot be created or destroyed, and the second states that some useful energy is always lost when one form of energy is transformed into another form.

2. In keeping with these laws, living things need an outside source of energy. Plants capture the energy of sunlight to make their own food, and animals eat either plants or other animals to obtain food.

3. Metabolism consists of 2 kinds of reactions, degradative and synthetic. During degradation, organic molecules derived from food are broken down to release energy; oxidation occurs and hydrogen atoms are removed from substances. During synthesis, an energy-requiring process, molecules derived from food accept hydrogen atoms and are reduced.

4. There are degradative metabolic pathways and synthetic metabolic pathways. Each reaction in a pathway requires its own enzyme. Enzymes speed up chemical reactions because they lower the energy of activation. They can do this because they form a complex with their substrate(s) at the active site.

5. Various factors affect the yield of enzymatic reactions, such as the concentration of the substrate(s), the temperature, and the pH. A high temperature or a pH outside the preferred range for that enzyme can lead to denaturation, a change in structure that prevents the enzyme from functioning.

6. The activity of most enzymes is regulated by inhibition. In competitive inhibition, a molecule competes with the substrate for the active site; during noncompetitive inhibition, a molecule binds to an allosteric site. During feedback inhibition, the product of an enzymatic reaction binds to the enzyme; for metabolic pathways, the end product usually binds to an allosteric site on the first enzyme in the pathway.

7. Many enzymes have cofactors or coenzymes that help them carry out a reaction. Coenzymes are nonprotein, organic molecules often derived at least in part from vitamins. $NAD^+$ and FAD are coenzymes that carry high-energy electrons to an electron transport system that is usually located in mitochondria. Here oxidation occurs each time one carrier passes electrons to the next one. At the same time, energy is released and is used to produce ATP molecules. Chloroplasts also contain an electron transport system that receives energized electrons taken from water.

8. NADPH is also a carrier of electrons, but it carries them to various synthetic reactions. These are reduction reactions that require hydrogen atoms and energy supplied by ATP.

9. Degradative pathways drive synthetic pathways. Degradative pathways supply the ATP and the NADPH that is needed to bring about synthesis. We must never forget, however, that in order for this to occur, the cell needs an outside source of energy and matter. Only a flow of energy through the organism allows it to sustain itself and to grow.

10. ATP is a high-energy molecule that serves as a common carrier of energy in cells. When ATP becomes ADP + $\circledP$, the right amount of energy is provided for many reactions. In order to produce ATP, the energy from carbohydrate breakdown is coupled to ATP production, and in order to pass this energy to synthetic reactions, coupling again occurs.

11. The chemiosmotic theory explains just how the electron transport system produces ATP. The carriers of this system deposit hydrogen ions ($H^+$) on one side of a membrane. When the ions flow down an electrochemical gradient in special channels, an enzyme utilizes the release of energy to make ATP from ADP and $\circledP$.

## Writing Across the Curriculum

In order to practice writing skills, students should write out the answers to any or all of the study questions and the critical thinking questions. The study questions are sequenced in the same order as the text. Suggested answers to the critical thinking questions are in appendix D.

---

# Study Questions

1. Explain the first energy law using the terms *system* and *surroundings* in your explanation. Explain the second energy law using the term *useful energy* in your explanation.

2. Discuss the importance of solar energy, not only to plants but to all living things.

3. Contrast a degradative reaction with a synthetic reaction.

4. Diagram a metabolic pathway and use this to explain the terms *substrate* and *product* and the need for various enzymes.

5. In what way do enzymes lower the energy of activation? What does the induced-fit model say about formation of the enzyme-substrate complex?

6. In what ways do substrate concentration, temperature, and pH affect the yield of enzymatic reactions?

7. Explain competitive inhibition, non-competitive inhibition, and feedback inhibition of enzyme activity.

8. Contrast the function of $NAD^+$ and FAD to that of $NADP^+$.

9. In what way do degradative reactions drive synthetic reactions? Why is there still a need for an outside source of energy and matter?

10. Discuss the structure and the function of ATP, and outline the formation of ATP by chemiosmotic phosphorylation.

## Objective Questions

1. Degradative reactions
   a. are often oxidation reactions.
   b. are the same as synthetic reactions.
   c. require a supply of NADPH molecules.
   d. Both a and c.

2. Enzymes
   a. make it possible for cells to escape the need for energy.
   b. are nonprotein molecules that help coenzymes.
   c. are not affected by a change in pH.
   d. lower the energy of activation.

3. An allosteric site on an enzyme is
   a. the same as the active site.
   b. where ATP attaches and gives up its energy.
   c. often involved in feedback inhibition.
   d. All of these.

4. A high temperature
   a. can affect the shape of an enzyme.
   b. lowers the energy of activation.
   c. makes cells less susceptible to disease.
   d. Both a and c.

5. Electron transport systems
   a. are found in both mitochondria and chloroplasts.
   b. contain oxidation reactions.
   c. are involved in the production of ATP.
   d. All of these.

6. The difference between NAD$^+$ and NADP$^+$ is
   a. only NAD$^+$ production requires niacin in the diet.
   b. one contains high-energy phosphate bonds and the other does not.
   c. one carries electrons to the electron transport system and the other carries them to synthetic reactions.
   d. All of these.

7. ATP
   a. is used only in animal cells and not in plant cells.
   b. carries energy between degradative pathways and synthetic pathways.
   c. is needed for chemical work, mechanical work, and transport work.
   d. Both b and c.

8. Chemiosmotic phosphorylation is dependent upon
   a. the diffusion of water across a selectively permeable membrane.
   b. an outside supply of phosphate and other chemicals.
   c. the establishment of an electrochemical H$^+$ gradient.
   d. the ability of ADP to join with ℗ even in the absence of a supply of energy.

9. Label this diagram that gives an overview of metabolism. Label the arrows either degradative reactions or synthetic reactions. Label the long lines either macromolecules or small molecules. Label the short lines ADP + ℗, ATP, NADP, NADPH.

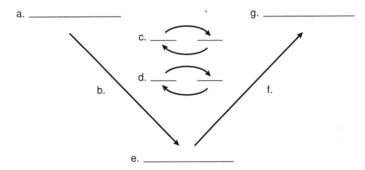

10. Label this diagram describing chemiosmotic phosphorylation.

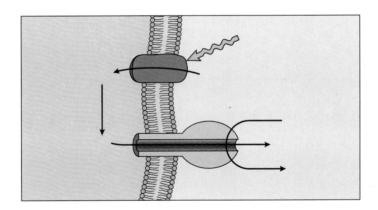

## Concepts and Critical Thinking

1. *A constant input of energy is required to maintain the organization of the cell.*

   In reference to work performed by the rough ER and the Golgi apparatus, explain specifically why a cell needs energy to maintain its organization.

2. *In keeping with the first and second laws of thermodynamics, a cell takes in useful energy and gives off heat.*

   Explain why you would expect both parts of the ATP cycle to be responsible for the loss of useful energy through heat.

3. *The actions of enzymes allow the cell to maintain itself and grow.*

   Explain why no reaction occurs in a cell unless a specific enzyme is present.

## Selected Key Terms

**metabolism** (mĕ-tab'o-lizm) 107
**substrate** (sub'strāt) 108
**enzyme** (en'zīm) 108
**active site** (ak'tiv sīt) 109
**denatured** (de-na'tūrd) 111
**feedback inhibition** (fēd'bak in"hī-bish'un) 112
**coenzyme** (ko-en'zīm) 112
**vitamin** (vi'tah-min) 112
**NAD⁺ (nicotinamide adenine dinucle-otide)** (nik"o-tin'ah-mīd ad'ē-nīn di-nu'kle-o-tīd) 113

**FAD (flavin adenine dinucleotide)** (fla'vin ad'ē-nīn di-nu'kle-o-tīd) 114
**cellular respiration** (sel'u-lar res"pī-ra'shun) 114
**electron transport system** (e-lek'tron trans'port sis'tem) 114
**NADP⁺ (nicotinamide adenine dinucle-otide phosphate)** (nik"o-tin'ah-mīd ad'ē-nīn di-nu'kle-o-tīd fos'fāt) 114
**photosynthesis** (fo"to-sin'thē-sis) 114

**ATP (adenosine triphosphate)** (ah-den'o-sēn tri-fos'fāt) 115
**ADP (adenosine diphosphate)** (ah-den'o-sēn di-fos'fāt) 116
**substrate-level phosphorylation** (sub'strāt lĕv-al fos"fōr-ī-la'shun) 117
**chemiosmotic phosphorylation** (kem"ī-os-mot'ik fos"fōr'ī-la'shun) 117

# 8

# Photosynthesis

Green leaves are a site for photosynthesis. They are situated to absorb energy from the sun and carbon dioxide from the air. Solar energy is used to convert carbon dioxide to a carbohydrate, which serves as a food not only for plants but for the entire biosphere.

Your study of this chapter will be complete when you can

1. give at least 3 examples of the importance of photosynthesis to living things;
2. relate the visible light range to photosynthesis, and describe the role of chlorophyll;
3. describe the structure of chloroplasts, and give the function of the various parts;
4. explain the terms *light-dependent reactions* and *light-independent reactions,* and describe how the light-dependent reactions drive the light-independent reactions;
5. trace the noncyclic electron pathway, pointing out significant events as the electrons move;
6. trace the cyclic electron pathway, pointing out significant events;
7. describe the process of chemiosmotic ATP synthesis in chloroplasts;
8. describe the 3 stages of the Calvin cycle;
9. contrast $C_3$ and $C_4$ plants, and relate the phenomenon of photorespiration to the success of $C_4$ plants in hot, dry climates;
10. contrast a CAM plant with a $C_4$ plant, using the terms *partitioning in space* and *partitioning in time.*

AVE YOU THANKED A GREEN PLANT TODAY? Plants dominate our environment, but most of us spend little time thinking about the various services they perform for us and for all living things. Chief among these is their ability to carry on photosynthesis, during which they use sunlight (*photo*) as a source of solar energy to produce food (*synthesis*):

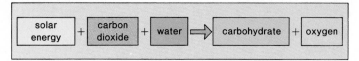

Other organisms, called algae, are also photosynthetic. Algae (chap. 24) are a diverse group, but many are water dwelling, microscopic organisms related to plants.

The organic food produced by photosynthesis is not only used by plants themselves, it is also the ultimate source of food for all other living things (fig. 8.1). Also, when organisms convert carbohydrate energy into ATP energy they make use of the oxygen ($O_2$) given off by photosynthesis. Nearly all living things are dependent on atmospheric oxygen derived from photosynthesis.

At one time in the distant past, plant and animal matter accumulated without decomposing. This matter became the fossil fuels (e.g., coal, oil, and gas) that we burn today for energy. This source of energy, too, is the product of photosynthesis.

Photosynthesis is absolutely essential for the continuance of life because it is the source of food and oxygen for nearly all living things.

## Sunlight

Photosynthesis is an energy transformation in which solar energy in the form of light is converted to chemical energy within carbohydrate molecules. Therefore, we will begin our discussion of photosynthesis with the energy source—sunlight.

Solar radiation can be described in terms of its energy content and its wavelength. The energy comes in discrete packets called **photons.** So, in other words, you can think of radiation as photons that travel in waves:

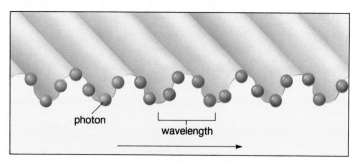

Figure 8.2*a* illustrates that solar radiation, or the **electromagnetic spectrum,** can be divided on the basis of wavelength—gamma rays have the shortest wavelength and radio waves have the longest

**Figure 8.1**
This squirrel is a herbivore. It feeds directly on plant material produced by a photosynthesizer. Carnivores, such as a hawk that may feed on this squirrel, are also dependent, although indirectly, on food produced by photosynthesizers.

wavelength. The energy content of photons is inversely proportional to the wavelength of the particular type of radiation; that is, short-wavelength radiation has photons of a higher energy content than long-wavelength radiation. High-energy photons, such as those of short-wavelength ultraviolet radiation, are dangerous to cells because they can break down organic molecules. Low-energy photons, such as those of infrared radiation, do not damage cells because they only increase the vibrational or rotational energy of molecules; they do not break bonds. But photosynthesis utilizes only the portion of the electromagnetic spectrum known as **visible light.** (It is called visible light because it is the part of the spectrum that allows us to see.) Photons of visible light have just the right amount of energy to promote electrons to a higher electron shell in atoms without harming cells.

We have mentioned previously that only about 42% of the solar radiation that hits the earth's atmosphere ever reaches the surface, and most of this radiation is within the visible-light range.

**Figure 8.2**

The electromagnetic spectrum, the components of visible light, and the absorption spectra of chlorophylls (**a**) and (**b**). Visible (white) light is actually made up of a number of different wavelengths of radiation; when it is passed through a prism (or through raindrops), we see these wavelengths as different colors of light. The chlorophylls, like other pigments, absorb only certain wavelengths; for example, the chlorophylls primarily absorb light in the blue-violet range and in the red-orange range. The chlorophylls do not absorb green light to any extent; therefore, light of this color is reflected to our eyes and we see chlorophyll as a green pigment. There are other pigments in plants, such as the carotenoids, that are yellow-orange and are able to absorb light in the violet-blue-green range. These pigments become noticeable in the fall when chlorophyll breaks down. To identify the absorption spectrum of a particular pigment, a purified sample is exposed to different wavelengths of light inside an instrument called a spectrophotometer. A spectrophotometer measures the amount of light that passes through the sample, but from this it can be calculated how much was absorbed. The amount of light absorbed at each wavelength is plotted on a graph, and the result is what we call the absorption spectrum in (**b**). When light is not absorbed, it is reflected of transmitted (**c**). How do we know that the peaks shown in chlorophyll's absorption spectrum indicate the wavelengths used for photosynthesis? Photosynthesis, of course, produces oxygen, and therefore we can use the rate of production of this gas as a means to measure the rate of photosynthesis at each wavelength of light. When such data are plotted, the resulting graph (called the action spectrum) is very similar to the absorption spectrum of the chlorophylls. Therefore, we are confident that the light absorbed by the chlorophylls does contribute extensively to photosynthesis.

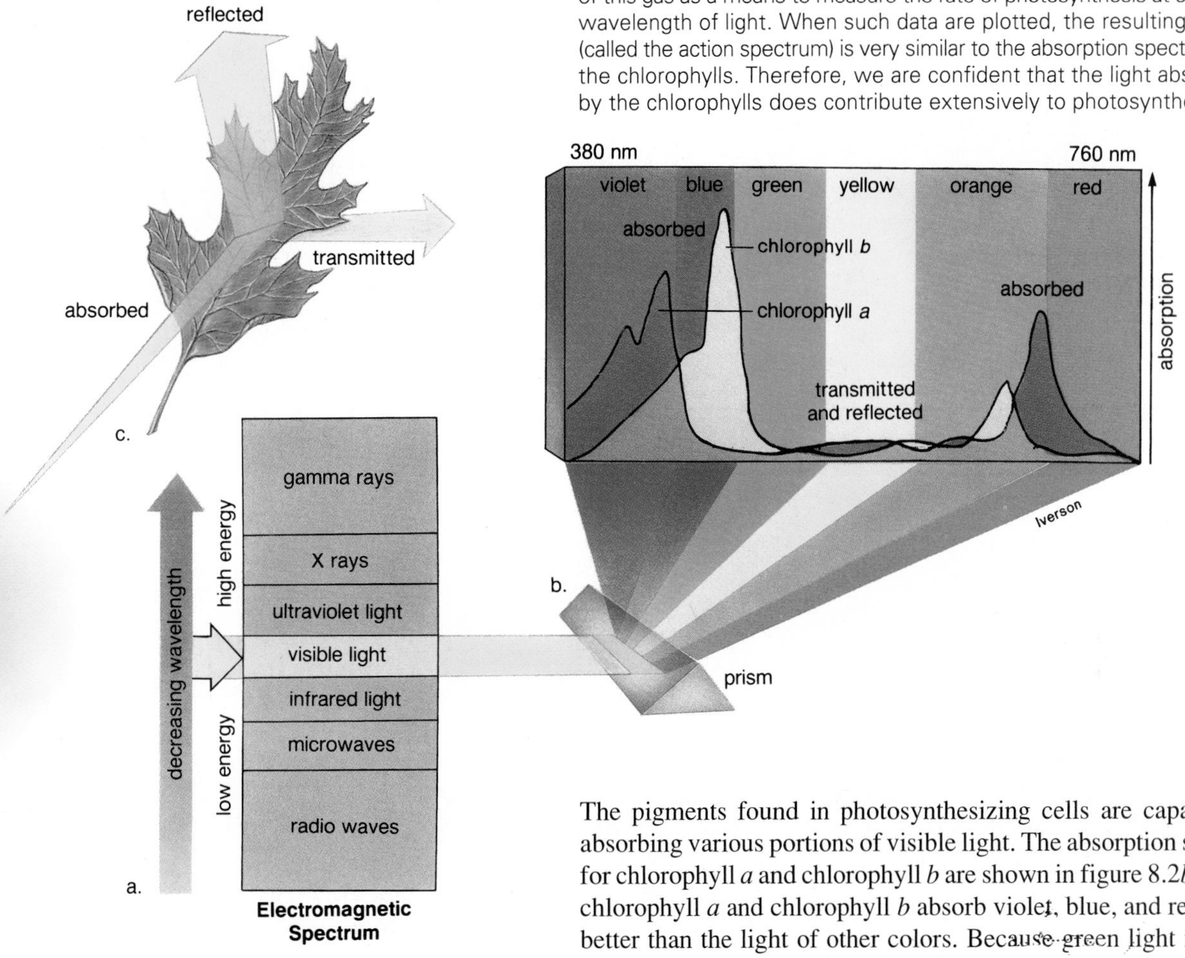

Higher energy wavelengths are screened out by the ozone layer in the atmosphere, and lower energy wavelengths are screened out by water vapor and carbon dioxide before they reach the earth's surface. The conclusion is, then, that both the organic molecules within organisms and certain life processes, such as vision and photosynthesis, are chemically adapted to the radiation that is most prevalent in the environment.

Photosynthesis utilizes the portion of the electromagnetic spectrum (solar radiation) known as visible light.

The pigments found in photosynthesizing cells are capable of absorbing various portions of visible light. The absorption spectra for chlorophyll *a* and chlorophyll *b* are shown in figure 8.2*b*. Both chlorophyll *a* and chlorophyll *b* absorb violet, blue, and red light better than the light of other colors. Because green light is only minimally absorbed and is primarily transmitted or reflected (fig. 8.2*c*), leaves appear green to us.

## Chloroplast Structure and Function

The site of photosynthesis is the **chloroplast,** an organelle found in the cells of green plants and certain algae (fig. 8.3). In the chloroplast, a double membrane, or envelope, surrounds a large central space called the **stroma.** A membrane system within the stroma forms the **grana,** so called because they looked like piles of seeds (*grana*—seeds) to early microscopists.

## Figure 8.3

Chloroplast structure. ***a.*** A chloroplast is bounded by a double membrane. Within, the grana contain pigments (e.g., chlorophyll) that absorb the energy of sunlight. The stroma contains enzymes that reduce carbon dioxide ($CO_2$). ***b.*** Electron micrograph of a chloroplast cross section showing the grana, their connecting stroma lamellae, and the stroma. Magnification, $\times 12,000$. ***c.*** A granum is a stack of membranous sacs called thylakoids. Each thylakoid has a thylakoid space.

## Figure 8.4

Photosynthesis overview. ***a.*** This overall equation for photosynthesis shows the reactants and the products. ***b.*** In the thylakoids, water is split and oxygen is released. ATP and NADPH are made. ***c.*** In the stroma, carbon dioxide ($R—CO_2$) is reduced to $R—CH_2O$ using the ATP and the NADPH from the thylakoids.

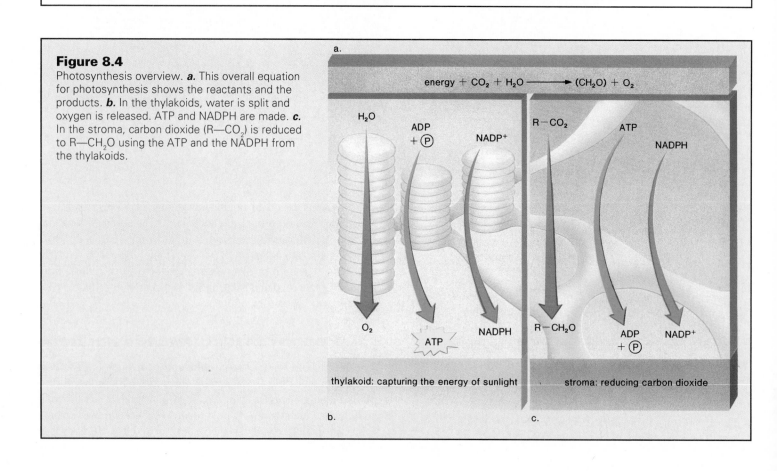

energy + $CO_2$ + $H_2O$ ⟶ ($CH_2O$) + $O_2$

thylakoid: capturing the energy of sunlight

stroma: reducing carbon dioxide

## Grana

The electron microscope reveals that each granum is actually a stack of flattened sacs or disks, now called **thylakoids.** The grana are connected to one another by stroma lamellae, and therefore it is believed that all the thylakoids in a chloroplast are continuous with one another and surround a common interior space.

**Chlorophyll** and the yellow-orange pigments called carotenoids are found within the membranes of the thylakoids. These pigments are able to absorb the energy of the sun (fig. 8.2b); therefore, they make photosynthesis possible. Perhaps we should think of these pigments as the most important biomolecules there are because nearly all living things are dependent on the food produced by photosynthesis.

Thylakoids are the energy-generating system of chloroplasts not only because they contain chlorophyll but also because the cytochrome (electron transport) system associated with the production of ATP and NADPH is also present in the thylakoid membrane.

## Stroma

The stroma contains an enzyme-rich solution. These enzymes incorporate carbon dioxide ($CO_2$) into organic molecules and reduce it to $CH_2O$. In other words, the carbohydrate produced by photosynthesis is formed within the stroma.

> Chloroplasts carry on photosynthesis, an energy-requiring reaction that results in carbohydrate synthesis. First the solar energy must be captured, and then carbon dioxide must be reduced to carbohydrate.

An overall equation for photosynthesis does not indicate that the process actually involves 2 sets of reactions (fig. 8.4). As described in this chapter's reading, the probability of 2 sets of reactions was suggested by F. F. Blackman in 1905 after he discovered that when light is being absorbed maximally, a rise in temperature still increases the rate of photosynthesis.

The first set of reactions, which takes place in the thylakoid, where chlorophyll and other pigments are located, is called the **light-dependent reactions** because they do not take place unless solar energy is available. As water is split and oxygen ($O_2$) is released, NADPH and ATP are made (fig. 8.4b).

The second set of reactions, which takes place in the stroma, where many enzymes are located, is called the **light-independent reactions** because solar energy is not required; that is, the reactions can occur in the dark as well as in the light. The light-independent reactions use the NADPH and ATP produced by the light-dependent reactions to reduce carbon dioxide (R—$CO_2$) to carbohydrate (R—$CH_2O$) (fig. 8.4c).

> The overall equation for photosynthesis does not indicate that the process involves the light-dependent reactions and the light-independent reactions. The light-dependent reactions, located in the thylakoid, capture the energy of sunlight needed by the light-independent reactions, located in the stroma, to reduce carbon dioxide to carbohydrate.

**Figure 8.5**
Components of a photosystem. An antenna consists of several hundred chlorophyll and carotenoid molecules plus a reaction-center chlorophyll *a* molecule. Light energy is absorbed in the form of photons and then passed from molecule to molecule (shown in yellow) until it reaches a reaction-center chlorophyll *a* molecule. Thereafter, energized electrons are passed to an acceptor molecule.

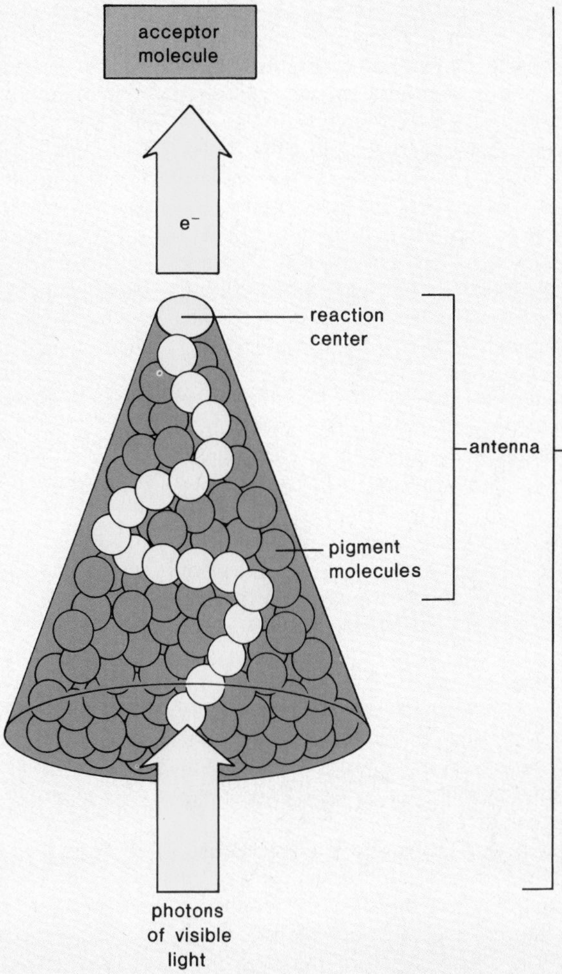

## Light-Dependent Reactions

The light-dependent reactions require the participation of 2 **photosystems,** called Photosystem I and Photosystem II. Each photosystem contains pigment molecules that are part of a light-harvesting antenna (fig. 8.5). Like television antennae aimed to pick up signals, the leaves of a plant turn so the photosystem antennae can collect as much solar energy as possible. The pigment molecules in an antenna are chlorophyll molecules and carotenoid molecules. One chlorophyll *a* molecule is special; it is the reaction-center chlorophyll molecule.

Photosynthesis begins when the antenna of each photosystem absorbs photons of visible light and funnels the energy to their respective reaction center. The reaction-center chloro-

**W**ith the rise of the scientific method in the seventeenth century, investigators began to perform experiments that eventually led to our present-day knowledge of photosynthesis. The ancient Greeks believed that plants were "soil eaters" that somehow converted soil into plant material. Apparently to test this hypothesis, a seventeenth-century Dutchman named Jean-Baptiste Van Helmont planted a willow tree weighing 5 lb in a large pot containing 200 lb of soil. He watered the tree regularly for 5 years and then reweighed both the tree and the soil. The tree weighed 170 lb and the soil weighed only a few ounces less than the original 200 lb. Van Helmont concluded that the increase in weight of the tree was due primarily to the addition of *water.* He did not consider the possibility that air had anything to do with it.

Later, a number of other investigators did consider the role of the air. Joseph Priestley (English, 1733-1804) put a plant and a lighted candle under a bell jar that had trapped a certain amount of air. He found that after the candle had gone out, the plant could "renew" the air in such a way that the candle would burn if lighted again. The science of chemistry was just emerging at this time, and Jan Ingenhousz (Dutch, 1730-99) identified the gas given off by plants as *oxygen.* He also observed that the amount of oxygen given off is proportional to the amount of sunlight reaching the plant; in the shade, plants "injure" air and release carbon dioxide. It was Nicholas de Saussure (Swiss, 1767-1845) who found that an increase in the dry weight of a plant was dependent upon the presence of *carbon dioxide.*

Microscopes were increasingly used during those days, and Matthias Schleiden, a German, is credited with the conclusion in 1838 that plants are composed of cells. Julius Sachs (German, 1832-97) was a plant physiologist who studied plant cells under the microscope. He noted that the green substance named *chlorophyll* is confined to the chloroplasts. In one experiment with a whole plant, he coated a few leaves with wax. He observed that after exposure to sunlight, only the uncoated leaves, which could take in carbon dioxide, increase in *starch* content. (A similar experiment is found in the lab manual that accompanies this text.) Sachs must have been an excellent microscopist because he followed up this experiment by observing that the starch grains in chloroplasts increase in size when a leaf is photosynthesizing.

By the end of the nineteenth century, scientists knew that photosynthesis occurs in chloroplasts and that the basic reaction is:

---

carbon dioxide + water

sunlight $\downarrow$

carbohydrate + oxygen

---

In 1905, F. F. Blackman performed experiments in which he steadily increased the amount of light energy while the temperature was held constant; once the rate of photosynthesis was maximum, he increased the temperature. Now the rate of photosynthesis increased. This led him to conclude that only one part of photosynthesis has to do with light, and that there is a second part that requires temperature-sensitive enzymes. Today these 2 parts of photosynthesis are called the light-dependent reaction and the light-independent reaction.

By the 1940s, cell fractionation allowed investigators to study certain parts of cells. Robert Hill isolated chloroplasts from cells and discovered that oxygen is given off in the absence of $CO_2$. He suggested, therefore, that the oxygen released during photosynthesis comes not from carbon dioxide but from water. This was substantiated in 1941 by Samuel Ruben and Martin Kamen, who performed a study that utilized heavy oxygen ($^{18}O_2$) as a tracer. A plant was exposed first to carbon dioxide that contained heavy oxygen and then to water that contained heavy oxygen. Only in the latter instance did the oxygen given off contain this isotope. To balance the equation, it is necessary to show water on both sides of the equation for photosynthesis (fig. 8.A).

Cell fractionation, radioactive tracers, and electron microscopy continue to add to our knowledge of photosynthesis; some of the more recent experiments using these techniques are covered elsewhere in this chapter.

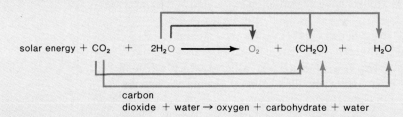

solar energy + $CO_2$ + $2H_2O \longrightarrow O_2$ + $(CH_2O)$ + $H_2O$

carbon dioxide + water → oxygen + carbohydrate + water

**Figure 8.A**

Overall equation for photosynthesis. The arrows show the relationship between the reactants and the products in the equation.

---

phyll *a* molecule of the Photosystem I antenna has an absorption spectrum that peak at a wavelength of around 700 nm and therefore is called *P700* (P—pigment). The reaction-center chlorophyll *a* molecule of the Photosystem II antenna has an absorption spectrum that peaks at a slightly shorter wavelength; it is called *P680.* The light energy received at a reaction center energizes electrons within the chlorophyll *a* molecule, and these electrons then pass to an acceptor molecule.

> In a photosystem, the light-harvesting antenna absorbs solar energy and funnels it to a reaction-center chlorophyll *a* molecule, which then sends energized electrons to an acceptor molecule.

## Two Electron Pathways

There are 2 possible pathways that electrons can take during the first phase of photosynthesis, but only one pathway, the *noncyclic electron pathway,* produces both ATP and NADPH. The other, the *cyclic electron pathway,* generates only ATP. Both pathways are diagrammed in figure 8.6.

## Figure 8.6

Photophosphorylation (ATP formation) occurs in thylakoid membranes. In noncyclic photophosphorylation, electrons move from water to P680 to an acceptor molecule that passes them down a transport system to P700, which sends them to another acceptor molecule before they are finally sent to NADP⁺. Then NADP⁺ combines with a hydrogen ($H^+$) ion from stroma and becomes NADPH. In cyclic photophosphorylation (dotted lines), electrons pass from P700 to an acceptor molecule that sends them down the electron transport system before they return to P700.

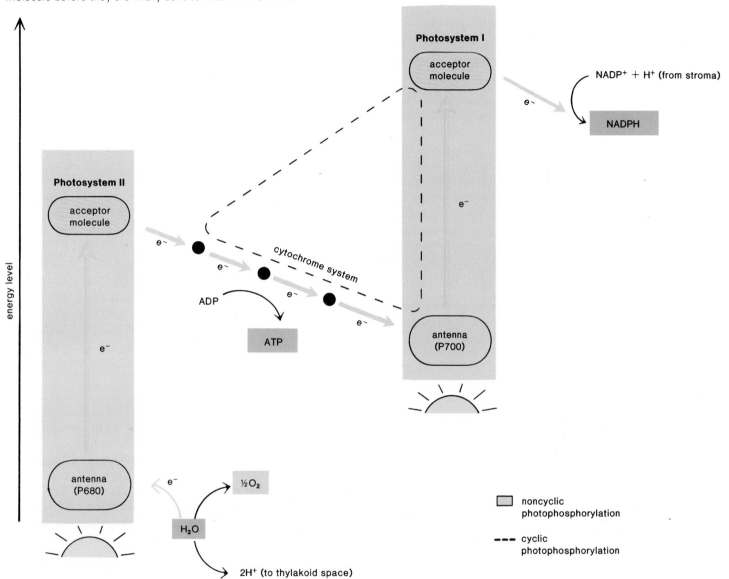

The light-dependent reactions of photosynthesis located in the thylakoid membrane include a noncyclic electron pathway (producing both ATP and NADPH) and a cyclic electron pathway (producing only ATP).

### The Noncyclic Electron Pathway

From figure 8.6, you can see that each photosystem absorbs sunlight at the same time; however, it is easier to describe the events as if they occur in a sequential manner and as if they begin with Photosystem II.

1. As P680 absorbs solar energy (*lower left*), electrons (e⁻) become so highly energized that they can leave the thylakoid-bound chlorophyll *a* molecule. The "hole" left in the molecule is filled by electrons that come from the splitting of water:

$$H_2O \longrightarrow 2H^+ + 2e^- + \tfrac{1}{2}O_2$$

The freed oxygen ($O_2$) is the oxygen gas given off during photosynthesis.

2. The electrons that leave P680 are received by an acceptor molecule that sends them to an electron transport system consisting of a series of thylakoid-bound carriers, some of which are cytochrome molecules. For this reason, this electron transport system often is called a **cytochrome system.** High-energy electrons are delivered to the system and low-energy electrons leave it. As the electrons pass from one cytochrome molecule to another, energy is made available for ATP formation:

$$ADP + \textcircled{P} + energy \longrightarrow ATP$$

3. Electrons leaving the cytochrome system are picked up by P700, the reaction-center chlorophyll *a* molecule of Photosystem I. As this antenna receives solar energy, the electrons are boosted to their highest energy level yet, and they move from the reaction center to an acceptor that passes them on to NADP$^+$, which also combines with hydrogen (H$^+$) ions from the stroma. In the end, NADPH results:

$$NADP^+ + 2e^- + H^+ \longrightarrow NADPH$$

These events make up the process of **noncyclic photophosphorylation** because it is possible to trace electrons in a one-way (noncyclic) direction from water (H$_2$O) to NADPH and because the energy from sunlight is used to produce ATP (photophosphorylation).

> The noncyclic electron pathway produces both ATP and NADPH, which are used in the stroma to reduce carbon dioxide. Oxygen is released as a by-product.

### The Cyclic Electron Pathway

The other electron pathway in chloroplasts is the cyclic electron pathway. In this pathway, electrons leave P700 and eventually return to it (dotted line in figure 8.6) instead of passing to NADP$^+$. Before they return to P700, they pass down the cytochrome system, and ATP is produced as previously described. This pathway sometimes is called **cyclic photophosphorylation** because it is possible to trace electrons in a cycle from P700 to P700 and because solar energy is used to produce ATP (photophosphorylation). Only ATP is produced during cyclic phosphorylation, and there is no concomitant formation of NADPH.

It is believed that the cyclic electron pathway, and therefore Photosystem I, evolved very early (before Photosystem II) in the history of the earth. There are some photosynthetic bacteria even today that use the cyclic electron pathway only. These bacteria do not produce oxygen because they do not split water. Instead, they get the hydrogen atoms from other sources; for example, the photosynthetic sulfur bacteria take hydrogen atoms from hydrogen sulfide (H$_2$S). These bacteria release sulfur and not oxygen (O$_2$). There is no release of oxygen during cyclic photophosphorylation.

**Table 8.1**
Light-Dependent Reactions

**Noncyclic Electron Pathway**

| Participant | Function | Result |
|---|---|---|
| Water | Splits to give O$_2$ H$^+$ Electrons | Becomes $\boxed{O_2}$ in atmosphere Collects in thylakoid space Passes to Photosystem II |
| Photosystem II | Absorbs solar energy | Supplies energized electrons to cytochrome system |
| Cytochrome system | Transports electrons | Releases energy for $\boxed{ATP}$ production |
| Photosystem I | Absorbs solar energy | Supplies energized electrons to NADP$^+$ |
| NADP$^+$ | Accepts electrons (final acceptor) | Becomes $\boxed{NADPH}$ |

**Cyclic Electron Pathway**

| Participant | Function | Result |
|---|---|---|
| Photosystem I | Absorbs solar energy | Supplies energized electrons to cytochrome system |
| Cytochrome system | Transports electrons | Releases energy for $\boxed{ATP}$ production |

$\boxed{\phantom{xxx}}$ = products of light-dependent reactions

From this discussion, you may think that the cyclic electron pathway is vestigial and of little use to water-splitting photosynthesizers. The cyclic electron pathway, however, may be utilized whenever CO$_2$ is in such limited supply that carbohydrate is not being produced. At these times, all available NADP$^+$ is already NADPH and there is no more NADP$^+$ to receive electrons. Obviously, the energized electrons from Photosystem I have to go somewhere, and under these circumstances, the cyclic electron pathway acts as a sort of safety valve. Finally, the cyclic electron pathway provides another way to establish a chemiosmotic gradient for ATP production.

> The cyclic electron pathway utilizes only Photosystem I and produces only ATP. It probably evolved before Photosystem II.

The characteristics of the light-dependent reactions are summarized in table 8.1.

### Chemiosmotic ATP Synthesis

The energy made available by the passage of electrons down the cytochrome system is coupled to the buildup of ATP in an indirect manner. Some of the carriers of the cytochrome system pump

## Figure 8.7

Chloroplast structure and function. **a.** Photosynthesis consists of light-dependent and light-independent reactions. The light-dependent reactions occur in the grana, where sunlight energy is captured, water ($H_2O$) is split and gives off oxygen ($O_2$), and ATP and NADPH are produced. The light-independent reactions occur in the stroma, where carbon dioxide ($CO_2$) is fixed and reduced after being incorporated into the Calvin cycle. Reduction uses the ATP and the NADPH from the light-dependent reactions. **b.** Chemiosmotic ATP synthesis. When water is split, the hydrogen ($H^+$) ions remain in the thylakoid space. Also, some of the cytochrome system carriers pump hydrogen ($H^+$) ions into the thylakoid space. When these flow out of the thylakoid space and down their electrochemical gradient through a special channel protein of the ATP synthase complex, ATP is produced.

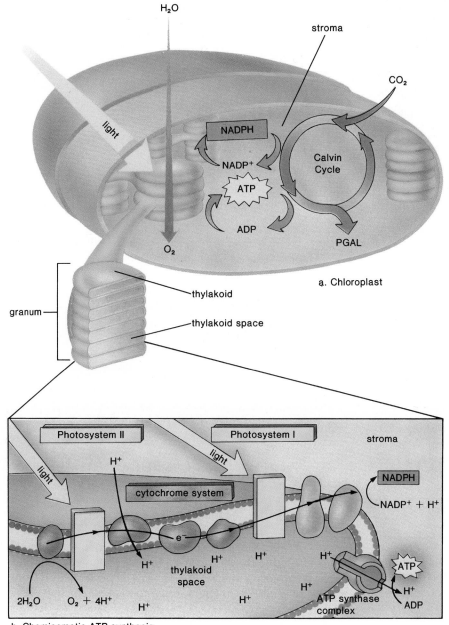

a. Chloroplast

b. Chemiosmotic ATP synthesis

hydrogen ions ($H^+$) from the stroma into the thylakoid space. The thylakoid space acts as a reservoir for hydrogen ions also because for every water molecule that is split at the beginning of noncyclic photophosphorylation, 2 hydrogen ions stay behind in the thylakoid space. The hydrogen ion taken up by $NADP^+$ comes from the stroma, not the thylakoid space.

Because of the large number of hydrogen ions in the thylakoid space compared to the stroma, an extreme electrochemical gradient is present. As previously discussed on page 117, when these hydrogen ions flow out of the thylakoid space by way of a channel protein present in a particle, called the *ATP synthase complex,* energy is provided for the ATP synthase enzyme to produce ATP from $ADP + \text{P}$ (fig. 8.7). This is called *chemiosmotic ATP synthesis* because chemical and osmotic events join to permit ATP synthesis.

## Light-Independent Reactions

The light-independent reactions are the second stage of photosynthesis. They take their name from the fact that light is not directly required for these reactions to proceed. In this stage of photosynthesis, the NADPH and ATP produced by the noncyclic electron pathway during the first stage are used to reduce carbon dioxide: $CO_2$ becomes $CH_2O$ within a carbohydrate molecule. This reduction process is a building-up or synthetic process because it requires the formation of new bonds. Hydrogen atoms ($e- + H^+$) and energy are needed for reduction synthesis, and these are supplied by NADPH and ATP.

> The light-independent reactions use the ATP and NADPH from the light-dependent reactions to reduce carbon dioxide.

### Calvin Cycle

The reduction of carbon dioxide occurs in the stroma of the chloroplast by means of a series of reactions known as the **Calvin cycle.** Melvin Calvin received a Nobel Prize in 1961 for his part in determining these reactions (fig. 8.8). The Calvin cycle includes carbon dioxide fixation, carbon dioxide reduction, and regeneration of RuBP.

## Figure 8.8

**a.** To discover the reactions involved in carbon dioxide fixation and reduction, Calvin and his colleagues used the apparatus shown here. Algae (*Chlorella*) were placed in the flat flask, which was illuminated by 2 lamps. Radioactive carbon dioxide ($^{14}CO_2$) was added, and the algae were killed by transferring them into boiling alcohol (in the beaker below the flat flask). This treatment instantly stopped all chemical reactions in the cells. Carbon-containing compounds were then extracted from the cells and analyzed. **b.** By tracing the radioactive carbon, Calvin and his colleagues were able to trace the reaction series, or pathway, through which $CO_2$ was incorporated. They found that the radioactive carbon (shown in boxes) is first incorporated into RuBP and that the resulting molecule immediately splits to form two molecules of PGA. By gradually increasing the exposure of the algae to $^{14}CO_2$ before killing, they were able to identify the rest of the molecules in the cycle that is now called the Calvin cycle (see fig. 8.9).

a.

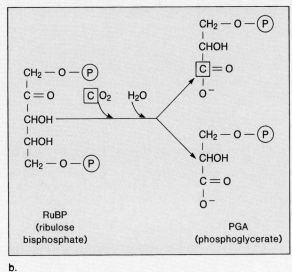

b.

### Carbon Dioxide Fixation

**Carbon dioxide ($CO_2$) fixation,** the attachment of carbon dioxide to an organic compound, is the first event in the Calvin cycle (figs. 8.9 and 8.10). Carbon dioxide fixation occurs when a 5-carbon molecule called **RuBP** (ribulose bisphosphate) combines with carbon dioxide. The enzyme that speeds up this reaction is called RuBP carboxylase, or *rubisco*. Rubisco makes up about 20%-50% of the protein content in chloroplasts, and some say it is the most abundant protein in the world. The reason for its abundance may be that it is unusually slow (it processes only about 3 molecules of substrate per second compared to about 1,000 per second for a typical enzyme), and so there has to be a lot of it to keep the Calvin cycle going.

---

Carbon dioxide fixation occurs when carbon dioxide combines with RuBP. The enzyme for this reaction is rubisco.

---

After $CO_2$ combines with RuBP, the resulting 6-carbon molecule immediately splits to give 2 copies of the same 3-carbon ($C_3$) molecule.

### Carbon Dioxide Reduction

During carbon dioxide reduction, $R—CO_2$ becomes $R—CH_2O$. (The R stands for the rest of the molecule.) The reduction process requires the NADPH and some of the ATP formed in the thylakoid membrane during the light-dependent reactions:

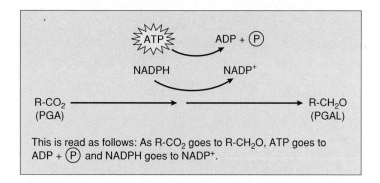

This is read as follows: As $R-CO_2$ goes to $R-CH_2O$, ATP goes to ADP + Ⓟ and NADPH goes to $NADP^+$.

The molecule *PGAL* (phosphoglyceraldehyde) is the end product of the Calvin cycle. For every 3 turns of the Calvin cycle (fig. 8.10), there is a net gain of one PGAL molecule. Two PGAL molecules combine to form glucose phosphate; therefore, glucose is often called the end product of photosynthesis.

Within the leaves of higher plants, glucose molecules are converted to the disaccharide sucrose for transport to other parts of the plant where glucose is used for the synthesis of macromolecules like cellulose and starch. In all organisms, glucose is also used as an energy source, as will be discussed in the next chapter.

---

Carbon dioxide reduction uses the NADPH and some of the ATP produced by the light-dependent reactions (fig. 8.7*b*).

---

## Figure 8.9

The Calvin cycle (simplified). RuBP combines with carbon dioxide ($CO_2$), forming a 6-carbon molecule ($C_6$), which immediately breaks down to 2 PGA (phosphoglycerate) molecules. PGA is reduced to PGAL (phosphoglyceraldehyde), the end product of this cycle. Outside the cycle, 2 PGA molecules can combine to give glucose-6-phosphate, a molecule that can be metabolized to other organic molecules. In the cycle, 5 out of every 6 PGAL are used to reform 3 RuBP.

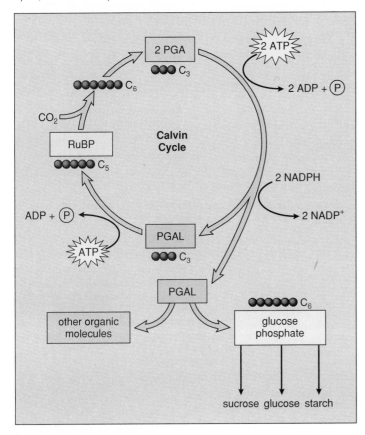

### Regeneration of RuBP

For every 3 turns of the Calvin cycle, 5 molecules of PGAL are used to reform 3 molecules of ribulose bisphosphate (RuBP), the molecule the cycle started with:

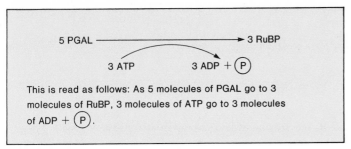

This reaction also utilizes some of the ATP produced by the light-dependent reactions. Altogether, the ATP and the NADPH consumed in producing glucose represents about a 30% energy-conversion rate. In other words, 30% of the energy available in ATP and NADPH is finally found in glucose.

Five out of 6 PGAL molecules of the Calvin cycle are used to regenerate 3 RuBP molecules so the cycle can start again. This regeneration process also uses up some of the ATP molecules produced by the light-dependent reactions.

Table 8.2 summarizes the light-independent reactions.

## $C_4$ Plants

Some plants that live in hot, dry environments do not take up $CO_2$ from the atmosphere by attaching it to RuBP. Instead they use the enzyme PEP carboxylase (*pepco*) to fix $CO_2$ to PEP (phosphoenolpyruvate). These plants are called **$C_4$ plants** because the product of carbon dioxide fixation is oxaloacetate, a $C_4$ or 4-carbon molecule. The plants that do use the enzyme rubisco to fix $CO_2$ to RuBP are called **$C_3$ plants** because the first detected molecule following fixation is PGA, a $C_3$ molecule:

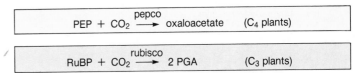

Since we have mentioned repeatedly that there is usually a close correlation between structure and function, you are probably not surprised that the structure of a leaf from a $C_3$ plant is different from that of a $C_4$ plant. In a $C_3$ plant, the *mesophyll cells* contain well-formed chloroplasts and are arranged in layers. In a $C_4$ leaf, not only the mesophyll cells but also the *bundle sheath cells* contain chloroplasts. Further, the mesophyll cells are arranged concentrically around the bundle sheath cells:

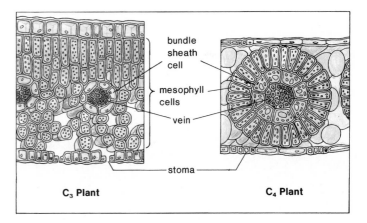

In a $C_3$ plant, all the mesophyll cells both fix $CO_2$ and produce glucose. In a $C_4$ plant, the mesophyll cells with chloroplasts only fix $CO_2$; the bundle sheath cells produce glucose by utilizing the Calvin cycle (fig. 8.11).

In a $C_4$ plant, PEP takes up $CO_2$ to give oxaloacetate in mesophyll cells and then a reduced molecule (malate) is pumped into the bundle sheath cells. After malate gives $CO_2$ to the Calvin cycle, the molecule pyruvate is returned to mesophyll cells. It takes 2 ATP molecules to change pyruvate to PEP once more. You would think that these "extra reactions" of the $C_4$ pathway

## Figure 8.10

Calvin cycle (detailed). Per 3 turns of the cycle 5 molecules of PGAL (15 carbons) are used to reform 3 molecules of RuBP (15 carbons). Each RuBP molecule combines with one molecule of carbon dioxide ($CO_2$), forming a 6-carbon molecule ($C_6$), which immediately breaks down to 2 PGA molecules. Each molecule is reduced in 2 steps to PGAL, the end product of this cycle. Per 3 turns, one PGAL molecule leaves the cycle.

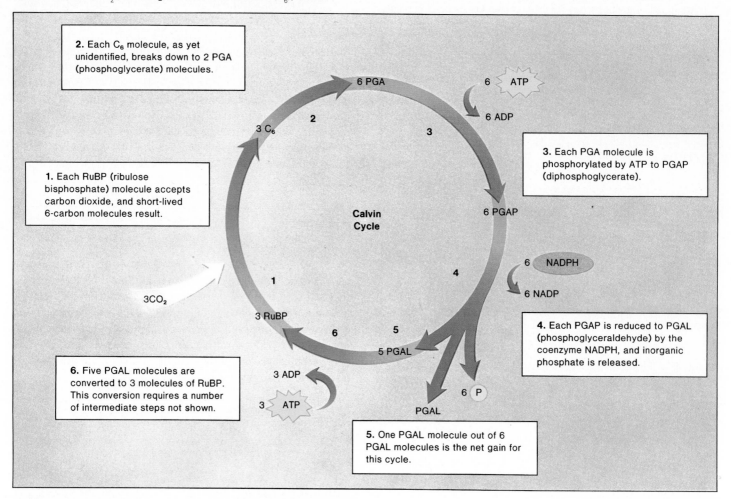

2. Each $C_6$ molecule, as yet unidentified, breaks down to 2 PGA (phosphoglycerate) molecules.

1. Each RuBP (ribulose bisphosphate) molecule accepts carbon dioxide, and short-lived 6-carbon molecules result.

6. Five PGAL molecules are converted to 3 molecules of RuBP. This conversion requires a number of intermediate steps not shown.

3. Each PGA molecule is phosphorylated by ATP to PGAP (diphosphoglycerate).

4. Each PGAP is reduced to PGAL (phosphoglyceraldehyde) by the coenzyme NADPH, and inorganic phosphate is released.

5. One PGAL molecule out of 6 PGAL molecules is the net gain for this cycle.

Calvin Cycle

6 PGA    6 ATP    6 ADP    3 $C_6$    2    3    6 PGAP    6 NADPH    6 NADP    4    5    5 PGAL    3 RuBP    1    6    3CO$_2$    3 ADP    3 ATP    6 P    PGAL

---

### Table 8.2
Light-Independent Reactions

| Participant | Function | Result |
|---|---|---|
| RuBP | Takes up $CO_2$ | $CO_2$ fixation |
| $CO_2$ | Provides carbon atoms | Reduced to $CH_2O$ |
| ATP | Provides energy for reduction of $CO_2$ and generation of RuBP | Broken down to ADP + Ⓟ |
| NADPH | Provides hydrogen atoms for reduction | Oxidized $NADP^+$ |
| PGAL | End product of photosynthesis | Glucose phosphate from 2 PGAL |

would represent an energy drain for these plants, yet in hot, dry climates, the net photosynthetic rate of $C_4$ plants such as sugarcane, corn, and Bermuda grass is about 2 to 3 times that of $C_3$ plants such as wheat, rice, and oats (fig. 8.12).

### Photorespiration

The explanation for the higher energy efficiency of $C_4$ plants in hot, dry climates compared to $C_3$ plants lies in the difference between *rubisco* and *pepco*, the respective enzymes for carbon dioxide fixation in $C_3$ and $C_4$ plants. Oxygen competes with $CO_2$ for the active site of rubisco but not for the active site of pepco. If there is an adequate supply of $CO_2$ inside the leaf, then most of RuBP is converted to 2 PGA molecules, and the $C_3$ pathway operates efficiently. But if the supply of $CO_2$ inside the leaf is inadequate, most of RuBP combines with $O_2$, giving one mol-

## Figure 8.11

The $C_4$ pathway is superimposed on an electron micrograph of the cells of the mesophyll and the bundle sheath in corn. During the $C_4$ pathway, the enzyme called pepco fixes $CO_2$ to PEP forming oxaloacetate, which is reduced to a molecule (malate) that carries $CO_2$ to the Calvin cycle in the bundle sheath cells. Pyruvate then returns to the mesophyll, where ATP is used to reform PEP. Magnification, X8,300.

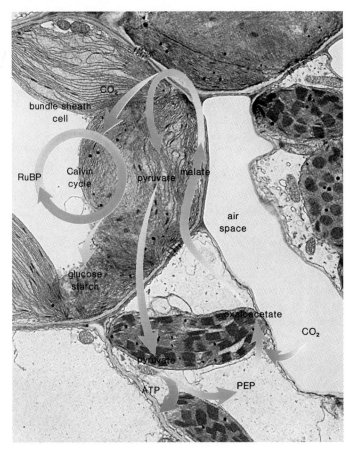

## Figure 8.12

Percentage of grasses that have $C_4$ photosynthesis in areas of North America. The higher temperatures in southern locations apparently offer some advantage to these plants because they are more prevalent in these locations.

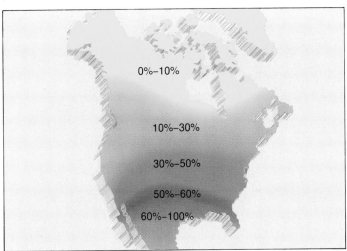

ecule of PGA and one molecule of phosphoglycolate; the latter rapidly breaks down to release $CO_2$:

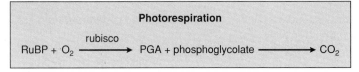

**Photorespiration**

$$RuBP + O_2 \xrightarrow{\text{rubisco}} PGA + \text{phosphoglycolate} \longrightarrow CO_2$$

This process is named photorespiration because in the presence of light (*photo*), oxygen is taken up and $CO_2$ is evolved (*respiration*). Obviously, this is just the reverse of the photosynthetic process: photorespiration wastes energy and does nothing to serve the needs of the plant.

Notice in the diagram on page 131 that there are little openings called stomata (stoma, sing.) in the surfaces of leaves through which water can leave and carbon dioxide can enter. If the weather is hot and dry, these openings close in order to conserve water. Therefore, this is also the time when the concentration of $CO_2$ is the lowest in the leaf and when photorespiration will predominate in a $C_3$ plant. $C_4$ photosynthesis evolved as a way to circumvent this problem since pepco never binds to $O_2$—regardless of the $CO_2$ concentration in the leaf. The energy-requiring "extra reactions" of $C_4$ photosynthesis that occur in the mesophyll cells can deliver a constant supply of $CO_2$ to the Calvin cycle in the bundle sheath cells. This keeps the $CO_2$ concentration high in bundle sheath cells and prevents photorespiration from occurring to any extent, even if the weather is hot and dry.

When the weather is moderate, $C_3$ plants are not at a disadvantage, but when the weather becomes hot and dry, $C_4$ plants have a distinct advantage, and we can expect them to predominate (fig. 8.12). In the early summer, $C_3$ plants such as Kentucky bluegrass and creeping bentgrass predominate in lawns in the cooler parts of the United States, but by midsummer, crabgrass, a $C_4$ plant, begins to take over.

$C_4$ plants have an advantage over $C_3$ plants when the weather is hot and dry because their method of $CO_2$ fixation and their related structure prevent photorespiration from occurring to any extent.

### CAM Plants

**CAM plants** use pepco (PEP carboxylase) to fix carbon dioxide at *night*, forming malate, which is stored in large vacuoles in their mesophyll cells until the next day. CAM stands for crassulacean-acid metabolism; the Crassulaceae are a family of flowering succulent plants, including the stonecrops—low-lying plants that live in warm, arid regions of the world (fig. 8.13). CAM was first

**Figure 8.13**

**Figure 8.13**
Most CAM plants, such as this stonecrop, are succulents that have fleshy stems or leaves. They conserve water by keeping their stomata closed during the day and can live under very arid conditions.

**Table 8.3**
The 3 Pathways of Carbon Dioxide Fixation

| Feature | $C_3$ | $C_4$ | CAM |
|---|---|---|---|
| Leaf structure | Bundle sheath cells lacking chloroplasts | Bundle sheath cells having chloroplasts | Large vacuoles in mesophyll cells |
| Enzyme utilized | Rubisco | Pepco | Pepco |
| Optimum temperature | 15°C–25°C | 30°C–40°C | 35°C |
| Productivity rate (tons/hectare/year) | 22 ± 0.3 | 39 ± 17 | Low—variable |

Modified with permission from D. K. Northington and J. R. Goodin, *The Botanical World.* St. Louis, 1984, Times Mirror/Mosby College Publishing.

discovered in these plants, but now it is known to be prevalent among most succulent plants that grow in desert environments, including the cacti.

Whereas a $C_4$ plant represents partitioning in space—carbon dioxide fixation occurs in mesophyll cells and the Calvin cycle occurs in bundle sheath cells—CAM is an example of partitioning carbon dioxide fixation in time. The malate formed at night releases $CO_2$ during the day to the Calvin cycle within the same cell, which now has NADPH and ATP available to it from the light-dependent reactions. The primary reason for this partitioning again has to do with the conservation of water. CAM plants open their stomata only at night, and therefore only at that time is atmospheric $CO_2$ available to pepco. When sunlight reaches the plant, the stomata close to conserve water and $CO_2$ cannot enter the plant.

Photosynthesis in a CAM plant is usually not as efficient as in a $C_3$ plant or a $C_4$ plant, but it does allow CAM plants to live under stressful conditions. Each of the 3 forms of carbon dioxide fixation does have an advantage. CAM plants can at least live under arid conditions; $C_4$ plants are more efficient photosynthesizers under hot, dry conditions; and $C_3$ plants are more efficient photosynthesizers below a temperature of 25°C.

CAM plants use pepco to carry on carbon dioxide fixation at night when the stomata are open. During the day, the stored carbon dioxide can enter the Calvin cycle.

Table 8.3 contrasts the features of these 3 types of plants.

## Summary

1. Photosynthesis (a) provides food either directly or indirectly for most living things, (b) produces oxygen, and (c) provides the energy present today in fossil fuels.

2. Photosynthesis uses solar energy in the visible-light range; the photons of this range contain the right amount of energy to energize the electrons of chlorophyll molecules. Specifically, chlorophylls *a* and *b* absorb violet-blue and red-orange wavelengths best. This causes chlorophyll to appear green to us.

3. A chloroplast is bounded by a double membrane and contains 2 main portions: the liquid stroma and the membranous grana made up of thylakoid sacs. The light-dependent reactions take place in the thylakoids, and the light-independent reactions take place in the stroma.

4. The noncyclic electron pathway of the light-dependent reactions begins when solar energy enters Photosystem II, where it activates the reaction center P680. Energized electrons leave P680 and go to an electron acceptor. The splitting of water replaces these electrons in P680, releases oxygen to the atmosphere, and gives hydrogen ions ($H^+$) to the thylakoid space. The acceptor molecule passes electrons to Photosystem I by way of a cytochrome (electron transport) system. Light energy absorbed by Photosystem I is received by its reaction center P700. The energized electrons that leave this molecule are ultimately received by $NADP^+$, which also combines with $H^+$ from the stroma to become NADPH.

5. In the cyclic electron pathway, the energized electrons that leave Photosystem I are not sent to $NADP^+$. Instead, they pass down the cytochrome system back to Photosystem I again. The cyclic pathway may occur only if there is no free $NADP^+$ to receive the electrons; this is a way to prevent an energy overload of Photosystem I. It serves to generate ATP,

The Cell

and in keeping with its name, it evolved first and is the only photosynthetic pathway present in certain bacteria today.

6. The energy made available by the passage of electrons down the cytochrome system allows carriers to pump hydrogen ($H^+$) ions into the thylakoid space. The buildup of hydrogen ions establishes an electrochemical gradient. When hydrogen ions flow down this gradient through the channel present in ATP synthase complexes, ATP is synthesized from ADP and $\textcircled{P}$.

7. The energy yield of the light-dependent reactions is stored in ATP and NADPH. These molecules are now used by the light-independent reactions to reduce carbon dioxide to carbohydrate.

8. During the light-independent reactions, the enzyme rubisco fixes carbon dioxide to RuBP giving a 6-carbon molecule that immediately breaks down to 2 $C_3$ molecules. This is the first stage of the Calvin cycle. During the second stage, $R—CO_2$ is reduced to $R—CH_2O$, and this step requires the NADPH and some of the ATP from the light-dependent reactions. For every 3 turns of the Calvin cycle, the net gain is one PGAL molecule; the other 5 PGAL molecules are used to reform 3 molecules of RuBP. This step also requires ATP for energy.

9. In $C_4$ plants, as opposed to the $C_3$ plants just described, the enzyme pepco fixes carbon dioxide to PEP to form a 4-carbon molecule, oxaloacetate, within mesophyll cells. A reduced form of this molecule is pumped into bundle sheath cells where carbon dioxide is released to the Calvin cycle. The resulting pyruvate returns to mesophyll cells, where ATP must be used to regenerate PEP. $C_4$ plants avoid photorespiration by a partitioning of pathways in space—carbon dioxide fixation in mesophyll cells and Calvin cycle in bundle sheath cells.

10. During CAM photosynthesis, pepco fixes carbon dioxide to PEP at night to produce malate. The next day, carbon dioxide is released and enters the Calvin cycle within the same cells. This represents a partitioning of pathways in time: carbon dioxide fixation at night and the Calvin cycle during the day. The plants that carry on CAM are desert plants, in which the stomata only open at night in order to conserve water.

### Writing Across the Curriculum

In order to practice writing skills, students should write out the answers to any or all of the study questions and the critical thinking questions. The study questions are sequenced in the same order as the text. Suggested answers to the critical thinking questions are in appendix D.

## Study Questions

1. Why is it proper to say that all living things are dependent on solar energy?
2. Contrast the electromagnetic spectrum with the absorption spectrum of chlorophyll. Why is chlorophyll a green pigment?
3. What are the inputs and outputs for the light-dependent reactions and for the light-independent reactions? Where do these reactions take place in the chloroplast?
4. Describe the structure of a photosystem. Explain the terms *P700* and *P680*.
5. Using figure 8.6, trace noncyclic and cyclic electron pathways, explaining the events as they occur.
6. Contrast the noncyclic and cyclic electron pathways and explain the differences noted.
7. Describe the chemiosmotic process of ATP production, mentioning the thylakoid membrane, the electron transport system, ATP synthase complex, and ATP synthase.
8. Describe the 3 stages of the Calvin cycle. What happens during the stage that utilizes ATP and NADPH from the light-dependent reactions? What specific molecules are involved?
9. Explain $C_4$ photosynthesis, contrasting the actions of rubisco and pepco.
10. Explain CAM photosynthesis, contrasting it to $C_4$ photosynthesis in terms of partitioning of a pathway.

## Objective Questions

1. The absorption spectrum of chlorophyll
   a. approximates the action spectrum of photosynthesis.
   b. explains why chlorophyll is a green pigment.
   c. shows that some colors of light are absorbed more than others.
   d. All of these.
2. The final acceptor of electrons during the noncyclic electron pathway is
   a. Photosystem I.
   b. Photosystem II.
   c. ATP.
   d. $NADP^+$.
3. A photosystem contains
   a. pigments, a reaction center, and an electron acceptor.
   b. ADP, $\textcircled{P}$, and $H^+$.
   c. proto ns, photons, and pigments.
   d. Both b and c.
4. Which of these should not be associated with the electron transport system?
   a. cytochromes
   b. movement of $H^+$ into the thylakoid space
   c. photophosphorylation
   d. absorption of solar energy
5. Pepco has an advantage compared to rubisco. The advantage is that
   a. pepco is present in both mesophyll and bundle sheath cells, rubisco is not.
   b. rubisco fixes carbon dioxide only in $C_4$ plants, whereas pepco does it in both $C_3$ and $C_4$ plants.
   c. rubisco is subject to photorespiration but pepco is not.
   d. Both b and c.

**6.** The NADPH and ATP from the light-dependent reactions are used to
  **a.** cause rubisco to fix carbon dioxide.
  **b.** reform the photosystems.
  **c.** cause electrons to move along their pathways.
  **d.** convert PGA to PGAL.
**7.** CAM photosynthesis
  **a.** is the same as $C_4$ photosynthesis.
  **b.** is an adaptation to cold environments in the Southern Hemisphere only.
  **c.** is prevalent in desert plants that close their stomata during the day.
  **d.** occurs in plants that live in marshy areas.
**8.** Chemiosmotic phosphorylation depends on
  **a.** an electrochemical gradient.
  **b.** a difference in $H^+$ concentration between the thylakoid space and the stroma.
  **c.** ATP breaking down to ADP + Ⓟ .
  **d.** Both a and b.
**9.** Label this diagram of a chloroplast.

**a.** The light-dependent reactions occur in which part of a chloroplast?
**b.** The light-independent reactions occur in which part of a chloroplast?
**10.** Label this diagram using these labels: water, carbohydrate, carbon dioxide, oxygen, ATP, ADP +Ⓟ, NADPH, and NADP.

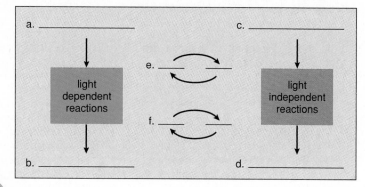

a. _____   c. _____

light dependent reactions

e. _____

light independent reactions

f. _____

b. _____   d. _____

---

## Concepts and Critical Thinking

**1.** *Virtually all living things are dependent on solar energy.*

What data would you use to convince a friend that the lives of human beings are dependent on solar energy?

**2.** *Photosynthesis makes solar energy available to living things.*

Tell how photosynthesis makes solar energy available by explaining the roles of chlorophyll, ATP, NADP$^+$, and $CO_2$ in photosynthesis.

**3.** *The cell and the organelles themselves are compartmentalized, and this assists their proper functioning.*

Name 2 major compartments (separate areas) of a chloroplast, and tell how photosynthesis is dependent on keeping the areas separate.

---

## Selected Key Terms

**photon** (fo'ton) 122
**electromagnetic spectrum** (e-lek"tro-mag-net'ik spek'trum) 122
**visible light** (viz'ĭ-b'l līt) 122
**chloroplast** (klo'ro-plast) 123
**stroma** (stro'mah) 123
**granum** (gra'num) 123
**thylakoid** (thi'lah-koid) 125
**chlorophyll** (klo'ro-fil) 125

**light-dependent reaction** (līt de-pen'dent re-ak'shun) 125
**light-independent reaction** (līt in"de-pen'dent re-ak'shun) 125
**photosystem** (fo"to-sis'tem) 125
**noncyclic photophosphorylation** (non-sik'lik fo"to-fos"for-ĭ-la'shun) 128
**cyclic photophosphorylation** (sik'lik fo"to-fos"for-ĭ-la'shun) 128
**ATP synthase complex** (sin'thase kom'pleks) 129

**Calvin cycle** (kal'vin ben'sun si'k'l) 129
**carbon dioxide (CO$_2$) fixation** (kar'bon di-ok'sīd fix-sa'shun) 130
**RuBP** 130
**C$_4$ plant** 131
**C$_3$ plant** 131
**photo respiration** (fo"to res"pi-ra'shun) 133
**CAM** 133

# 9

# Glycolysis and Cellular Respiration

The powerful muscles of this cheetah allow it to dart forward to capture prey. Muscles are powered by ATP energy, and mitochondria are the organelles that convert the energy of other organic compounds to the energy of ATP.

Your study of this chapter will be complete when you can

1. describe the general function of cellular respiration;
2. describe both the anaerobic and aerobic fate of pyruvate;
3. discuss glycolysis, explaining the process of substrate-level phosphorylation and giving the inputs and outputs of the pathway;
4. describe lactate fermentation and alcoholic fermentation and the evolutionary significance of the fermentation process;
5. discuss the transition reaction, explaining the role of acetyl CoA and giving the inputs and outputs of the reaction;
6. discuss the Krebs cycle, explaining how it cycles and giving the inputs and outputs of the cycle;
7. discuss the electron transport system, and explain how it contributes to the formation of ATP;
8. describe the arrangement of electron carriers in the inner membrane of mitochondria, and explain the process of chemiosmotic ATP synthesis;
9. calculate the yield of ATP molecules per glucose molecule for glycolysis and cellular respiration;
10. discuss the concept of a metabolic pool and how the breakdown of carbohydrate, protein, and fat contributes to the pool.

t is common knowledge that human beings eat food, which they digest to nutrient molecules, but you may not be aware that most other organisms also use organic molecules as a source of building blocks and energy. Carbohydrate molecules (e.g., glucose) are a primary energy source because they are often broken down to acquire a supply of ATP molecules. We can write the overall equation for carbohydrate breakdown in this manner:

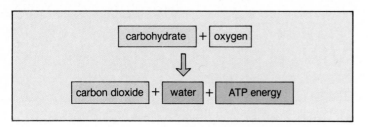

ATP, the energy currency of cells (p. 115), provides the energy that cells need for transport work, mechanical work, and synthetic work. Why do living things convert carbohydrate energy to ATP energy? Why not, for example, use glucose energy directly, thereby bypassing mitochondria? The answer is that the energy content of glucose is too great for individual cellular reactions; ATP contains the right amount of energy, and enzymes are adapted through evolution to couple ATP breakdown with energy-requiring cellular processes (fig. 7.12).

---

In cells, the energy of carbohydrates is converted to that of ATP molecules, the energy currency of cells.

---

You know from chapter 5 that chloroplasts produce the energy-rich carbohydrates that typically undergo cellular respiration in mitochondria. Cellular respiration, then, is the link between the energy captured by plants and the energy utilized by both plants and animals (fig. 9.1a). It is the means by which energy flows from the sun through all living things. This is true because, for example, plants produce the food eaten by animals.

As we discussed in chapter 7, there is always loss of usable energy when one form of energy is transformed into another form of energy. Eventually, the energy content of food is completely dissipated, but the chemicals themselves cycle. Note that when

**Figure 9.1**
*a.* Which of these 2 organisms is carrying on cellular respiration—the giraffe or the acacia tree? Both are, because both animals and plants require a source of ATP energy. *b.* The energy-rich carbohydrate molecules, broken down largely within the mitochondria, are originally derived from photosynthesis carried out within chloroplasts. The flow of *energy* among organisms (e.g., from plants to animals) goes one way only, but the *chemicals* can recycle because the carbon dioxide and water given off by the mitochondria are reused by the chloroplasts.

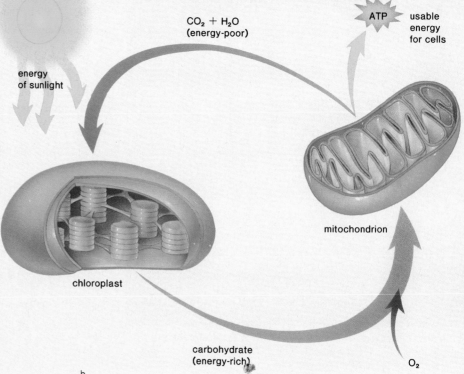

carbohydrate is broken down, the low-energy molecules carbon dioxide and water are given off. These are the raw materials for photosynthesis (fig. 9.1*b*).

> The energy content of carbohydrate was originally provided by solar energy.

## Overview of Glucose Metabolism

We will especially be concerned with the breakdown of glucose because this molecule is the primary carbohydrate that cells break down for a constant supply of ATP molecules. The energy in the bonds of glucose is not released all at once; instead the energy is released slowly, step by step. *Metabolic pathways* (p. 108) are a series of enzymatic reactions that in this case allow the step-by-step breakdown of glucose. This gradual breakdown allows a concomitant gradual buildup of ATP molecules. If glucose is metabolized completely to carbon dioxide ($CO_2$) and water ($H_2O$), the cell realizes a maximum yield of 36 ATP molecules. The energy in 36 ATP molecules is equivalent to 39% of the energy that was available in glucose.

### Glycolysis and Fermentation

**Glycolysis** is a metabolic pathway that breaks down glucose ($C_6$) to 2 molecules of **pyruvate** ($C_3$) (fig. 9.2). Glycolysis takes place in the cytosol outside mitochondria and is anaerobic—it does not require oxygen. Oxidation by removal of hydrogen atoms ($e^- + H^+$) does occur, though, and the energy of oxidation is used to generate 2 ATP molecules.

     Pyruvate is a pivotal metabolite. What happens to pyruvate depends on whether oxygen is available to the cell. If oxygen is not available, anaerobic metabolism continues and fermentation occurs. **Fermentation** includes glycolysis plus the reduction of pyruvate to alcohol or to lactate. Yeast cells carry on alcoholic fermentation when they ferment sugars in fruits and grains to the alcohol in wine and beer. You may be aware that, in the absence of oxygen, our muscle cells carry on lactate fermentation. A cramping of muscles may occur when lactate builds up. With aerobic exercise, a common endeavor today, lactate fermentation is not expected and there should be no cramping of the muscles.

> **Glycolysis:** anaerobic process in cytosol; breakdown of glucose to pyruvate; net gain of 2 ATP per glucose molecule
> **Fermentation:** anaerobic process in cytosol; breakdown of glucose to alcohol or lactate; net gain of 2 ATP per glucose molecule

### Glycolysis and Cellular Respiration

Pyruvate from glycolysis enters a mitochondrion when oxygen is available. **Cellular respiration** is the complete breakdown of pyruvate to carbon dioxide and water. Cellular respiration occurs in mitochondria and is aerobic—oxygen is required. Cellular respiration involves the conversion of pyruvate to an acetyl group, and 2 other

**Figure 9.2**

Overview of glucose metabolism. Glycolysis in the cytosol produces pyruvate, a key intermediary metabolite. If oxygen is not available, anaerobic metabolism continues and pyruvate is reduced to either alcohol or lactate depending on the type of cell. If oxygen is available, aerobic respiration begins in a mitochondrion and pyruvate is metabolized completely to carbon dioxide and water. The net gain of ATP for glycolysis (and fermentation) is 2 ATP; the net gain of ATP for cellular respiration is 34 ATP; the net gain of ATP for the complete breakdown of glucose is 36 ATP.

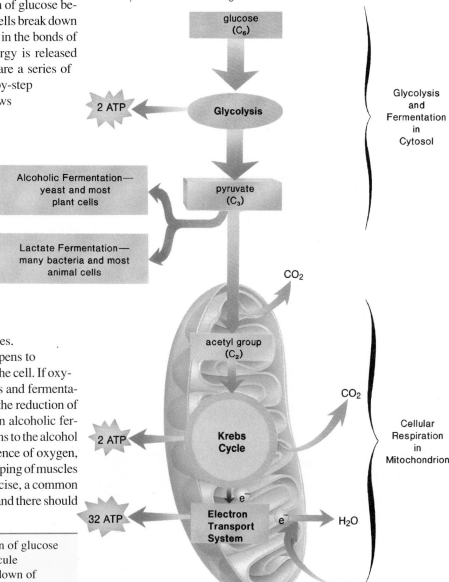

metabolic pathways shown in figure 9.2, namely the Krebs cycle and an electron transport system. At the completion of cellular respiration, glucose has been broken down completely to carbon dioxide and water. It is possible to write an overall equation for the complete breakdown of glucose as is done in figure 9.3, that is:

$$C_6H_{12}O_6 + 6O_2 \rightarrow 6CO_2 + 6H_2O$$

## Figure 9.3

Complete glucose breakdown. Glucose enters a cell from the bloodstream and is metabolized to pyruvate which enters a mitochondrion. Oxygen from the bloodstream also enters a mitochondrion. Here pyruvate is metabolized completely to carbon dioxide and water, which exit the cell. The resulting ATP molecules stay behind in the cell to enable the cell to perform all sorts of energy-requiring processes.

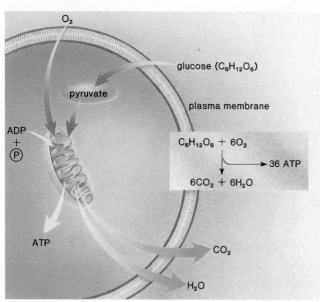

Unfortunately, the overall equation has limited usefulness because it does not remind us that complete glucose breakdown involves glycolysis, the conversion of pyruvate to an acetyl group, the Krebs cycle, and the electron transport system. The equation, however, allows us to discuss how the human body provides the reactants and deals with the end products of glucose breakdown. The air we breathe contains *oxygen* ($O_2$) and the food we eat contains *glucose* ($C_6H_{12}O_6$). The oxygen and the glucose enter the bloodstream, which carries them to the body's cells, where they diffuse into each and every cell. During cellular respiration in a mitochondrion, pyruvate is broken down to *carbon dioxide* ($CO_2$) and *water* ($H_2O$) as ATP is produced. All 3 of these then leave the mitochondrion. The ATP is utilized inside the cell for energy-requiring processes. Carbon dioxide diffuses out of the cell into the bloodstream. The bloodstream takes the carbon dioxide to the lungs, where it is exhaled. The water molecules produced, called metabolic water, play little if any role in humans. In other organisms, metabolic water helps prevent dehydration.

Even though the overall equation for glucose breakdown seems to suggest that oxygen combines directly with glucose, this is not the case at all (fig. 9.3). Instead, as expected, oxidation occurs by the removal of hydrogen atoms ($e^- + H^+$) from metabolites. We will see that these electrons ($e^-$) are carried by coenzyme $NAD^+$ (and FAD) to the electron transport system (p. 146), where they are eventually transferred to oxygen molecules. Reduced oxygen then takes up hydrogen ($H^+$) ions and becomes water.

---

**Cellular respiration:** aerobic respiration in mitochondria; breakdown of pyruvate to carbon dioxide and water; pyruvate → acetyl group, Krebs cycle, electron transport system; gain of 34 ATP per glucose molecule

---

Net maximum gain of ATP per glucose molecule breakdown to $CO_2$ and $H_2O$:

| | |
|---|---|
| Glycolysis | 2 ATP |
| Cellular respiration | 34 ATP |
| | 36 ATP |

## Glycolysis

Figure 9.4 illustrates the main reactions of glycolysis, the breakdown of glucose to 2 molecules of pyruvate. First, 2 ATP molecules are used to phosphorylate metabolites in order to activate them (steps 1 and 2). The term *glycolysis,* meaning the splitting of sugar, is appropriate because glucose is split into 2 smaller molecules (step 3). All reactions after this step are multiplied by 2.

Hydrogen atoms ($e^- + H^+$) are removed from a metabolite of the pathway and are picked up by $NAD^+$, a coenzyme of dehydrogenases (p. 114):

$$NAD^+ \xrightarrow{\text{2H}} NADH + H^+$$

Altogether, 2 NADH molecules result because step 4 occurs twice. The energy of oxidation results in high-energy phosphate bonds attached to certain substrates. These substrates are used to produce 4 ATP molecules (steps 5 and 7). Since 2 ATP molecules were used to get started, glycolysis yields a net gain of 2 ATP molecules per glucose molecule.

---

Glycolysis provides 2 NADH molecules and a net gain of 2 ATP per glucose molecule.

---

ATP production during glycolysis is called **substrate-level phosphorylation** because substrates have transferred high-energy phosphate bonds to ADP. This process is illustrated in figure 9.5, where the molecule PGAP gives up a high-energy phosphate to ADP and becomes PGA. ADP becomes ATP (Step 5 in fig. 9.4).

The end product of glycolysis is 2 molecules of pyruvate, a $C_3$ molecule (step 8 in fig. 9.4).

Altogether the inputs and outputs of glycolysis are as follows:

| Glycolysis | |
|---|---|
| inputs | outputs |
| glucose | 2 pyruvate |
| 2 $NAD^+$ | 2 NADH |
| 2 ADP + $P$ | 2 ATP (net) |

## Figure 9.4

Glycolysis is a metabolic pathway that begins with glucose and ends with pyruvate. Net gain of 2 ATP molecules can be calculated by subtracting those expended from those produced. Text in boxes to the right explains the reactions.

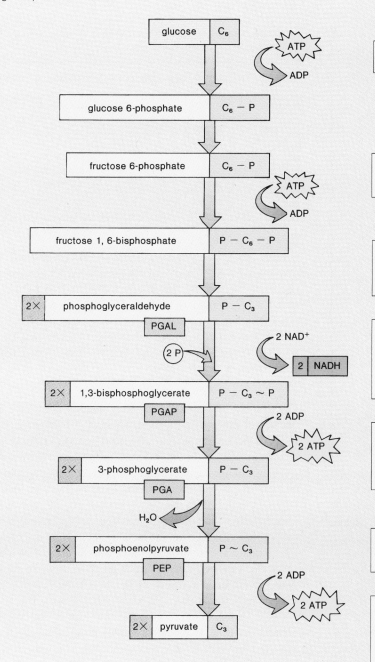

1. Phosphorylation of glucose by ATP produces an activated molecule.

2. Rearrangement, followed by a second ATP phosphorylation, gives fructose bisphosphate.

3. The 6-carbon molecule is split into 2 3-carbon PGAL. After this, each reaction has to occur twice per each glucose molecule.

4. Oxidation, followed by phosphorylation, produces 2 NADH molecules and gives 2 PGAP molecules—each with one high-energy phosphate bond, represented by a wavy line.

5. Removal of high-energy phosphate by 2 ADP molecules produces 2 ATP molecules and gives 2 PGA molecules. This pays back the original investment of 2 ATP used in steps 1 and 3.

6. Oxidation by removal of water gives 2 PEP molecules, each with a high-energy phosphate bond.

7. Removal of high-energy phosphate by 2 ADP molecules produces 2 ATP molecules. The net energy yield for the glycolysis pathway is 2 molecules of ATP.

8. Pyruvate is the end product of the glycolytic pathway. If cellular respiration occurs, pyruvate enters the mitochondria for further breakdown.

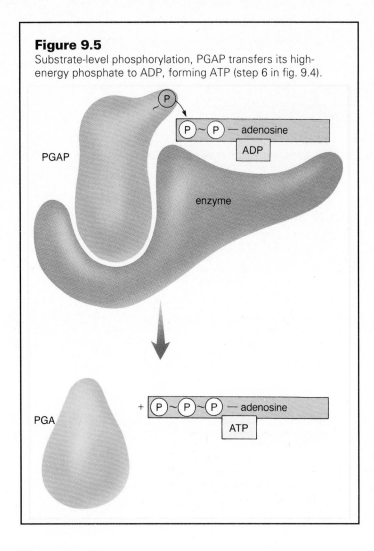

**Figure 9.5**
Substrate-level phosphorylation, PGAP transfers its high-energy phosphate to ADP, forming ATP (step 6 in fig. 9.4).

## Fermentation

*Fermentation* includes glycolysis plus the reduction of pyruvate to either lactate or alcohol. During fermentation, pyruvate accepts the 2 hydrogen atoms that were removed from glycolytic metabolites earlier. In the 2 most familiar forms of fermentation, lactate fermentation and alcoholic fermentation, the end products are lactate (in many bacteria and in animal cells) or ethyl alcohol and carbon dioxide (in yeasts and in most plant cells):

$$\text{pyruvate} \xrightarrow{\text{2H}} \text{lactate}$$

$$\text{pyruvate} \xrightarrow{\text{2H}} \text{alcohol} + CO_2$$

Recall that step 4 in glycolysis (fig. 9.4) requires the coenzyme NAD$^+$ and that following this step, ATP production occurs. Glycolysis is an **anaerobic** process, and if oxygen is not available, glycolysis is the only means of producing ATP. Therefore, it is extremely important to the life of a cell that glycolysis

continue when oxygen is not available. For glycolysis to continue, there must be a source of "free" NAD$^+$, that is, NAD$^+$ that is not carrying hydrogen atoms. During fermentation, NADH from step 4 of glycolysis (fig. 9.4) becomes free NAD$^+$ by passing on hydrogen atoms to pyruvate. Now glycolysis can continue and ATP production can occur (fig. 9.6).

> The purpose of fermentation is to supply the glycolytic pathway with free NAD$^+$ when oxygen is not available.

The inputs and outputs of fermentation are as follows:

| Fermentation | |
|---|---|
| **inputs** | **outputs** |
| glucose | 2 lactate or |
| | 2 alcohol and 2 $CO_2$ |
| 2 ADP | 2 ATP (net) |

### Energy Yield of Fermentation

A high-energy phosphate bond in an ATP molecule has an energy content of 7.3 Kcal, and 2 ATP are produced per glucose molecule during fermentation. This is equivalent to 14.6 Kcal. Complete glucose breakdown to carbon dioxide and water represents a possible energy yield of 686 Kcal per molecule. Therefore, the energy yield for fermentation is only 14.6/686 = 2.1%. This is much less than the energy yield for glycolysis followed by aerobic cellular respiration (p. 143).

### Usefulness of Fermentation

Despite its low yield of ATP, fermentation does have its place, because it can provide a *rapid burst* of ATP energy. When our muscles work vigorously over a short period of time, as when we run, fermentation provides a way to produce ATP even though oxygen is temporarily in limited supply. At first, the blood carries away all the lactate formed in the muscles, but eventually lactate begins to build up in the muscles, changing the pH and causing muscle fatigue and discomfort. When we stop running, our body has a chance to catch up; the lactate is converted back to pyruvate by the removal of hydrogen atoms, which can now be transported to the electron transport system. In the meantime, we are said to be in **oxygen debt,** as signified by the fact that we keep on breathing very heavily for a time.

Fermentation allows yeast cells to grow and divide anaerobically for a time. Eventually, though, if the initial glucose level is high, they are killed by the very alcohol they produce. Presumably, human beings were delighted to discover this form of fermentation, as the ethyl alcohol produced by the process has been consumed in great quantity for thousands of years. In addition, we use the $CO_2$ produced to make bread rise.

## Evolutionary Perspective

Glycolysis and fermentation have probably been present as long as there have been living things on earth. At first, there was no oxygen

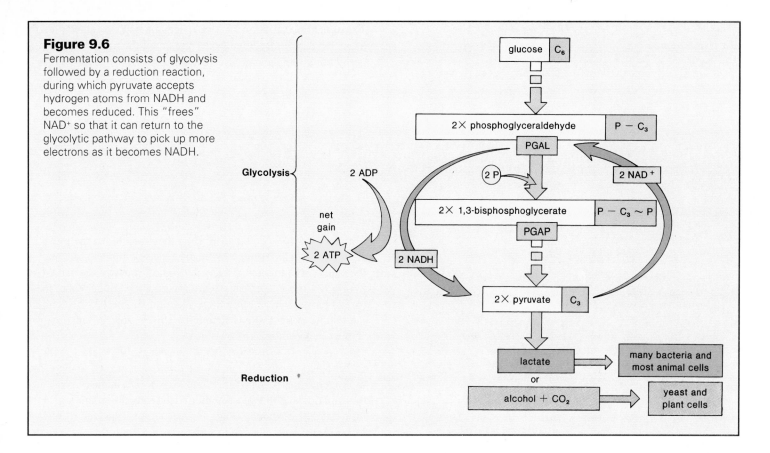

**Figure 9.6**
Fermentation consists of glycolysis followed by a reduction reaction, during which pyruvate accepts hydrogen atoms from NADH and becomes reduced. This "frees" NAD$^+$ so that it can return to the glycolytic pathway to pick up more electrons as it becomes NADH.

in the atmosphere, but there were organic molecules in the oceans where the first cell(s) arose. Any cell that could utilize the energy of these molecules anaerobically had a competitive advantage, even if the method was not the most efficient. The presence of some form of fermentation in nearly all organisms shows how important this ancient competitive edge was to the continuance of life. Apparently, in the ancient past, only cells capable of fermentation survived, passing on the trait to future organisms.

Since fermentation is very inefficient compared to aerobic respiration, you might think it would have simply disappeared sometime during the history of life on earth. As with the 2 photosystems involved in photosynthesis, however, it appears that evolution simply builds on what has already been accomplished. In this instance, we see that pyruvate, the end product of glycolysis, can be metabolized either anaerobically or aerobically. Glycolysis, then, may have initially evolved as a part of fermentation, but today it is usually preliminary to aerobic respiration.

# Cellular Respiration

Cellular respiration is aerobic respiration that occurs in mitochondria. During cellular respiration pyruvate from glycolysis is broken down completely to carbon dioxide and water. Altogether there is a gain of 36 ATP for the complete breakdown of glucose (fig. 9.7). This is far more than the 2 ATP that result from fermentation, an anaerobic process.

During cellular respiration, organic molecules are oxidized by the removal of hydrogen atoms (e$^-$ + H$^+$). Most often, these hydrogen atoms are accepted by NAD$^+$, which then becomes NADH. NADH delivers electrons (e$^-$) to an electron transport system, which passes them on to oxygen, which then becomes water (H$_2$O). Oxygen is the final acceptor for electrons during cellular respiration.

Cellular respiration requires 3 different events: the oxidation of pyruvate to an acetyl group (called the transition reaction), the Krebs cycle, and an electron transport system.

## Transition Reaction

The **transition reaction** is so called because it connects glycolysis to the Krebs cycle (fig. 9.8). In this reaction, pyruvate (C$_3$) is oxidized and broken down to an acetyl group (C$_2$) and one molecule of CO$_2$. The acetyl group is transferred to coenzyme A (CoA)—a large molecule that contains a nucleotide and a portion of one of the B vitamins. This combination of acetyl group and CoA is called **acetyl CoA**. Acetyl CoA contains a high-energy bond, and therefore the molecule is called active acetate, AA (fig. 9.7). Oxidation of pyruvate results in an NADH molecule, which delivers electrons to the electron transport system.

During the transition reaction, pyruvate is broken down to an acetyl group and CO$_2$ is released. Oxidation of pyruvate results in an NADH molecule.

## Figure 9.7

The overall equation for glucose breakdown shown does not indicate that the process requires these metabolic pathways plus the transition reaction.

Glycolysis: Glucose is broken down to 2 molecules of pyruvate (PYR) for a net gain of 2 ATP molecules. Electrons exit the pathways and are taken to the electron transport system by NADH.

Transition Reaction: Carbon dioxide ($CO_2$) and electrons are removed from pyruvate (PYR). Acetyl CoA (AA) results. The electrons are taken to the electron transport system by NADH.

Krebs Cycle: Acetyl CoA enters the cycle; carbon dioxide ($CO_2$) and electrons exit the cycle. There is a gain of 2 ATP per glucose molecule.

Electron Transport System: The system receives electrons (usually carried by NADH), and as they pass down the system from carrier to carrier (cytochromes are carriers), 32 ATP molecules are built up. Oxygen ($O_2$) is the final acceptor for electrons and water ($H_2O$) results.

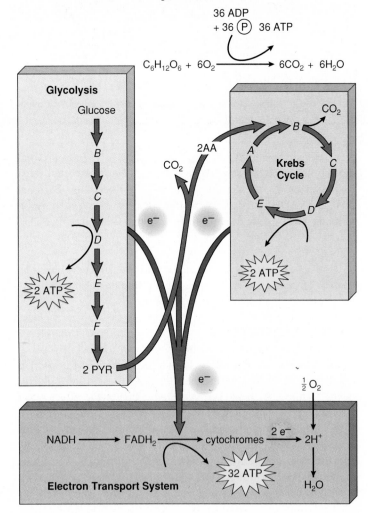

The transition reaction occurs twice for each original glucose molecule. Therefore, inputs and outputs of the transition reaction per glucose molecule are as follows:

| Transition Reaction | |
|---|---|
| inputs | outputs |
| 2 pyruvate<br>2 CoA<br>2 NAD$^+$ | 2 $CO_2$<br>2 acetyl CoA<br>2 NADH |

## Krebs Cycle

The acetyl CoA produced during the transition reaction then enters the **Krebs cycle,** a cyclical metabolic pathway located in the matrix of mitochondria (fig. 9.9). The Krebs cycle is named for Sir Hans Krebs, a British scientist who received the Nobel Prize for his part in determining these reactions in the 1930s. The first metabolite of the cycle is citrate; for this reason, it is also known as the **citric acid cycle.**

The $C_2$ acetyl group that enters the Krebs cycle is oxidized to 2 molecules of $CO_2$. During the oxidation process most of the hydrogen atoms ($e^-$ + $H^+$) are donated to NAD$^+$, but in one instance they are donated to FAD, another coenzyme of oxidation in cells. Some of the energy of oxidation is used immediately to form ATP by substrate-level phosphorylation, as in glycolysis.

> During the Krebs cycle, the $C_2$ acetyl group is oxidized to 2 molecules of $CO_2$. Oxidation results in NADH and FADH$_2$ molecules. Substrate phosphorylation yields an ATP molecule.

The Krebs cycle turns twice for each original glucose molecule. Therefore, the inputs and outputs of the Krebs cycle per glucose molecule are as follows:

| Krebs Cycle | |
|---|---|
| inputs | outputs |
| 2 acetyl groups<br>2 ADP + 2 P<br>6 NAD$^+$<br>2 FAD | 4 $CO_2$<br>2 ATP<br>6 NADH<br>2 FADH$_2$ |

## Electron Transport System

The electrons that were removed from the substrates of glycolysis and the Krebs cycle are donated by NADH and FADH$_2$ to an **electron transport system** located on the cristae of mitochondria. Some of the electron carriers of the system are cytochrome molecules; therefore, the system is also called the **cytochrome system.**

Figure 9.10 of the electron transport system illustrates that high-energy electrons are delivered to the top of the system and low-energy electrons leave at the bottom. As a pair of electrons is passed from carrier to carrier, oxidation occurs, and the energy

## Figure 9.8

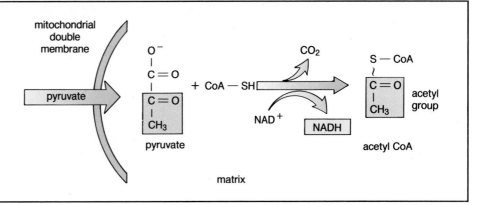

Transition reaction. Inside mitochondria, the oxidation of pyruvate by $NAD^+$ is accompanied by the release of $CO_2$. The remaining 2-carbon (acetyl) group binds to coenzyme A (CoA) forming acetyl CoA. This reaction occurs twice per molecule of glucose broken down.

## Figure 9.9

Krebs cycle. The net result of this cycle is the oxidation of an acetyl group to 2 molecules of $CO_2$. The energy of oxidation allows the immediate formation of one molecule of ATP. Most of the energy, however, resides in high-energy electrons, which are accepted by 3 molecules of $NAD^+$ and one molecule of FAD.

NADH and $FADH_2$ then take the electrons to the electron transport system. (Keep in mind that the Krebs cycle turns twice per glucose molecule.)

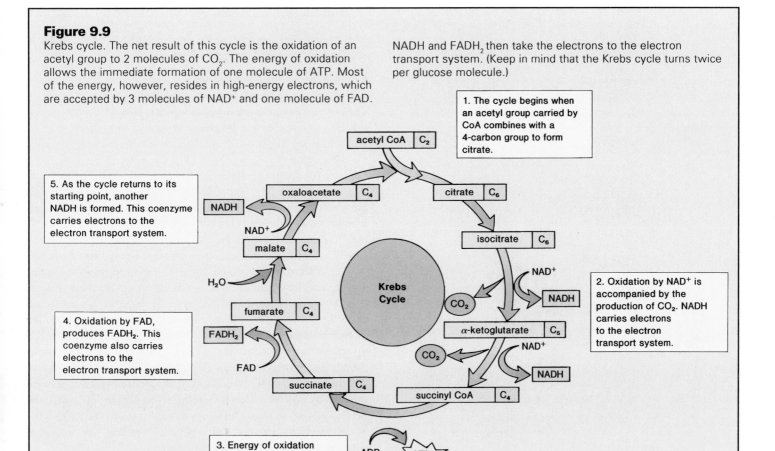

released is used to form ATP molecules at 3 different sites. This process is sometimes called **oxidative phosphorylation** because oxygen receives the energy-spent electrons from the last of the carriers. An enzyme called cytochrome oxidase splits and reduces molecular oxygen ($O_2$) to water:

$$\frac{1}{2}O_2 + 2e^- + 2H^+ \longrightarrow H_2O$$

Oxygen is, then, the final acceptor for electrons during cellular respiration, an **aerobic** process.

Electrons removed by $NAD^+$ and FAD from the metabolites of glycolysis and the Krebs cycle are passed from one carrier to the next in the electron transport system. Each pair of electrons releases enough energy to allow the production of 3 ATP molecules.

## Figure 9.10

The electron transport system. NADH and FADH₂ bring electrons to the electron transport system. As the electrons move down the system energy is released and used to form ATP. For every pair of electrons that enters by way of NADH, 3 ATP result. For every pair of electrons that enters by way of FADH₂, 2 ATP result. Oxygen, the final acceptor of the electrons, becomes water.

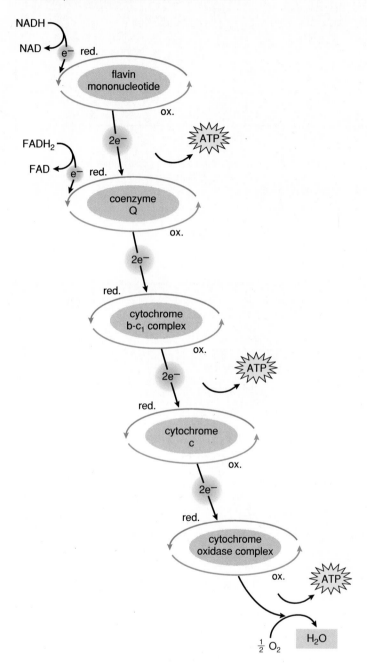

*Recycling of NAD⁺ and FAD*

### Recycling of NAD⁺ and FAD

The coenzymes NAD⁺ and FAD are constantly being reduced and oxidized inside the cell:

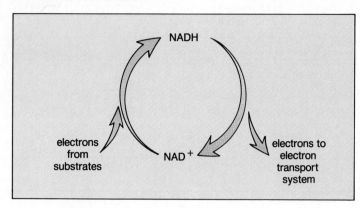

This means that the cell needs only a limited supply of these coenzymes because they are used repeatedly.

### Chemiosmotic ATP Synthesis

ATP synthesis occurs as a result of the electron transport system. The *chemiosmotic theory* suggests how it is possible for the energy released by the electron transport system to be coupled to ATP buildup. Coupling is only possible because of the structure of a mitochondrion.

As discussed in chapter 5, a mitochondrion has a double membrane (fig. 9.11). The inner membrane forms cristae, which are shelflike projections that jut out into the matrix, the innermost compartment, which is filled with a gel-like substance. The transition reaction and the Krebs cycle occur within the matrix; the carriers of the electron transport system are located on the cristae.

Electrons are passed down the electron transport system and certain carriers of the electron transport system pump hydrogen (H⁺) ions from the matrix into the intermembrane space between the outer and inner membrane. As hydrogen ions enter the intermembrane space, an extreme electrochemical gradient builds up across the inner membrane. The reduction of oxygen to water in the matrix removes hydrogen ions from the matrix and also adds to the gradient. When hydrogen ions flow down an electrochemical gradient by way of a channel protein present in a particle, called the ATP synthase complex, ATP is synthesized from ADP + ℗. This is called chemiosmotic ATP synthesis because chemical and osmotic events join to permit ATP synthesis.

## Energy Yield From Glucose Metabolism

As we have suggested, it is possible to calculate the ATP yield for the complete breakdown of glucose to carbon dioxide and water. Table 9.1 shows the number of molecules of ATP produced by each part of the process.

## Figure 9.11

Mitochondrion structure and function. **a.** Mitochondria have a double membrane with a space between the membranes. The inner membrane forms cristae, which project into the matrix, the innermost compartment. **b.** The transition reaction (PYR→ AA) and the Krebs cycle are located in the matrix. The carriers of the electron transport system are located on the cristae. The chemiosmotic theory says that some of the carriers pump H$^+$ into the intermembrane space. When these hydrogen ions flow down their electrochemical gradient through ATP synthase complexes, ATP is produced. Notice that the reduction of O$_2$ in the matrix contributes to the electrochemical gradient by lowering the number of H$^+$ in the matrix.

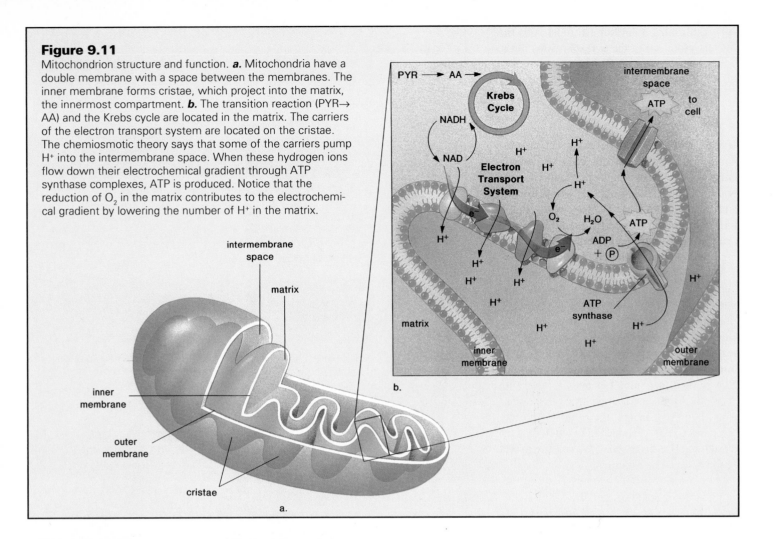

## Table 9.1
Energy Yield of Glycolysis and Cellular Respiration per Glucose Molecule

| Pathway | ATP Yield per Glucose Molecule | | Total ATP |
| | Substrate-Level Phosphorylation | Oxidative Phosphorylation | |
| --- | --- | --- | --- |
| Glycolysis | 2 ATP (net)* | 2 NADH = 4 ATP+ | 6 ATP |
| Transition reaction | | 2 NADH = 6 ATP | 6 ATP |
| Krebs cycle | 2 ATP | 6 NADH = 18 ATP | |
| | | 2 FADH$_2$ = 4 ATP | 24 ATP |
| Total | 4 ATP | 32 ATP | 36 ATP |

*The only amount that is produced under anaerobic conditions.

+In most cells each NADH produced in the cytosol results in only 2 rather than 3 ATP because the electrons are shuttled to an FAD coenzyme. In the heart and the liver, a shuttle system delivers the electrons to an NAD$^+$ coenzyme, and in these cells each NADH does result in the production of 3 ATP.

## Substrate-Level Phosphorylation

Per glucose molecule, 4 ATP molecules are formed directly by substrate-level phosphorylation: 2 during glycolysis and 2 during 2 turns of the Krebs cycle.

## Oxidative Phosphorylation

Per glucose molecule 10 NADH molecules and 2 FADH$_2$ molecules take electrons to the electron transport system. For each molecule of NADH produced *inside* the mitochondria by the Krebs cycle, 3 ATP are produced by the electron transport system, but for each FADH$_2$, there are only 2 ATP. Figure 9.10 explains the reason for this difference: FADH$_2$ delivers its hydrogens to the transport system below the level of the first site for ATP production.

What about the ATP yield of NADH generated *outside* the mitochondria by the glycolytic pathway? NADH cannot cross mitochondrial membranes, but there is a "shuttle" mechanism, which allows its electrons to be delivered to the electron transport system inside the mitochondria. In most types of cells, the shuttle consists of an organic molecule, which can cross the outer membrane, accept the electrons, and deliver them to an FAD molecule in the inner membrane. This FAD, however, can produce only 2

ATP molecules; therefore, in most cells the NADH produced in the cytosol results in the production of only 2 ATP instead of 3 ATP.

## Efficiency of Energy Transformation

It is interesting to consider how much of the energy in a glucose molecule eventually becomes available to the cell. The difference in energy content between the reactants (glucose and oxygen) and the products (carbon dioxide and water) is 686 Kcal. In other words, this is the total amount of energy available for the production of high-energy phosphate bonds. A high-energy phosphate bond has an energy content of 7.3 Kcal, and 36 of these are produced during glucose breakdown, so 36 phosphates are equivalent to a total of 263 Kcal. Therefore, 263/686, or 39%, of the available energy is transferred from glucose to ATP. The rest of the energy is lost in the form of heat. In birds and mammals, in particular, some of this heat is used to maintain a body temperature above that of the environment. In other organisms, including plants, there are few if any measures to conserve heat, and the body temperature generally fluctuates according to that of the external environment.

---

The energy yield of 36 ATP molecules represents 39% of the total energy that is available in a glucose molecule.

---

## The Metabolic Pool and Biosynthesis

Figure 9.12 shows how the primary organic compounds of cells make up a metabolic pool. Cells can oxidize, or degrade, other molecules besides glucose to release energy. For example, a fat molecule is hydrolyzed to glycerol and 3 fatty acids when it is used as an energy source. Inside the cell, glycerol is easily broken down to PGAL and thereby can enter the glycolytic pathway. The fatty acids are degraded to acetyl groups, which can enter the Krebs cycle. If a fatty acid contains 18 carbons, it breaks down to give 9 acetyl groups. Calculation shows that these 9 groups would produce 108 ATP molecules, a very large number. Fats are a very efficient form of stored energy, especially since there are 3 long-chain fatty acids per fat molecule.

The energy yield of protein is equivalent to carbohydrates. Every protein is broken down to amino acids, which undergo deamination, the removal of the amino group. The number of carbons remaining now determines just where the molecule enters the Krebs cycle (fig. 9.12). The amino group is converted to ammonia ($NH_3$) and excreted.

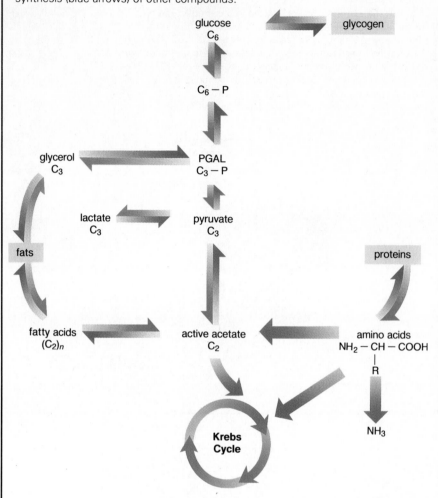

**Figure 9.12**
The metabolic pool concept. Carbohydrates, fats, and proteins can be used as energy sources, and they enter degradative pathways at specific points. Degradation (green arrows) produces metabolites that can also be used for synthesis (blue arrows) of other compounds.

The substrates making up the pathways in figure 9.12 can also be used as starting materials for synthetic reactions. In other words, compounds that enter the pathways are oxidized to substrates that can be used for biosynthesis. This is the cell's **metabolic pool,** in which one type of molecule can be converted to another. In this way, carbohydrate intake can result in the formation of fat. PGAL molecules can be converted to glycerol molecules, and acetyl groups can be joined together to form fatty acids. Fat synthesis follows. This explains why you gain weight from eating too much candy, ice cream, and cake.

Some metabolites of the Krebs cycle can be converted to amino acids. Plants are able to synthesize all of the amino acids they need. Animals, however, lack some of the enzymes necessary for synthesis of all amino acids. Humans (and white rats), for example, can synthesize 9 of the common amino acids but cannot synthesize the other 11. The amino acids that cannot be synthesized must be supplied by the diet; they are called the essential amino acids. The nonessential amino acids are those that can be synthesized. It is quite possible for animals to eat large quantities of protein and still suffer from protein deficiency if their diet does not contain adequate quantities of the essential amino acids.

---

All the reactions involved in cellular respiration are part of a metabolic pool; the metabolites from the pool can be used either as energy sources or as substrates for various synthetic reactions.

---

There are other degradative pathways, aside from those discussed in this chapter or depicted in figure 9.12, that are needed to supply substrates for biosynthesis. For example, a pathway called the pentose phosphate shunt converts glucose to pentose (a 5-carbon sugar), rather than to pyruvate. NADPH is a by-product of the pentose phosphate shunt. The pentose sugars are needed for nucleotide formation, and the NADPH is used in synthetic reactions. Therefore, this pathway is also a very important one.

## Summary

1. In cells, the energy of carbohydrate molecules, in particular, is converted to ATP energy.

2. Glucose metabolism often provides energy. Glycolysis anaerobically produces pyruvate, which either becomes reduced within the cytosol (fermentation) or becomes fully oxidized in mitochondria. Many bacteria and most animal cells carry on lactate fermentation; yeast and plant cells carry on alcoholic fermentation. Both kinds of fermentation only give a net yield of 2 ATP molecules.

3. Glycolysis is the breakdown of glucose to 2 molecules of pyruvate. This pathway is a series of enzymatic reactions that occurs in the cytosol. Oxidation by NAD$^+$ releases enough energy immediately to give a net gain of 2 ATP molecules by substrate-level phosphorylation. NADH either gives electrons to pyruvate or takes them to a mitochondrion.

4. Fermentation involves glycolysis, followed by the reduction of pyruvate by NADH to either lactate or alcohol and $CO_2$. The reduction process "frees" NAD$^+$ so that it can accept more hydrogen atoms from the glycolysis.

5. Although fermentation results in only 2 ATP molecules, it still serves a purpose: In vertebrates, it provides a quick burst of ATP energy for short-term, strenuous muscular activity. The accumulation of lactate puts the individual in oxygen debt because oxygen is needed when lactate is completely metabolized to $CO_2$ and $H_2O$. Fermentation evolved before aerobic respiration, and its continued presence exemplifies how evolution usually works—the newly evolved process is added onto those that have already evolved.

6. When glucose is completely oxidized, at least 36 ATP molecules are produced. Three subpathways are required: glycolysis, the Krebs cycle, and the electron transport system. Oxidation involves the removal of hydrogen atoms (e$^-$ + H$^+$) from substrate molecules, usually by the coenzyme NAD$^+$.

7. Pyruvate from glycolysis can enter the mitochondrion, where the transition reaction takes place. During this reaction, oxidation occurs as $CO_2$ is removed. NAD$^+$ is reduced, and CoA receives the $C_2$ acetyl group that remains. Since the reaction must take place twice per glucose molecule, 2 NADH result.

8. The acetyl group enters the Krebs cycle, a cyclical series of reactions located in the mitochondrial matrix. Complete oxidation follows, as 2 $CO_2$ molecules, 3 NADH molecules, and one FADH$_2$ molecule are formed. The cycle also produces one ATP molecule. The entire cycle must turn twice per glucose molecule.

9. The final stage of pyruvate breakdown involves the electron transport system located in the inner membrane of mitochondria. The electrons received from NADH and FADH$_2$ are passed down a chain of carriers until they are finally received by oxygen, which combines with hydrogen ions to produce water. As the electrons pass down the chain, ATP is produced. The term *oxidative phosphorylation* is sometimes used for ATP production by the electron transport system.

10. Some of the electron carriers pump H$^+$ into the intermembrane space setting up an electrochemical gradient. H$^+$ flow down this gradient, and the energy released is used to form ATP molecules from ADP and (P).

11. In the mitochondria, for each NADH molecule donating electrons to the electron transport system, 3 ATP molecules are produced. In the cytosol, however, each NADH formed results in only 2 ATP molecules. This is because its hydrogen atoms must be shuttled across the mitochondrial outer membrane by a molecule that can cross it. In most cells, these hydrogen atoms are then taken up by FAD. Each molecule of FADH$_2$ results in the formation of only 2 ATP because the electrons enter the electron transport system at a lower level.

12. Once NAD$^+$ and FAD have delivered electrons to the electron transport system, they are "free" to turn around and accept more hydrogens. Therefore, the cell only needs a limited supply of these coenzymes because they are used repeatedly.

13. Of the 36 ATP formed by aerobic cellular respiration, 4 are the result of substrate-level phosphorylation and the rest are produced by oxidative phosphorylation. The energy for the latter comes from the electron transport system.

14. Carbohydrate, protein, and fat can be broken down by entering the degradative pathways at different locations. These pathways also provide metabolites needed for the synthesis of various important substances. Degradation and synthesis, therefore, both utilize the same metabolic pool of reactants.

---

### Writing Across the Curriculum

In order to practice writing skills, students should write out the answers to any or all of the study questions and the critical thinking questions. The study questions are sequenced in the same order as the text. Suggested answers to the critical thinking questions are in appendix D.

---

# Study Questions

1. What types of organisms carry on cellular respiration? What is the function of this process?
2. Glycolysis results in pyruvate molecules. Contrast the fate of pyruvate under anaerobic conditions and aerobic conditions.
3. What is the overall chemical equation for the complete breakdown of glucose to carbon dioxide and water? How does a human cell acquire the needed substrates, and what happens to the products?
4. What are the 3 pathways involved in the complete breakdown of glucose to carbon dioxide and water? What reaction is needed to join 2 of these pathways?
5. Outline the main reactions of glycolysis, emphasizing those that produce energy.
6. Contrast glycolysis with fermentation. What is the benefit of pyruvate reduction during fermentation? What types of organisms carry out lactate fermentation, and what types carry out alcoholic fermentation?
7. Give the substrates and products of the transition reaction. Where does it take place?
8. What happens to the acetyl group that enters the Krebs cycle? What are the other steps in this cycle?
9. What is the electron transport system, and what are its functions? Relate these functions to the location of this system.
10. Explain how, when, and where chemiosmotic phosphorylation takes place in mitochondria.
11. Calculate the energy yield of glycolysis and cellular respiration per glucose molecule. Distinguish between substrate-level phosphorylation and oxidative phosphorylation.
12. Give examples to support the concept of the metabolic pool.

# Objective Questions

For questions 1-8, identify the pathway involved and match to the terms in the key.

*Key:*

    **a.** glycolysis
    **b.** Krebs cycle
    **c.** electron transport system

1. carbon dioxide given off
2. water formed
3. PGAL
4. NADH becomes $NAD^+$
5. oxidative phosphorylation
6. cytochrome carriers
7. pyruvate
8. FAD becomes $FADH_2$
9. Which of these is *not* true of fermentation?
    **a.** net gain of only 2 ATP
    **b.** occurs in cytosol
    **c.** NADH donates electrons to electron transport system
    **d.** begins with glucose
10. The transition reaction
    **a.** connects glycolysis to the Krebs cycle.
    **b.** gives off $CO_2$.
    **c.** utilizes $NAD^+$.
    **d.** All of these.
11. The greatest contributor of electrons to the electron transport system is
    **a.** oxygen.
    **b.** glycolysis.
    **c.** the Krebs cycle.
    **d.** the transition reaction.
12. Substrate-level phosphorylation takes place in the
    **a.** glycolysis and the Krebs cycle.
    **b.** electron transport system and the transition reaction.
    **c.** glycolysis and the electron transport system.
    **d.** Krebs cycle and the transition reaction.
13. Fatty acids are broken down to
    **a.** pyruvate molecules, which take electrons to the electron transport system.
    **b.** acetyl groups, which enter the Krebs cycle.
    **c.** amino acids, which excrete ammonia.
    **d.** All of these.
14. Of the 36 ATP molecules that are produced during the complete breakdown of glucose, most are formed by
    **a.** chemiosmotic phosphorylation.
    **b.** the electron transport system.
    **c.** substrate-level phosphorylation.
    **d.** Both a and b.

For questions 15-20, match the items below to one of the locations named in the key.

*Key:*

    **a.** matrix of the mitochondria
    **b.** cristae of the mitochondria
    **c.** between the inner and outer membranes of the mitochondria
    **d.** in the cytosol
15. electron transport system
16. Krebs cycle
17. glycolysis
18. transition reaction
19. $H^+$ reservoir
20. ATP synthase complex
21. Label this diagram of a mitochondrian:

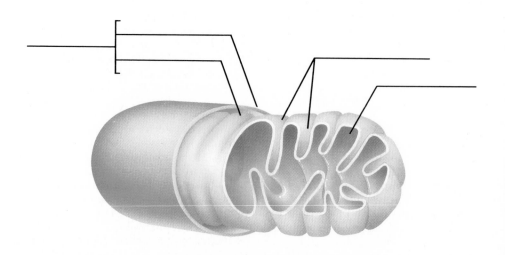

22. Label this diagram of glycolysis and cellular respiration. Place the labels lactate, electron transport system, acetyl CoA, pyruvate, and Krebs cycle on the appropriate lines. Label the arrows as either $CO_2$ or NADH. Indicate the correct number of ATP on the lines provided.

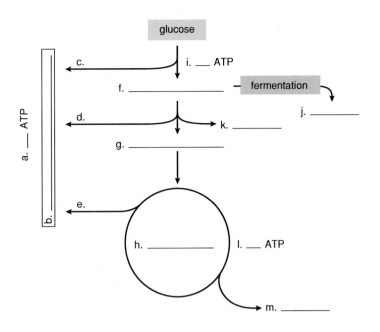

## Concepts and Critical Thinking

1. *Photosynthesis and cellular respiration permit a flow of energy and a recycling of matter.*

   Explain this concept by referring to figure 9.1.

2. *Solar energy first stored in nutrient molecules is transformed into a form that can be "spent" by the cell.*

   How are nutrient molecules utilized by cells as a source of energy?

3. *Certain metabolic pathways are virtually universal and thereby demonstrate the unity of living things.*

   In what way does glycolysis demonstrate the unity of living things and support the theory of evolution?

## Selected Key Terms

**cellular respiration** (sel'u-lar res"pǐ-ra'shun) 138
**glycolysis** (gli-kol i-sis) 139
**pyruvate** (pi'roo-vāt) 139
**fermentation** (fer"men-ta'shun) 142
**substrate-level phosphorylation** (sub'strāt lěv-al fos"fōr-i-la'shun) 142

**oxygen debt** (ok'sǐ-jen det) 142
**transition reaction** (tran-zish'un re-ak'shun) 143
**acetyl CoA** (as"ě-til-ko-en'zīm A) 143
**Krebs cycle** (Krebz si'k'l) 144
**citric acid cycle** (sit'rik as'id si'kl) 144

**cytochrome system** (si'to-krōme sis'tem) 144
**electron transport system** (e-lek'tron trans'port sis'tem) 144
**oxidative phosphorylation** (ok"si-da'tiv fos"for-i-la'shun) 145
**metabolic pool** (mět'a-bol-ic pōōl) 148

# CRITICAL THINKING CASE STUDY

### The Processing of Insulin

An active insulin molecule is composed of 2 polypeptides linked by disulfide chemical bonds. It is known that the original protein, produced at the ribosomes, is a larger molecule than active insulin, which is secreted into the blood. Therefore, it is hypothesized that *protein processing occurs within the organelles of the cell before active insulin is secreted.*

To test the hypothesis, pancreatic tissue, which produces insulin, was subjected to autoradiography. Pancreatic tissue was immersed *for a short time* in a solution containing radioactive amino acids and then, at regular intervals, tissue was sectioned and placed on slides for autoradiography (see p. 55 for an explanation of this technique). The autoradiographs showed the location of the radioactivity over time. Where do you predict the radioactivity will be as time elapses?

**Prediction 1** Radioactive protein will appear in different organelles of the cell as time elapses.

**Result 1** Within 5 min, the labeled protein moved to the inside of the endoplasmic reticulum. After 15 min, the labeled protein appeared in the Golgi apparatus, and after 60 min, it had moved to the secretory vesicles between the Golgi apparatus and the plasma membrane. These data indicate that the protein moved from the endoplasmic reticulum to the Golgi apparatus and then to the secretory vesicles.

---

I. In keeping with these results, would an autoradiograph produced after 60 min show any sign of remaining radioactivity? Why or why not?

If the tissue had been exposed to radioactive amino acids for a considerable length of time (instead of a short time), how might the results differ from those described?

---

These results tell us the pathway of protein movement, but they do not indicate where protein processing occurred. Where do you predict protein processing might occur?

**Prediction 2** Protein processing might occur inside the endoplasmic reticulum or in the Golgi apparatus. Most likely active insulin is in the secretory vesicles.

**Result 2** Chemical analysis of organelle contents showed that a small amino acid chain is removed from the protein as soon as it enters the endoplasmic reticulum. The remaining molecule is called proinsulin.

---

II. Molecules that cross membranes usually combine with receptor sites in a lock-and-key manner. What might have been the purpose of the small amino acid chain that was removed in the endoplasmic reticulum?

---

Researchers then turned their attention to the vesicles. Using electron microscopy they observed vesicles coated with protein bristles close to the Golgi saccules, and uncoated vesicles were some distance away. Then they continued the autoradiography experiment. What do you predict regarding the movement of protein via these vesicles?

**Prediction 3** Protein will be found first in the coated vesicles near the Golgi and then later in the uncoated vesicles some distance away.

**Result 3** The results showed radioactivity first in the coated vesicles and then about 30 min later in the uncoated vesicles.

---

III. Is it possible that the protein might move from one type vesicle to the other or that the coated vesicles might become uncoated vesicles? Why does the latter possibility seem more logical?

---

Another series of experiments was done to identify whether the protein in the vesicles was proinsulin or active insulin.

Antibodies, molecules that bind to specific molecules, were prepared for proinsulin or insulin. Fine gold particles, which can be detected by electron microscopy, were linked to these antibodies. Pancreatic tissue slices were exposed *either* to proinsulin antibody or insulin antibody and then examined using the electron microscope. What do you predict in regard to binding of the antibodies to the vesicles?

**Prediction 4** Since coated vesicles precede uncoated vesicles and proinsulin precedes active insulin, proinsulin antibody will bind to coated vesicles and insulin antibody will bind to uncoated vesicles.

**Result 4** Results showed that proinsulin antibody binds only to the coated vesicles. Active insulin antibody binds only to the uncoated vesicles. These results strongly suggest that proinsulin is converted to insulin as coated vesicles develop into smooth vesicles.

---

IV. Why did the researchers prepare antibodies to proinsulin and insulin? Why not just continue with radiography?

Why did the researchers choose to apply the 2 different types of antibodies separately? Why not apply them together?

---

The results of these experiments provide strong support for the hypothesis that protein processing occurs within the organelles of the cell before active insulin is secreted. Understanding the complete pathway of the formation of the insulin molecule required a variety of techniques, each of which made an essential contribution to understanding the process.

## Other Questions

1. Cell fractionation could be done following brief exposure of pancreatic tissue to radioactive amino acids. With reference to figure 5.C, what would be the results of this experiment? Assume three fractions: (a) endoplasmic reticulum; (b) Golgi apparatus; and (c) vesicles.
2. The researchers were careful to use the beta cells of the Isles of Langerhans in their experiment and not the entire pancreas? Why? What would it mean if they got similar results using the alpha cells?

Orci, L., J. D. Vassalli, and A. Perrelet. September 1988. The insulin factory. *Scientific American*, 85-94.

## Suggested Readings for Part 1

Alberts, B., et al. 1989. *Molecular biology of the cell.* 2d ed. New York: Garland Publishing.

Allen, R. D. February 1987. The microtubule as an intracellular engine. *Scientific American.*

Baker, J. J. W., and Allen, G. E. 1981. *Matter, energy, and life.* 4th ed. Reading, Mass.: Addison-Wesley.

Berns, M. W. 1983. *Cells.* 3d ed. New York: Holt, Rinehart & Winston.

Cronquist, A. 1982. *Basic botany.* 2d ed. New York: Harper & Row.

Dautry-Varsat, A., and Lodish, H. F. May 1984. How receptors bring proteins and particles into cells. *Scientific American.*

de Duve, C. May 1983. Microbodies in the living cell. *Scientific American.*

Dickerson, E. March 1980. Cytochrome and the evolution of energy metabolism. *Scientific American.*

Dustin, P. August 1980. Microtubules. *Scientific American.*

Govindjee, and Coleman, William J. February 1990. How plants make oxygen. *Scientific American.*

Grivell, L. A. March 1983. Mitochondrial DNA. *Scientific American.*

Karplus, M., and McCammon, J. A. April 1986. The dynamics of proteins. *Scientific American.*

Lake, J. A. July 1981. The ribosome. *Scientific American.*

Miller, K. R. October 1979. The photosynthetic membrane. *Scientific American.*

Ostro, M. J. January 1987. Liposomes. *Scientific American.*

Porter, K. R., and Bonneville, M. A. 1973. *Fine structure of cells and tissues.* 4th ed. Philadelphia: Lea and Febiger.

Raven, H., et al. 1986. *Biology of plants.* 4th ed. New York: Worth.

Rothman, J. September 1985. The Compartmental organization of the Golgi apparatus. *Scientific American.*

*Scientific American.* October 1985. The molecules of life.

Sharon, N. November 1980. Carbohydrates. *Scientific American.*

Sheeler, P., and Bianchi, D. E. 1980. *Cell biology: Structure, biochemistry, and function.* New York: John Wiley and Sons.

Stryer, L. 1988. *Biochemistry.* 3d ed. San Francisco: W. H. Freeman.

Unwin, N. February 1984. The structure of proteins in biological membranes. *Scientific American.*

Wickramasinghe, H. Kumar. October 1989. Scanned-probe microscopes. *Scientific American.*

Yougan, D. C., and Mars, B. L. June 1987. Molecular mechanisms of photosynthesis. *Scientific American.*

# Classification of Organisms

The classification system given here is a simplified one, containing all the major kingdoms, as well as the major divisions (called phyla in the kingdom Protista and the kingdom Animalia).

## Kingdom Monera

Prokaryotic, unicellular organisms. Nutrition principally by absorption, but some are photosynthetic or chemosynthetic.
    Division Archaebacteria: methanogens, halophiles, and thermoacidophiles
    Division Eubacteria: all other bacteria, including cyanobacteria (formerly called blue-green algae)

## Kingdom Protista

Eukaryotic, unicellular organisms (and the most closely related multicellular forms). Nutrition by photosynthesis, absorption, or ingestion.
    Phylum Sarcodina: amoeboid protozoans
    Phylum Ciliophora: ciliated protozoans
    Phylum Zoomastigina: flagellated protozoans
    Phylum Sporozoa: parasitic protozoans
    Phylum Chlorophyta: green algae
    Phylum Pyrrophyta: dinoflagellates
    Phylum Euglenophyta: *Euglena* and relatives
    Phylum Chrysophyta: diatoms
    Phylum Rhodophyta: red algae
    Phylum Phaeophyta: brown algae
    Phylum Myxomycota: slime molds
    Phylum Oomycota: water molds

## Kingdom Fungi

Eukaryotic organisms, usually having haploid or multinucleated hyphal filaments. Spore formation during both asexual and sexual reproduction. Nutrition principally by absorption.
    Division Zygomycota: black bread molds
    Division Ascomycota: sac fungi
    Division Basidiomycota: club fungi
    Division Deuteromycota: imperfect fungi (means of sexual reproduction not known).

## Kingdom Plantae

Eukaryotic, terrestrial, multicellular organisms with rigid cellulose cell walls and chlorophylls *a* and *b*. Nutrition principally by photosynthesis. Starch is the food reserve.
    Division Bryophyta: mosses and liverworts
    Division Psilophyta: whisk ferns
    Division Lycophyta: club mosses
    Division Sphenophyta: horsetails
    Division Pterophyta: ferns
    Division Cycadophyta: cycads
    Division Ginkgophyta: ginkgo
    Division Gnetophyta: gnetae
    Division Coniferophyta: conifers
    Division Anthophyta: flowering plants
        Class Dicotyledonae: dicots
        Class Monocotyledonae: monocots

## Kingdom Animalia

Eukaryotic, usually motile, multicellular organisms without cell walls or chlorophyll. Nutrition principally ingestive, with digestion in an internal cavity.
    Phylum Porifera: sponges
    Phylum Cnidaria: radially symmetrical aquatic animals
        Class Hydrozoa: hydras, Portuguese man-of-war
        Class Scyphozoa: jellyfishes
        Class Anthozoa: sea anemones and corals
    Phylum Platyhelminthes: flatworms
        Class Turbellaria: free-living flatworms
        Class Trematoda: parasitic flukes
        Class Cestoda: parasitic tapeworms
    Phylum Nematoda: roundworms
    Phylum Rotifera: rotifers
    Phylum Mollusca: soft-bodied, unsegmented animals
        Class Polyplacophora: chitons
        Class Monoplacophora: *Neopilina*
        Class Gastropoda: snails and slugs
        Class Bivalvia: clams and mussels
        Class Cephalopoda: squids and octopuses
    Phylum Annelida: segmented worms
        Class Polychaeta: sandworms
        Class Oligochaeta: earthworms
        Class Hirudinea: leeches

Phylum Arthropoda: chiton exoskeleton, jointed appendages
    Class Crustacea: lobsters, crabs, barnacles
    Class Arachnida: spiders, scorpions, ticks
    Class Chilopoda: centipedes
    Class Diplopoda: millipedes
    Class Insecta: grasshoppers, termites, beetles
Phylum Onychophora: small, sluglike animals with legs; internal features of both annelids and arthropods
Phylum Echinodermata: marine; spiny, radially symmetrical animals
    Class Crinoidea: sea lilies and feather stars
    Class Asteroidea: sea stars
    Class Ophiuroidea: brittle stars
    Class Echinoidea: sea urchins and sand dollars
    Class Holothuroidea: sea cucumbers
Phylum Chordata: dorsal supporting rod (notochord) at some stage; dorsal hollow nerve cord; pharyngeal pouches or gill slits
    Subphylum Urochordata: tunicates
    Subphylum Cephalochordata: lancelets
    Subphylum Vertebrata: vertebrates
        Class Agnatha: jawless fishes (lampreys, hagfishes)
        Class Chondrichthyes: cartilaginous fishes (sharks, rays)
        Class Osteichthyes: bony fishes
            Subclass Dipnoi: lungfishes
            Subclass Crossopterygii: lobe-finned fishes
            Subclass Actinopterygii: ray-finned fishes
        Class Amphibia: frogs, toads, salamanders
        Class Reptilia: snakes, lizards, turtles
        Class Aves: birds
        Class Mammalia: mammals

Subclass Prototheria: egg-laying mammals
    Order Monotremata: duckbilled platypus, spiny anteater
Subclass Metatheria: marsupial mammals
    Order Marsupialia: opossums, kangaroos
Subclass Eutheria: placental mammals
    Order Insectivora: shrews, moles
    Order Chiroptera: bats
    Order Edentata: anteaters, armadillos
    Order Rodentia: rats, mice, squirrels
    Order Lagomorpha: rabbits, hares
    Order Cetacea: whales, dolphins, porpoises
    Order Carnivora: dogs, bears, weasels, cats, skunks
    Order Proboscidea: elephants
    Order Sirenia: manatees
    Order Perissodactyla: horses, hippopotamuses, zebras
    Order Artiodactyla: pigs, deer, cattle
    Order Primates: lemurs, monkeys, apes, humans
        Suborder Prosimii: lemurs, tree shrews, tarsiers, lorises, pottos
        Suborder Anthropoidea: monkeys, apes, humans
            Superfamily Ceboidea: New World monkeys
            Superfamily Cercopithecoidea: Old World monkeys
            Superfamily Hominoidea: apes and humans
                Family Hylobatidae: gibbons
                Family Pongidae: chimpanzees, gorillas, orangutans
                Family Hominidae: *Australopithecus,* * *Homo erectus,* * *Homo sapiens sapiens*

*extinct

## Table of Chemical Elements

Atomic number → 1 ← Atomic weight
**H** ← Chemical symbol
hydrogen

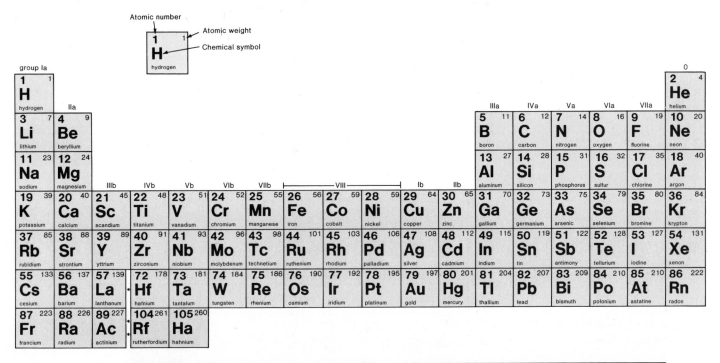

## Metric System

### Think Metric
### Length

1. The speed of a car is 60 miles/hr or 100 km/hr.
2. A man who is 6 feet tall is 180 centimeters tall.
3. A 6-inch ruler is 15 centimeters long.
4. One yard is almost a meter (0.9 m).

### Metric System

| Standard Metric Units | | Abbreviations |
|---|---|---|
| Standard unit of mass | gram | g |
| Standard unit of length | meter | m |
| Standard unit of volume | liter | l |

| Common Prefixes | | Examples |
|---|---|---|
| kilo (k) | 1,000 | a kilogram is 1,000 grams |
| centi (c) | 0.01 | a centimeter is 0.01 of a meter |
| milli (m) | 0.001 | a milliliter is 0.001 of a liter |
| micro (μ) | one-millionth | a micrometer is 0.000001 (one-millionth) of a meter |
| nano (n) | one-billionth | a nanogram is $10^{-9}$ (one billionth) of a gram |
| pico (p) | one-trillionth | a picogram is $10^{-12}$ (one trillionth) of a gram |

### Units of Length

| Unit | Abbreviation | Equivalent |
|---|---|---|
| meter | m | approximately 39 in |
| centimeter | cm | $10^{-2}$ m |
| millimeter | mm | $10^{-3}$ m |
| micrometer | μm | $10^{-6}$ m |
| nanometer | nm | $10^{-9}$ m |
| angstrom | Å | $10^{-10}$ m |

**Length Conversions**

| | | |
|---|---|---|
| 1 in = 2.5 cm | 1 mm = 0.039 in | 1 m = 1.094 yd |
| 1 ft = 30 cm | 1 cm = 0.39 in | 1 km = 0.6 mi |
| 1 yd = 0.9 m | 1 m = 39 in | |
| 1 mi = 1.6 km | | |

| To Convert | Multiply By | To Obtain |
|---|---|---|
| inches | 2.54 | centimeters |
| feet | 30 | centimeters |
| centimeters | 0.39 | inches |
| millimeters | 0.039 | inches |

*Think Metric*
*Volume*

1. One can of beer (12 oz) contains 360 milliliters.
2. The average human body contains between 10-12 pints of blood or between 4.7-5.6 liters.
3. One cubic foot of water (7.48 gal) is 28.426 liters.
4. If a gallon of unleaded gasoline costs $1.00, a liter costs 26¢.

## Units of Volume

| Unit | Abbreviation | Equivalent |
|---|---|---|
| liter | l | approximately 1.06 qt |
| milliliter | ml | $10^{-3}$ l (1 ml = 1 cm³ = 1 cc) |
| microliter | µl | $10^{-6}$ l |

### Volume Conversions

| | | |
|---|---|---|
| 1 tsp = 5 ml | 1 pt = 0.47 l | 1 ml = 0.03 fl oz |
| 1 tbsp = 15 ml | 1 qt = 0.95 l | 1 l = 2.1 pt |
| 1 fl oz = 30 ml | 1 gal = 3.8 l | 1 l = 1.06 qt |
| 1 cup = 0.241 | | 1 l = 0.26 gal |

| To Convert | Multiply by | To Obtain |
|---|---|---|
| fluid ounces | 30 | milliliters |
| quarts | 0.95 | liters |
| milliliters | 0.03 | fluid ounces |
| liters | 1.06 | quarts |

*Think Metric*

*Weight*

1. One pound of hamburger is 448 grams.
2. The average human male brain weighs 1.4 kilograms (3 lb 1.7 oz).
3. A person who weighs 154 pounds weighs 70 kilograms.
4. Lucia Zarate weighed 5.85 kilograms (13 lbs) at age 20.

## Units of Weight

| Unit | Abbreviation | Equivalent |
|---|---|---|
| kilogram | kg | $10^3$ g (approximately 2.2 lb) |
| gram | g | approximately 0.035 oz |
| milligram | mg | $10^{-3}$ g |
| microgram | µg | $10^{-6}$ g |
| nanogram | ng | $10^{-9}$ g |
| picogram | pg | $10^{-12}$ g |

### Weight Conversions

| | |
|---|---|
| 1 oz = 28.3 g | 1 g = 0.035 oz |
| 1 lb = 453.6 g | 1 kg = 2.2 lb |
| 1 lb = 0.45 kg | |

| To Convert | Multiply By | To Obtain |
|---|---|---|
| ounces | 28.3 | grams |
| pounds | 453.6 | grams |
| pounds | 0.45 | kilograms |
| grams | 0.035 | ounces |
| kilograms | 2.2 | pounds |

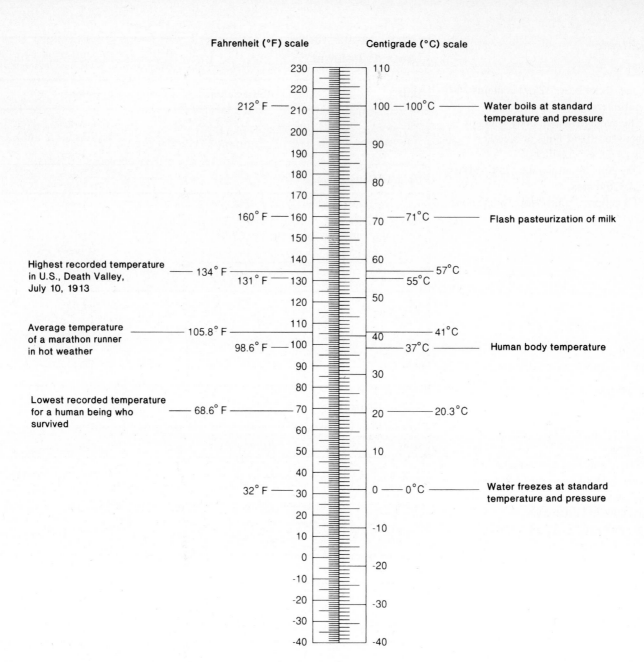

230 — 110

220

212° F — 210    100 — 100°C ——— Water boils at standard
                                   temperature and pressure
200

190    90

180

170    80

160° F — 160    70 — 71°C ——— Flash pasteurization of milk

150

Highest recorded temperature    140    60 ——— 57°C
in U.S., Death Valley,    — 134° F
July 10, 1913    131° F — 130    55°C
                                   ——— 55°C
120    50

Average temperature    — 105.8° F    110
of a marathon runner    40 ——— 41°C
in hot weather    98.6° F — 100    37°C ——— Human body temperature

90    30

80

Lowest recorded temperature
for a human being who    — 68.6° F ——— 70    20 ——— 20.3°C
survived
60

50    10

40

32° F — 30    0 — 0°C ——— Water freezes at standard
                                   temperature and pressure
20

10    -10

0

-10    -20

-20    -30

-30

-40    -40

# APPENDIX D

## Answers

### Chapter 1
#### Objective Questions

1. d
2. c
3. a
4. b
5. e
6. c
7. b
8. c
9. a. plants; b. animals; c. nutrients for plants; d. death and decay.

#### Concepts and Critical Thinking

1. A common ancestor has passed on to all living things common characteristics, such as using DNA genes.
2. Ways of life are diverse; therefore, organisms are diverse.
3. Nonliving things may have levels of organization, but they do not demonstrate emergent properties, where the whole is greater than the sum of the parts. For example, a cell is alive, but the structures, which make up a cell, are not alive.

### Chapter 2
#### Objective Questions

1. d
2. a
3. d
4. a
5. b
6. c
7. b
8. d
9. a. Bacteria exposed to sunlight do not die if medium contains dye. b. Bacteria can be protected from UV light by dye. c. Experimental plate contains bacteria and dye; control contains only bacteria; exposure to UV light causes all bacteria to die. d. Hypothesis refuted.

#### Concepts and Critical Thinking

1. Even though the results of experimentation or observation support a hypothesis, it could be that the further studies will prove it false.
2. In the everyday sense, a theory means a supposition; in the scientific sense, a theory means a hypothesis that has been supported by many experiments and observations.
3. A scientist does studies to understand the natural world; citizens make decisions how these results should be used.

### Chapter 3
#### Objective Questions

1. c
2. b
3. c
4. d
5. d
6. d
7. b
8. c
9. a
10. 7 p and 7 n in nucleus; 2 electrons in inner shell, and 5 electrons in outer shell. This means that nitrogen needs 3 more electrons to complete its outer shell; therefore the formula for ammonia is $NH_3$.

#### Concepts and Critical Thinking

1. For example, people take supplemental calcium to keep their bones strong.
2. Cells have a complex structure that is built upon the atoms and molecules in cells.
3. For example, if water was not slow to heat up and slow to cool down, living things might be subject to the killing effects of rapid heating and cooling.

### Chapter 4
#### Objective Questions

1. c
2. a
3. d
4. b
5. c
6. b
7. c
8. c
9. d
10. c
11. a
12. a
13. a. monomer; b. condensation; c. polymer; d. hydrolysis. The diagram shows the manner in which macromolecules are synthesized and degraded in cells.
14. a. primary level; b. secondary level; c. tertiary level; d. quaternary level.
15. AGTTCGGCATGC

#### Concepts and Critical Thinking

1. They are unified in the use of carbohydrate as a structural molecule, but they are diversified as to the particular carbohydrate.
2. Butter is solid, and an oil is a liquid at room temperature because a fat containing saturated hydrocarbon chains melts at a higher temperature than one containing unsaturated chains.

3. Phospholipids have a polar head and nonpolar tails; they make up the plasma membrane, which is selectively permeable. Keratin, a fibrous protein that contains only helix polypeptides, is found in tough structures such as hair and nails.

## Chapter 5
### Objective Questions

1. c
2. c
3. d
4. a
5. c
6. c
7. a
8. d
9. d
10. d
11. a. Golgi apparatus further modifies; b. smooth ER modifies; c. chromatin—DNA directs; d. nucleolus—RNA helps; e. rough ER produces.
12. a. See text. b. Mitochondria and chloroplasts are a pair because they are both membranous structures involved in energy metabolism. c. Centrioles and flagella are a pair because they both contain microtubules; centrioles give rise to the basal bodies of flagella. d. Endoplasmic reticulum and ribosomes are a pair because together they are rough ER, which produces proteins.

### Concepts and Critical Thinking

1. Show them microscopic slides of all sorts of tissues from different organisms. You would have to find an organism whose tissues did not contain cells.
2. The various organelles listed in table 5.1 show that the cell is compartmentalized; each organelle has a separate function.
3. All the structures labeled in question 11 plus mitochondria supply ATP energy. All the other parts of a cell assist to some degree also.

## Chapter 6
### Objective Questions

1. b
2. b
3. a
4. c
5. c
6. d
7. d
8. b
9. b
10. See figure 6.3. Phospholipid tails have no polar groups to interact with water's polar groups.
11. a. hypertonic—cell shrinks due to loss of water; b. hypotonic—cell has swelled due to gain of water.

### Concepts and Critical Thinking

1. Structurally, the plasma membrane is the outer boundary of the cell. Functionally, the plasma membrane regulates what enters and leaves a cell.
2. A cell can die in either a severely hypertonic or severely hypotonic solution.
3. The plasma membrane secretes chemical messengers that communicate with distant cells; the plasma membrane participates in the structure of junctions between adjacently located cells.

## Chapter 7
### Objective Questions

1. a
2. d
3. c
4. a
5. d
6. c
7. d
8. c
9. a. macromolecules; b. degradative reactions; c. ADP + P → ATP; d. NADP → NADPH; e. small molecules; f. synthetic reactions; g. macromolecules.
10. See figure 7.13.

### Concepts and Critical Thinking

1. Both the rough ER and Golgi apparatus produce molecules that are used for structural purposes. It takes energy to produce molecules.
2. When glucose energy is converted to ATP, there is a loss, and the energy released when ATP is broken down eventually becomes heat.

3. The temperature of the body is not high, and enzymes are needed to bring reactants together.

## Chapter 8
### Objective Questions

1. d
2. d
3. a
4. d
5. c
6. d
7. c
8. d
9. See figure 8.3.
10. a. water; b. oxygen; c. carbon dioxide; d. carbohydrate; e. ADP + P → ATP; f. NADP → NADPH.

### Concepts and Critical Thinking

1. Show that solar energy is needed for photosynthesis, and photosynthetic organisms are the food for the biosphere, including humans. The bodies of plants became the fossil fuels, which we use to produce electricity and heat houses and convert to gasoline for cars.
2. Chlorophyll captures solar energy, which is converted to ATP energy and used to produce NADPH. ATP and NADPH are used to reduce $CO_2$ to a carbohydrate.
3. The thylakoid space is separated from the stroma by the thylakoid membrane; the build-up of hydrogen ions in the thylakoid space leads to ATP production.

## Chapter 9
### Objective Questions

1. b
2. c
3. a
4. c
5. c
6. c
7. a
8. b
9. c
10. d
11. c
12. a
13. b
14. d
15. b
16. a
17. d
18. a
19. c
20. b
21. See figure 9.11*a*.
22. a. 32; b. electron transport system; c. NADH; d. NADH; e. NADH; f. pyruvate; g. acetyl CoA; h. Krebs cycle; i. 2; j. lactate; k. $CO_2$; l. 2; m. $CO_2$.

## Concepts and Critical Thinking

1. Flow of energy; photosynthesis converts the energy of the sun into carbohydrates that are converted to ATP energy by cellular respiration. Recycling of matter: Carbohydrates and oxygen from photosynthesis participate in cellular respiration, whose end product carbon dioxide reenters plants again.

2. Nutrient molecules are broken down to molecules that participate in cellular respiration, with the concomitant build-up of ATP. It is the ATP that is used by cells.

3. Glycolysis is a metabolic pathway that is almost universally found in organisms; it must have evolved in an ancestor common to all other organisms.

# Chapter 10
## Objective Questions

| | | | |
|---|---|---|---|
| 1. | c | 2. | b |
| 3. | d | 4. | d |
| 5. | a | 6. | c |
| 7. | b | 8. | c |
| 9. | b | 10. | b |
| 11. | See figure 10.5 | | |

## Concepts and Critical Thinking

1. As cells grow, they become ready for reproduction by producing more plasma membrane, cytoplasm, and organelles.

2. The daughter cells inherit DNA along with its inherent regulation from the parent cell.

3. Mitosis involves the use of a spindle, a structure that ensures that each daughter cell receives a copy of each chromosome.

# Chapter 11
## Objective Questions

| | | | |
|---|---|---|---|
| 1. | b | 2. | c |
| 3. | c | 4. | b |
| 5. | a | 6. | c |
| 7. | d | 8. | d |
| 9. | c | | |
| 10. | a—It shows bivalents at equators. | | |

## Concepts and Critical Thinking

1. Asexual reproduction and sexual reproduction begin with cells, usually single cells. This shows how fundamental the cell is to the life of an organism.

2. Variation during asexual reproduction is limited to the occurrence of mutations. Variation during sexual reproduction is introduced due to mutations, crossing-over, and recombination (from independent assortment and fertilization).

3. Advantage: favorable mutations may be of immediate advantage. Disadvantage: unfavorable mutations may be of immediate disadvantage.

# Chapter 12
## Practice Problems 1

1. a. 100% $W$; b. 50% $W$, 50% $w$; c. 50% $T$, 50% $t$; d. 100% $T$

2. a. gamete; b. genotype; c. gamete

## Practice Problems 2

1. $bb$
2. $Tt \times tt$, $tt$
3. 3/4 or 75%
4. $Yy$ and $yy$

## Practice Problems 3

1. a. 100% $tG$; b. 50% $TG$, 50% $tG$; c. 25% $TG$, 25% $Tg$; 25% $tG$; 25% $tg$; d. 50% $TG$, 50% $Tg$

2. a. genotype; b. gamete; c. genotype; d. gamete

## Practice Problems 4

1. $BbTt$ only
2. a. $LlGg \times llgg$
   b. $LlGg \times LlGg$
3. 9/16

## Objective Questions

| | | | |
|---|---|---|---|
| 1. | b | 2. | a |
| 3. | c | 4. | d |
| 5. | c | 6. | d |
| 7. | a | 8. | b |
| 9. | b | 10. | b |
| 11. | c | 12. | d |

## Additional Genetics Problems

1. 100% chance for widow's peak and 0% chance for continuous hairline
2. $Ee$

3. 50%
4. 210 gray body and 70 black body; 140 = heterozygous; cross fly with recessive
5. $F_1$ = all black with short hair; $F_2$ = 9:3:3:1; offspring would be 1 brown long: 1 brown short: 1 black long: 1 black short
6. $Bbtt \times bbTt$ and $bbtt$
7. $GGLl$
8. 25%

## Concepts and Critical Thinking

1. You inherit chromosomes containing the same types of genes from parents, but each parent contributes one-half of the particular genes inherited.

2. Plants and animals both have chromosomes containing the genetic material, DNA.

3. The individual's genes were not modified by the accident.

# Chapter 13
## Practice Problems 1

1. $Cc^h$ (wild); $c^{ch}c^h$ (light gray); $Cc^h$ (wild); $c^hc^h$ (Himalaya)
2. 1 pink:1 white
3. 7
4. pleiotropy, epistasis

## Practice Problems 2

1. females: $X^RX^R$, $X^RX^r$, $X^rX^r$, males: $X^RY$ (gametes $X^R$, Y); $XrY$ (gametes $Xr$, Y)
2. b; 1:1
3. 100%, none, 100%
4. mother: $X^BX^bRr$; father: $X^BYRr$; son: $X^bYrr$

## Practice Problems 3

1. 9:3:3:1, linkage
2. $54.5 - 13.0 = 41.5\%$
3. 12
4. bar eye, scalloped wings, garnet eye

## Objective Questions

| | | | |
|---|---|---|---|
| 1. | b | 2. | c |
| 3. | e | 4. | a |
| 5. | b | 6. | b |
| 7. | d | 8. | b |
| 9. | c | 10. | d |

## Additional Genetics Problems

1. pleiotropy
2. codominance, $F^BF^B$, $F^WF^W$, $F^BF^W$
3. 50%
4. 6, yes
5. linkage, 10
6. Both males and females are 1:1.
7. $X^BX^B \times X^bY$ = both males and females bar-eyed; $X^bX^b \times X^BY$ = females bar-eyed; males all normal
8. $XbXbWw$
9. females—1 short-haired tortoise shell: 1 short-haired yellow; males—1 short-haired black: 1 short-haired yellow

### Concepts and Critical Thinking

1. Most human traits, such as height and color of skin, are examples of polygenic inheritance.
2. The genes behave similarly to the chromosomes (i.e., they come in pairs, they segregate during meiosis, and there is only one in the gametes).
3. Almost all males have an Y chromosome; XO individuals are females. This shows that most likely sex determining alleles on the Y chromosome have some degree of dominance to sex-determining alleles on the X chromosome.

## Chapter 14

### Practice Problem 1

1. 25%
2. 50%
3. heterozygous
4. heterozygous—unless a new mutation, which is a fairly common cause of NF.

### Practice Problem 2

1. $Hb^AHb^S$; either $Hb^SHb^S$ or $Hb^AHb^S$
2. light
3. white
4. Doe—baby #1
   Jones—baby #2

### Practice Problem #3

1. mother; mother $X^HX^b$, father $X^HY$, son $X^bY$

2. 0%, 0%, 100%
3. $rrX^bX^b$ and $RrX^BX^b$, $rrX^BX^b$, $RrX^bX^b$
4. colorblind

### Objective Questions

1. c          2. a
3. b          4. c
5. All are consistent.
6. c
7. b
8. d
9. c
10. a
11. autosomal dominant condition

### Additional Genetics Problems

1. 0%, 0%, 100%
2. 25%
3. 50%, 50%
4. AB, Yes, A, B, AB, or O
5. light, white
6. $X^cYww$, $X^CX^cWw$, $X^CYWw$; normal vision with widow's peak
7. **a.** recessive, Aa
   **b.** dominant, Aa
   **c.** sex-linked recessive, $X^AX^a$

### Concepts and Critical Thinking

1. Yes, the concept still holds, because any differences don't negate the basic similarities.
2. Divide the cells of an early embryo to produce as large a number of identical children as possible. (This solves the nature question.) Raise each child under entirely different circumstances and then see how similar the children are.
3. This is a question that each person must decide on their own. Genetic counseling and genetic testing can sometimes tell parents their chances of having a child with a genetic disease.

## Chapter 15

### Objective Questions

1. b          2. d
3. a          4. c
5. d          6. c
7. a          8. a
9. b          10. d

11. The parental helix is heavy-heavy, and each daughter helix is heavy-light. (The cell was supplied with $^{14}$N nucleotides as replication began.) This shows that each daughter helix is composed of one template strand and one new strand. This is consistent with semiconservative replication.

### Concepts and Critical Thinking

1. $^{35}$S would have been found with the bacterial cells, showing that it was needed for viral replication.
2. Replication is a part of duplication of the chromosomes. Replication before mitosis ensures that each daughter cell will have the diploid number of chromosomes, and replication before meiosis ensures that the sperm and egg will have the haploid number of chromosomes.
3. Mutation is the ultimate source of genetic variation. Without variation, new organisms would never evolve.

## Chapter 16

### Objective Questions

1. b          2. d
3. a          4. c
5. a          6. a
7. c          8. b
9. c          10. d
11. a. ACU CCU GAA UGC AAA;
    b. UGA GGA CUU ACG UUU;
    c. threonine-proline-glutamate-cysteine-lysine.

### Concepts and Critical Thinking

1. The genetic information stored by DNA is the sequence of amino acids in a protein; this is stored in the sequence of DNA bases.
2. A universal code means that the ancestry of all types of organisms can be traced to a common origin.
3. The sequence of bases of DNA ultimately determines the sequence of bases in a protein. Figure 16.5 shows how the molecules are colinear.

# Chapter 17

## Objective Questions

| | | | |
|---|---|---|---|
| 1. | c | 2. | a |
| 3. | d | 4. | a |
| 5. | b | 6. | b |
| 7. | b | 8. | d |
| 9. | d | 10. | d |

11. a. See figure 17.2*a*. b. For the trp operon to be in the off position, a corepressor has to be attached to the repressor.

## Concepts and Critical Thinking

1. Only certain mRNA molecules are present in cells at any particular time.
2. Gene expression necessitates a functioning protein product. Posttranslational control involves regulation of the activity of the protein product.
3. Checks and balances are often seen in biological systems. This is a safety feature that allows greater control.

# Chapter 18

## Objective Questions

| | | | |
|---|---|---|---|
| 1. | a | 2. | c |
| 3. | d | 4. | a |
| 5. | d | 6. | d |
| 7. | b | 8. | c |
| 9. | d | 10. | a |

11. a. AATT; b. TTAA.
12. 2. reverse transcription occurs; 4. most of the viral genes are removed, etc.; 5. transcription of DNA occurs; 6. recombinant RNA is repackaged; 8. reverse transcription occurs.

## Concepts and Critical Thinking

1. It is possible to make recombinant DNA, to modify natural DNA in the laboratory, to perform PCR analysis, to do DNA fingerprinting, and to determine the sequence of the bases of DNA. This is disturbing to people who resist thinking of humans as physical and chemical machines.
2. Human genes can be placed in a bacterium, where they function normally.

# Chapter 19

## Objective Questions

| | | | |
|---|---|---|---|
| 1. | d | 2. | b |
| 3. | b | 4. | d |
| 5. | d | 6. | d |
| 7. | d | 8. | b |

9. a. All the continents of today were one continent during the Triassic period, and the reptiles spread throughout the land. b. All vertebrates share a common ancestor, who had pharyngeal pouches during development. c. Two different continents can have similar environments and therefore unrelated organisms that are similarly adapted. d. This demonstrates that diversification has occurred in the nightshade family.

## Concepts and Critical Thinking

1. Both bacteria and humans use DNA as the hereditary material and have similar metabolic pathways.
2. If adaptation were purposeful, organisms wouldn't have structures that don't serve a current function.
3. It is possible to formulate a hypothesis about evolution and then make observations to see if the hypothesis is correct.

# Chapter 20

## Objective Questions

| | | | |
|---|---|---|---|
| 1. | c | 2. | b |
| 3. | c | 4. | c |
| 5. | c | 6. | c |
| 7. | d | 8. | c |
| 9. | b | | |

10. See figure 20.7*b*.

## Population Genetics Problems

1. 99%
2. recessive allele = 0.2; dominant allele = 0.8; homozygous recessive = 0.04; homozygous dominant = 0.64; heterozygous = 0.32

## Concepts and Critical Thinking

1. Traits are passed from one generation to the next via the gametes, and no mechanism has been found by which phenotype changes can affect the genes in the gametes.

2. The more fit organisms have more offspring, and this is the way by which adaptive traits accumulate in a population.
3. Increase variation between populations: genetic drift, natural selection, mutation, nonrandom mating. Decrease variation between populations: gene flow. These mechanisms alter gene frequencies in populations.

# Chapter 21

## Objective Questions

| | | | |
|---|---|---|---|
| 1. | d | 2. | d |
| 3. | c | 4. | b |
| 5. | b | 6. | d |
| 7. | a | 8. | d |
| 9. | b | | |

10. Do away with 5 middle arrows so that 2 sets of circles remain. The circles of each set are connected by arrows.

## Concepts and Critical Thinking

1. The higher taxa show how species are related. Species in different kingdoms are distantly related, and those in the same genus are closely related, for example.
2. All new species have evolved from previous species; therefore all organisms have a phylogenetic history.
3. One group gives rise to many groups, each adapted to a particular environment.

# Chapter 22

## Objective Questions

| | | | |
|---|---|---|---|
| 1. | d | 2. | c |
| 3. | b | 4. | b |
| 5. | a | 6. | c |
| 7. | c | 8. | b |
| 9. | b | 10. | d |

11. The diagram suggests that the eukaryotic cell acquired mitochondria and, if present, chloroplasts by engulfing particular prokaryotic cells. This hypothesis is called the endosymbiotic theory.

## Concepts and Critical Thinking

1. All 3 hypotheses suggest that a chemical evolution produced the first cell(s).
2. The increase in variability among the offspring of sexually reproducing organisms (as compared to asexual ones) provides more opportunity for change.
3. Under the conditions of the primitive earth, a chemical evolution produced the first cell(s). Under the conditions of today's earth, life comes only from life.

## Chapter 23
### Objective Questions

| | | | |
|---|---|---|---|
| 1. | b | 2. | b |
| 3. | c | 4. | b |
| 5. | c | 6. | c |
| 7. | d | 8. | c |
| 9. | b | 10. | d |
| 11. | d | | |

12. See figure 23.3.

### Concepts and Critical Thinking

1. Viruses are unable to carry on reproduction outside a living cell. They reproduce by taking over the machinery of the cell; therefore, they are obligate parasites.
2. Monerans are metabolically diverse, as witnessed by the fact that there is probably no single type of organic compound that some Moneran cannot digest.
3. This statement refers to structure, not biochemistry.

## Chapter 24
### Objective Questions

| | | | |
|---|---|---|---|
| 1. | d | 2. | a |
| 3. | b | 4. | c |
| 5. | d | 6. | a |
| 7. | a | 8. | d |

9. a. Protista, Chlorophyta (green algae); b. Fungi, Ascomycota (sac fungi); c. Protista, Sarcodina (amoeboid protozoans).
10. a. locomotion; b. pigmentation.
11. See figure 24.9b. Sexual reproduction involves the union of gametes.

## Concepts and Critical Thinking

1. The protozoans are heterotrophic; the algae are photosynthetic; and the slime molds are saprophytic.
2. In the haplontic cycle, the zygote undergoes meiosis, and therefore the adult is haploid. In alternation of generations, the sporophyte produces haploid spores that mature into the gametophyte. In the diplontic cycle, meiosis produces haploid gametes, the only part of the cycle that is haploid.
3. Fungi live on the organic material they digest. Plants produce their own organic food, and animals go out and find it.

## Chapter 25
### Objective Questions

| | | | |
|---|---|---|---|
| 1. | d | 2. | c |
| 3. | c | 4. | c |
| 5. | d | 6. | b |
| 7. | b | 8. | b |
| 9. | d | | |

10. See figure 25.3b.
11. a. leafy shoot; b. heart-shaped structure; c. stalk with sporangium; d. rhizome with fronds, roots. The fern has the sporophyte dominant. There are sporangia on the underside of the fronds, and this is the generation that persists longer.
12. a. underside of scales of male cone; b. in the anther of stamens; c. upper surface of scales of female cone; d. in ovary of pistil. The flowering plant produces fruit, which develops from the ovary.

### Concepts and Critical Thinking

1. The 5-kingdom system of classification is based in part on mode of nutrition. From this standpoint, it would seem that all photosynthesizers should be in the same kingdom. On the other hand, it would seem that we need some other criteria to go by; for example, all organisms in the plant kingdom are multicellular with some degree of complexity. In this text, all organisms in the plant kingdom protect the zygote.
2. Green algae at the base could lead to nonvascular plants (bryophytes) and vascular plants. The primitive vascular plants as a group could lead to gymnosperms (4 divisions) and angiosperms (2 classes). The tree goes from the first evolved to the latest evolved according to the fossil record.
3. Roots enable a plant to take water out of the soil; stems hold the leaves aloft to catch solar energy, and the leaves also take in carbon dioxide at stomata.

## Chapter 26
### Objective Questions

| | | | |
|---|---|---|---|
| 1. | d | 2. | c |
| 3. | d | 4. | d |
| 5. | b | 6. | b |
| 7. | a | | |

8. See figure 26.4.
9. Sponges (no items)
   Cnidarians
      radial symmetry
      tissue level of organization
   Flatworms
      3 germ layers
   Roundworms
      pseudocoelom
      3 germ layers
      tube-within-a-tube
10. Sponges
      collar cells
   Planarians
      eyespots
      gastrovascular cavity
   Tapeworms
      proglottids
   Roundworms
      uterus

### Concepts and Critical Thinking

1. For a contrast of the plant cell and the animal cell, see table 5.2. Note that animal cells cannot make their own food (lack chloroplasts) and are motile (have flagella), 2 distinct characteristics of animals.
2. A common ancestor would have to have those features shared by both types of organisms: bilateral symmetry, 3 germ layers, and organs.

3. Internal parasites need to have a means of protecting themselves from host attack, absorbing nutrients from the host, and reproducing that allows dispersal to new hosts.

## Chapter 27
### *Objective Questions*

| | | | |
|---|---|---|---|
| 1. | a | 2. | a |
| 3. | c | 4. | c |
| 5. | a | 6. | d |
| 7. | d | 8. | b |
| 9. | d | | |

10. Mollusks
    organ system level of
        organization
    coelom
    cephalization in some
        representatives
    complete gut
    Annelids
    organ system level of
        organization
    segmentation
    coelom
    cephalization in some
        representatives
    complete gut
    Arthropods
    organ system level of
        organization
    segmentation
    coelom
    cephalization in some
        representatives
    complete gut
    Chordates
    organ system level of
        organization
    segmentation
    coelom
    cephalization in some
        representatives
    complete gut
11. Clam
    Mollusk
    3 ganglia
    gills
    open circulatory system
    hatchet foot

Earthworm
    annelid
    ventral nerve cord
    closed circulatory system
    hydrostatic skeleton
    setae

12. Label wings (for flying in the air), spiracle (for breathing air), tympanum (for picking up sound waves), ovipositor (for laying eggs in earth), digestive system (for eating grass), Malpigian tubules (for excretion without loss of water), tracheae (for breathing air), vagina and seminal receptacle (for reception and storage of sperm so they do not dry out). See figure 27.12 for placement of labels.

### *Concepts and Critical Thinking*

1. Body rings indicate that the earthworm's body is segmented and there are setae on each segment. In the grasshopper, segmentation is not so obvious because there is a head, thorax with legs, and abdomen—that is, specialization of parts.
2. See table 27.2, which lists all the features that arthropods and chordates have in common. The fact that both lineages have these features must mean that these features increase the fitness of organisms having them.
3. See answer to objective question 12.

## Chapter 28
### *Objective Questions*

| | | | |
|---|---|---|---|
| 1. | a | 2. | d |
| 3. | a | 4. | b |
| 5. | b | 6. | d |
| 7. | d | 8. | b |
| 9. | b | 10. | c |

11. a. mammals; b. birds; c. reptiles; d. amphibians; e. modern bony fishes; f. cartilagenous fishes; g. modern jawless fishes.
12. Primate (all), Prosimian (only lemurs), Anthropoid (all but lemurs), Hominoid (all but lemurs and monkeys), Hominid (only Australopithecines and humans).

13. Kingdom Animalia, phylum Chordata, subphylum Vertebrata, class Mammalia, order Primate, superfamily Hominoidea, family Hominidae, genus and species *Homo sapiens.*

### *Concepts and Critical Thinking*

1. The number of species in varied habitats is a good criterion. Vertebrates are chordates; therefore table 27.2 lists the features they have in common. In addition, they both have a rigid, but jointed, skeleton.
2. Gametes are protected from drying out: reptiles pass sperm directly from male to female; angiosperms rely on the pollen grain. Embryo is protected from drying out: reptiles develop in a shelled egg; angiosperms produce seeds that protect the embryo until germination under favorable conditions occurs.
3. No living organism evolved from another living organism. Instead, 2 living organisms may have shared a common ancestor.

## Chapter 29
### *Objective Questions*

| | | | |
|---|---|---|---|
| 1. | c | 2. | b |
| 3. | c | 4. | c |
| 5. | b | 6. | b |
| 7. | b | 8. | c |
| 9. | c | 10. | b |

11. a. epidermis; b. cortex; c. endodermis; d. phloem; e. xylem.
12. a. cork; b. phloem; c. vascular cambium; d. xylem; e. pith; f. bark.
13. a. upper epidermis; b. palisade mesophyll; c. leaf vein; d. spongy mesophyll; e. lower epidermis.

### *Concepts and Critical Thinking*

1. A leaf that is broad and flat is well shaped to catch the sun's rays; the cells contain chloroplasts; the leaf veins bring water from the roots; the stomata admit carbon dioxide into the air spaces of the spongy layer.
2. Figure 29.1 and table 29.1 show how a plant is organized.

3. Wood helps a plant resist the pull of gravity. It serves as an internal skeleton and allows the plant to lift the leaves, exposing them to the rays of the sun.

## Chapter 30
### Objective Questions

1. d     2. a
3. c     4. b
5. c     6. c
7. c     8. d
9. The diagram shows that gravity pushing down on mercury in the pan can only raise a column of mercury to 76 cm. When water above the column is transpired, it pulls on the mercury and raises it higher. This suggests that transpiration would be able to raise water to the top of trees.
10. See figure 30.6*b* and *c*. After K$^+$ enters guard cells, water follows by osmosis and the stoma opens.
11. There is more solute in (1) than (2); therefore water enters (1); this creates a pressure that causes water, along with solute, to flow toward (2).

### Concepts and Critical Thinking

1. A physical process accounts for the transport of water in xylem; causes a stoma to open; and accounts for the transport of organic substances in phloem.
2. Because vessel cells are hollow and nonliving, they form a continuous pipeline from the roots to the leaves. This allows plants to transport water and minerals from the roots to the leaves.

## Chapter 31
### Objective Questions

1. c     2. a
3. b     4. a
5. a     6. b
7. a     8. d
9. c     10. c
11. a. sporophyte; b. microsporangium; c. megasporangium (in ovule); d. meiosis; e. microspore; f. megaspore; g. male gametophyte (pollen grain); h. female gametophyte; i. egg and sperm; j. fertilization; k. zygote in seed.

### Concepts and Critical Thinking

1. Flowering plants follow the alternation of generations life cycle, but the gametophyte is reduced to a microscopic size and is protected by the sporophyte. The gametes are also protected; the egg stays within the ovule within the ovary and awaits the sperm. A pollen grain is transported to the pistil by wind or animals and then germinates, allowing the sperm to move down the style to the egg. The zygote (new sporophyte) develops into an embryo within the ovule, which becomes the seed enclosed by fruit. Fruit is dispersed in various ways.
2. As stated, the gametophyte is protected by the sporophyte and the embryonic sporophyte is protected by the seed.
3. Coevolution results in one type of pollinator gathering food from a particular plant. This reduces competition and assures the pollinator of food from this particular source.

## Chapter 32
### Objective Questions

1. d     2. b
3. e     4. a
5. c     6. d
7. b     8. d
9. d     10. a
11. The arrows show the movement of auxin from the shady side to the sunny side of the stem. Auxin causes the cells on the shady side to elongate; therefore, the oat seedling bends toward the light.
12. The arrows (K$^+$ ions) leave guard cells under the influence of abscisic acid (ABA) when a plant is water stressed. Water osmotically follows the movement of K$^+$ and the stoma closes.

### Concepts and Critical Thinking

1. The best example is that the ratio of auxin to cytokinin determines whether a plant tissue will form an undifferentiated mass or form roots, vegetative shoots and leaves, or flowers.
2. It is adaptive, for example, for leaves to unfold and stomata to open during the day, when photosynthesis can occur. A biological clock system needs a receptor that is sensitive to light and dark; a timekeeper, that is, a biological clock; and a means of communication within the body of the organism. In regard to flowering, phytochrome is the receptor and hormones are probably the means of communication. The possible timekeeper is not known.

## Chapter 33
### Objective Questions

1. b     2. b
3. a     4. d
5. d     6. d
7. b     8. d
9. d     10. c
11. a. smooth muscular tissue; b. blood cells—connective tissue; c. nervous tissue; d. ciliated columnar epithelium.

### Concepts and Critical Thinking

1. The cell is alive, but the organelles making up a cell are not alive; a tissue performs functions that the individual cell in the tissue cannot perform and so forth from level to level.
2. Acquiring food is located to the right and left of the diagram. Cellular respiration is a degradative pathway requiring an exchange of gases. Excretion of waste accompanies degradation.
3. Acquiring food provides materials and energy; the musculoskeletal system helps us get food; exchange of gases gets rid of carbon dioxide and brings in oxygen; transporting materials brings nutrients to and takes wastes away from cells;

protecting the body from disease maintains its integrity; coordinating body activities makes sure that all the other functions are performed.

## Chapter 34
### Objective Questions

| | | | |
|---|---|---|---|
| 1. | b | 2. | b |
| 3. | a | 4. | d |
| 5. | d | 6. | c |
| 7. | d | 8. | b |
| 9. | b | 10. | b |
| 11. | d | | |
| 12. | a (See figure 34.6.) | | |

### Concepts and Critical Thinking

1. Tissue fluid remains relatively constant because materials are constantly being added and removed at the capillaries.
2. The cells are far removed from the exterior, and therefore they have to have nutrients brought to them and wastes removed from them.
3. ATP energy keeps the heart pumping (arterial flow) and the skeletal muscles contracting (venous flow). ATP energy is produced by mitochondria, where chemical energy is transformed into usable energy.

## Chapter 35
### Objective Questions

| | | | |
|---|---|---|---|
| 1. | d | 2. | d |
| 3. | a | 4. | b |
| 5. | c | 6. | b |
| 7. | a | 8. | a |
| 9. | b | 10. | d |

### Concepts and Critical Thinking

1. Death occurs when immunity fails to prevent microorganisms from taking over the body.
2. Immunity consists of immediate defense mechanisms and specific defense mechanisms, which work more slowly.
3. Bones are of course part of the skeletal system. Red bone marrow is the site of red and white blood cell formation; therefore, it is part of the circulatory and lymphatic systems.

## Chapter 36
### Objective Questions

| | | | |
|---|---|---|---|
| 1. | d | 2. | b |
| 3. | d | 4. | b |
| 5. | c | 6. | a |
| 7. | c | 8. | c |
| 9. | d | 10. | c |
| 11. | d | 12. | c |
| 13. | See figure 36.6. | | |
| 14. | Test tube 1: No digestion | | |

Test tube 1: No digestion
No enzyme and no HCl
Test tube 2: Some digestion
No HCl
Test tube 3: No digestion
No enzyme
Test tube 4: Digestion
Both enzyme and HCl are present

### Concepts and Critical Thinking

1. In humans, the digestive system takes in nutrients, excretes certain metabolites like heavy metals, and prevents bacterial invasion by the low pH of the stomach.
2. See figure 36.3.
3. Only autotrophs are able to produce their own food. Heterotrophs feed on autotrophs and acquire material and energy. Autotrophs use wastes from heterotrophs (fertilizer, $CO_2$).

## Chapter 37
### Objective Questions

| | | | |
|---|---|---|---|
| 1. | a | 2. | b |
| 3. | b | 4. | d |
| 5. | c | 6. | b |
| 7. | c | 8. | b |
| 9. | d | 10. | b |
| 11. | See figure 37.7. | | |

### Concepts and Critical Thinking

1. The respiratory system carries out exchange of gases and helps maintain the pH of the blood.
2. Gills are external extensions and lungs are internal cavities. Both are minutely divided and highly vascularized for the exchange of gases.
3. In humans, the lungs expand when the chest moves up and out and deflate when the chest moves down and in. The lungs contain many alveoli (air sacs) and are highly

vascularized. Oxygen enters the blood following inhalation, and carbon dioxide leaves the body upon exhalation.

## Chapter 38
### Objective Questions

| | | | |
|---|---|---|---|
| 1. | d | 2. | a |
| 3. | c | 4. | b |
| 5. | a | 6. | c |
| 7. | b | 8. | d |
| 9. | a | 10. | b |
| 11. | See figure 38.10. | | |

### Concepts and Critical Thinking

1. When molecules undergo degradation, wastes result. For example, following glucose metabolism, carbon dioxide must be excreted. Following amino acid breakdown, urea must be excreted.
2. Water contains a lower amount of oxygen than air. The countercurrent mechanism in the gills of fishes helps them extract oxygen out of water. Mammals evolved on land; a countercurrent mechanism in the kidneys helps them conserve water.
3. Reabsorption, in part by active transport, takes place in the proximal convoluted tubule. An increased surface area (microvilli) helps reabsorption, and mitochondria supply the ATP energy for active transport.

## Chapter 39
### Objective Questions

| | | | |
|---|---|---|---|
| 1. | c | 2. | b |
| 3. | d | 4. | c |
| 5. | a | 6. | a |
| 7. | b | 8. | d |
| 9. | b | 10. | b |
| 11. | a. receptor (initiates nerve impulse); b. sensory neuron (takes impulses to cord); c. interneuron (passes impulses to motor neuron and other interneurons in the cord); d. white matter (contains tracts that take impulses up and down the cord); e. motor neuron (takes impulses to effector); f. effector (brings about adaptive response). | | |

### Concepts and Critical Thinking

1. Controls internal organs like the beating of the heart and skeletal muscles, which allow an animal to seek environments compatible with a constant internal environment.
2. There is a spinal nerve in each segment, but the brain is an obvious specialization of part of the spinal cord.
3. A neuron has long processes that conduct nerve impulses. The postsynaptic membrane is part of a dendrite, and the presynaptic membrane is part of an axon. Impulses always flow in this direction because only an axon contains synaptic vesicles.

## Chapter 40
### Objective Questions

1. d
2. c
3. c
4. d
5. d
6. c
7. c
8. b
9. d
10. c
11. See figure 40.5 and table 40.1.

### Concepts and Critical Thinking

1. The sense organs help animals find favorable environments and avoid unfavorable ones; they help animals find food and avoid being eaten; they help animals communicate with others for the purpose of cooperation.
2. The sense organs are receptors for external and internal stimuli; they generate nerve impulses, which supply information to the central nervous system.
3. Relative fitness determines what structures characterize a species. Therefore, animals have those sense organs that are most adaptive.

## Chapter 41
### Objective Questions

1. b
2. f
3. c
4. e
5. b
6. b
7. b
8. c
9. a
10. d

11. See table 41.3.
12. See figure 41.10a.

### Concepts and Critical Thinking

1. Consult, for example, figure 21.19 and note how the skeleton of a horse was adapted at first to a forest and then later to a grassland habitat.
2. Animals have to go out and get their food; therefore, locomotion is essential to them.
3. Nutrient molecules from the digestive system travel in the blood to muscle cells. Here, glucose energy is converted to ATP energy, largely within mitochondria. ATP energy is used by myosin filaments to pull actin filaments, and muscle contraction occurs.

## Chapter 42
### Objective Questions

1. f
2. b
3. c
4. a
5. e
6. c
7. a
8. d
9. b
10. d
11. b
12. b
13. See figure 42.7. In a negative feedback system, the output (i.e., thyroxin) cancels the input (e.g., thyroid growth hormone from the pituitary); therefore, the production of thyroxin is shut off when it gets too high.

### Concepts and Critical Thinking

1. Neurotransmitters are released when needed and then reabsorbed by the presynaptic membrane or broken down within the synaptic cleft by an enzyme, for example. The level of a hormone is controlled typically by using negative feedback to regulate its production.
2. Specialized structures (i.e., neurons) deliver messages in the nervous system, whereas the endocrine system uses the blood system.
3. The human body undergoes a dramatic change following puberty.

## Chapter 43
### Objective Questions

1. b
2. d
3. c
4. d
5. c
6. c
7. c
8. c
9. a
10. a
11. See figure 43.3 and table 43.1. Testis, epididymis, vas deferens, urethra (in penis).

### Concepts and Critical Thinking

1. They are similarly adapted, except the embryos of reptiles develop in shelled eggs while the embryos of placental mammals develop within the uterus of the female. Even so, the same extraembryonic membranes surround the embryo.
2. Reproduction is a process that takes some time; it is not like movement, which takes place quickly.

## Chapter 44
### Objective Questions

1. b
2. b
3. b
4. a
5. b
6. b
7. d
8. a
9. d
10. d
11. See figure 44.8. Chorion exchanges wastes for nutrients with the mother; amnion protects and prevents desiccation; blood vessels of allantois become umbilical blood vessels; yolk sac is the first site of blood cell formation.

### Concepts and Critical Thinking

1. Senescence is a change that is part of the normal life cycle from fertilization to death.
2. They could be considered local hormones because they are messengers that are produced by one cell and act on another cell.
3. Perhaps as individuals age, genes are turned off (instead of on), and this leads to degenerative changes.

# Chapter 45

## Objective Questions

1. d     2. a
3. d     4. a
5. c     6. d
7. d     8. c
9. d     10. a
11. a. communication, chemical;
    b. learning, imprinting; c. commu-
    nication, sound; d. learning,
    operant conditioning; e. communi-
    cation, touch; f. communication,
    visual.

## Concepts and Critical Thinking

1. The genes control the development
   of the brain and therefore even
   insight learning has a genetic basis.
2. Feeding behavior patterns affect
   fitness because an animal needs a
   body in good working order for
   reproduction to be successful.
   Territoriality affects fitness because
   an animal needs enough food to
   nourish its offspring. Reproductive
   patterns affect fitness because an
   animal needs to have a mate to
   reproduce.
3. Altruistic acts can be shown to
   increase inclusive fitness.

# Chapter 46

## Objective Questions

1. b     2. c
3. d     4. d
5. d     6. c
7. b     8. c
9. See figure 46.3b. The r stands for
   rate of natural increase.

## Concepts and Critical Thinking

1. When there are unlimited resources,
   such as space, food, and shelter, and
   where there are no deaths due to
   natural disasters, population growth
   could be infinite.
2. Human life strategy is only like that
   of a K-strategist. See table 46.4.

# Chapter 47

## Objective Questions

1. b     2. d
3. c     4. d
5. c     6. b
7. d     8. c
9. d
10. a. mutualism; b. parasitism;
    c. commensalism.

## Concepts and Critical Thinking

1. The organisms themselves provide
   niches for other organisms. For
   example, several types of warblers
   find niches in spruce trees (fig. 47.3).
2. As prey populations decrease in
   size, it is less likely that a predator
   will find that type of prey, and prey
   populations have defenses against
   predation.
3. Parasites keep down population
   sizes when they weaken or directly
   kill off their hosts. Parasites
   increases diversity because, for
   example, even parasites have
   parasites. As mentioned in number
   1, organisms themselves provide
   niches for other organisms.

# Chapter 48

## Objective Questions

1. a     2. a
3. d     4. a
5. a     6. c
7. d     8. c
9. a
10. a. 4+ for both; b. 3+ for both; c. 2+
    for both; d. 1+ for both.

## Concepts and Critical Thinking

1. The neritic province is more
   productive because it receives both
   nutrients and sunlight. In contrast,
   only the epipelagic zone of the
   oceanic province receives adequate
   sunlight. This zone is not as
   nutrient rich as the neritic province.
2. Succession does not require a
   definite sequence, but there is a
   sequence of communities from the
   simple to the complex.
3. It gets colder as you go from the
   base of a mountain to the top;
   therefore, the communities will
   change as noted.

# Chapter 49

## Objective Questions

1. b     2. a
3. c     4. d
5. b     6. d
7. c     8. c
9. a     10. b
11. a. algae; b. zooplankton; c. small
    fishes; d. large fishes; e. humans.

## Concepts and Critical Thinking

1. For example, the atmosphere is the
   abiotic component of the nitrogen
   cycle, and the passage of nitrogen
   compounds from producer to
   consumers is the biotic component
   of the nitrogen cycle.
2. It is impossible to create energy to
   keep an ecosystem going (first
   law), and as energy is passed from
   one trophic level to the next, there
   is always some loss of energy
   (second law). Therefore, there is a
   flow of energy from the sun
   through an ecosystem.

# Chapter 50

## Objective Questions

1. a     2. d
3. c     4. b
5. b     6. c
7. d     8. c
9. a     10. c
11. a. acid deposition; b. ozone shield
    destruction; c. greenhouse effect; d.
    photochemical smog.

## Concepts and Critical Thinking

1. The primary productivity of the
   other terrestrial biomes might
   increase due to carbon dioxide
   buildup following tropical rain
   forest destruction.
2. The figure shows, for example, that
   nitrous dioxide ($NO_2$), nitrous
   oxide, carbon dioxide, methane,
   CFCs, and halons all contribute to
   the greenhouse effect, and of these
   CFCs and halons also caused ozone
   shield destruction.
3. Fossil fuel energy contributes to
   agricultural yield and therefore
   helps support the increased human
   population size of today.

# GLOSSARY

## A

**abscisic acid (ABA)**(ab-sis′ik) a plant hormone causing stomata to close and initiating and maintaining dormancy *528*

**abscission** (ab-sizh′un) the dropping of leaves, fruits, or flowers from a plant body *528*

**acetylcholine** (as″ĕ-til-ko′lēn) a neurotransmitter used within both the peripheral and central nervous systems *639*

**acetyl CoA** (as″ĕ-til-ko-A) molecule made up of a 2-carbon acetyl group attached to coenzyme A, which enters the Krebs cycle for further oxidation *143*

**acid** a compound tending to dissociate and yield hydrogen ions in a solution and to lower its pH numerically *36*

**acid deposition** the return to earth as rain or snow of the sulfate or nitrate salts of acids produced by commercial and industrial activities on earth *800*

**acoelomate** (a-sēl′o-māt) condition where an organism does not develop a coelom or true body cavity *424*

**ACTH** adrenocorticotrophic hormone, which is secreted by the anterior pituitary and stimulates activity in the adrenal cortex *683*

**actin** a muscle protein making up the thin filaments in a sarcomere; its movement shortens the sarcomere, yielding muscle contraction *671*

**action potential** change in the polarity of a neuron caused by opening and closing of ion channels in the axon plasma membrane *639*

**active site** that part of an enzyme molecule where the substrate fits and the chemical reaction occurs *109*

**active transport** use of a plasma membrane carrier molecule to move particles from a region of lower to higher concentration; it opposes an equilibrium and requires energy *94*

**adaptation** an organism's modification in structure, function, or behavior to increase the likelihood of continued existence *8, 307*

**adaptive radiation** formation of a large number of species from a common ancestor *336*

**adenine (A)** (ad′ē-nīn) one of 4 organic bases in the nucleotides composing the structure of DNA and RNA *54, 238*

**adenosine diphosphate (ADP)** (ah-den′o-sēn di-fos′fāt) one of the products of the hydrolysis of ATP, a process that liberates energy *116*

**adenosine triphosphate (ATP)** (ah-den′o-sēn tri-fos′fāt) a compound containing adenine, ribose, and 3 phosphates, 2 of which are high-energy phosphates. The breakdown of ATP to ADP makes energy available for energy-requiring processes in cells *115*

**adipose tissue** type of loose connective tissue in which fibroblasts enlarge and store fat *541*

**ADP** *see* adenosine diphosphate

**adrenal gland** lies atop a kidney and secretes the stress hormones epinephrine, norepinephrine, and the corticoid hormones *685*

**aerobic** process in which oxygen is required *145*

**age structure diagram** a representation of the number of individuals in each age group in a population *748*

**agglutination** (ah-glūt-en-ā′shun) clumping of red blood cells, as when incompatible red cell antigens and antibodies mix *568*

**AIDS** acquired immunodeficiency syndrome, caused by the HIV virus in humans; the virus attacks T4 lymphocytes, making them ineffective and leading to the development of other infections *707*

**aldosterone** (al″do-ster′on) hormone secreted by the adrenal gland that maintains the sodium and potassium balance of the blood *630, 686*

**alga (pl., algae)** aquatic organisms carrying out photosynthesis and belonging to the kingdom Protista *383*

**allantois** (ah-lan′to-is) an extraembryonic membrane that accumulates nitrogenous wastes in the embryo of birds and reptiles and contributes to the formation of umbilical blood vessels in mammals *719*

**allele** (ah-lēl′) alternative forms of a gene, for example, an allele for long wings and an allele for short wings in fruit flies *186*

**allopatric speciation** origin of new species in populations that are separated geographically *334*

**alternation of generations cycle** life cycle in which a haploid generation alternates with a diploid generation *384*

**alveolus** (al-ve′o-lus) terminal, microscopic, grapelike air sacs found in vertebrate lungs *612*

**altruism** form of social behavior found in insect societies, exemplified by the willingness of the daughters to help the queen rather than reproduce their own offspring *740*

**amino acid** organic subunits each having an amino group and an acid group, that covalently bond to produce protein molecules *50*

**amnion** an extraembryonic membrane of the bird, reptile, and mammal embryo that forms an enclosing fluid-filled sac *719*

**amphibian** class of terrestrial vertebrates that includes frogs, toads, and salamanders; still tied to a watery environment for reproduction *452*

**anaerobic** a process that does not require oxygen *142*

**angiosperm** flowering plant having double fertilization that results in development of seed-bearing fruit *404*

**annelid** (ah′nel-id) a phylum of segmented worms characterized by a tube-within-a-tube body plan, bilateral symmetry, and segmentation of some organ systems *434*

**anther** part of stamen where pollen grains develop *407, 510*

**antheridium** (an"ther-id'i-um) male structures in nonseed plants where swimming sperm are produced *398*

**antibody** a protein molecule usually formed naturally in an organism to combat antigens or foreign substances *563, 577*

**anticodon** 3 nucleotides on a tRNA molecule attracted to a complementary codon on mRNA *257*

**antidiuretic hormone (ADH)** (an"ti-di"u-ret'ik) hormone secreted by the posterior pituitary that increases the permeability of the collecting duct in a nephron *628, 681*

**antigen** substances, usually proteins, that are not normally found in the body *563, 577*

**aorta** largest systemic artery, which transports blood from the heart to all other systemic arteries *560*

**appendicular skeleton** part of the skeleton forming the upper appendages, shoulder girdle, lower appendages, and hip girdle *666*

**archaebacterium** (ar"ke-bak-te're-um) probably the earliest prokaryote; the cell wall and plasma membrane differ from other bacteria, and many live in extreme environments *374*

**archegonium** (ar"kē-go'ne-um) female structure in nonseed plants where an egg is produced *398*

**artery** a blood vessel that transports blood away from the heart toward arterioles *555*

**arthropod** a diverse phylum of animals that have jointed appendages; includes lobsters, insects, and spiders *437*

**asexual reproduction** reproduction involving only mitosis which produces offspring genetically identical to the parent *166*

**aster** short, radiating fibers produced by the centrioles; important during mitosis and meiosis *160*

**atom** the smallest particle of an element that displays its properties *26*

**ATP** *see* adenosine triphosphate

**autonomic nervous system** a branch of the peripheral nervous system administering motor control over internal organs *641*

**autosome** any chromosome other than the sex-determining pair *204*

**autotroph** self-nourishing organism; referring to producers starting food chains that make organic molecules from inorganic nutrients *356*

**auxin** plant hormone regulating growth, particularly cell elongation; also called indoleacetic acid *523*

**axial skeleton** part of the skeleton forming the vertical support or axis, including the skull, rib cage, and vertebral column *666*

**axon** the part of a neuron that conducts the impulse from cell body to synapse *635*

# B

**B lymphocyte** white blood cell that produces and secretes antibodies that can combine with antigens *577*

**bacteriophage** a virus that parasitizes a bacterial cell as its host, destroying it by lytic action *237*

**bacterium (pl., bacteria)** one-celled organism that lacks a nucleus and cytoplasmic organelles other than ribosomes; reproduces by binary fission and occurs in one of 3 shapes (rod, sphere, spiral) *66, 370*

**Barr body** X chromosome in females that is inactive *269*

**basal body** structure in the cell cytoplasm, located at the base of a cilium or flagellum *78*

**base** a compound tending to lower the hydrogen ion concentration in a solution and raise its pH numerically *36*

**behavior** all responses made by an organism to changes in the environment *5*

**benthic division** the ocean floor with a unique set of organisms in contrast to the pelagic division or open waters *768*

**bicarbonate ion** form in which most of the carbon dioxide is transported in the bloodstream *614*

**bilaterally symmetrical** a body having 2 corresponding or complementary halves *419*

**bile** a substance released from the gallbladder to small intestine to emulsify fats prior to chemical digestion *594*

**binary fission** splitting of a parent cell into 2 daughter cells; serves as an asexual form of reproduction in single-celled organisms *156, 371*

**biogeography** the study of the geographical distribution of organisms *303*

**biological clock** internal mechanism that maintains a biological rhythm in the absence of environmental stimuli *530, 732*

**biological magnification** process by which substances become more concentrated in organisms in the higher trophic levels of the food chain *784*

**biome** a major biotic community having well-recognized life forms and a typical climax species *766*

**biosphere** a thin shell of air, land, and water around the earth that supports life *8, 766*

**biotic potential** maximum rate of natural increase of a population that can occur under ideal circumstances *745*

**bivalent (tetrad)** (bi-va'lent) homologous chromosomes each having sister chromatids that are joined by a nucleoprotein lattice during meiosis *172*

**blastocoel** (blas'to-sēl) the fluid-filled cavity of a blastula in the early embryo *713*

**blastula** a hollow, fluid-filled ball of cells prior to gastrula formation of the early embryo *713*

**blood** type of connective tissue in which cells are separated by a liquid called plasma *541*

**blood pressure** force of flowing blood pushing against the inside wall of an artery *561*

**bone** type of connective tissue in which a matrix of calcium salts is deposited around protein fibers *541*

**bronchus (pl., bronchi)** one of 2 main branches of the trachea in vertebrates that have lungs *612*

**bronchiole** small tube that conducts air from a bronchus to the alveoli *612*

**bryophyte** a plant of a group that includes mosses and liverworts *398*

**budding** in animals an asexual form of reproduction whereby a new organism develops as an outgrowth of the body of the parent *695*

**buffer** a substance or group of substances that tend to resist pH changes in a solution, thus stabilizing its relative acidity *36*

**bundle sheath** cells that surround the xylem and phloem of leaf veins *131, 491*

# C

**calorie** 1/1,000 of a kilocalorie; a unit used to express the quantity of energy of carbohydrates, fats, and proteins *34, 109*

**$C_3$ plant** plant using the enzyme rubisco to fix carbon dioxide to RuBP; the first detected molecule after fixation is PGAL, a 3-carbon molecule *131*

**$C_4$ plant** plant that does not directly use the Calvin cycle and produces an immediate 4-carbon molecule during photosynthesis *131*

**Calvin cycle** series of photosynthetic reactions in which carbon dioxide is reduced in the chloroplast *129*

**cambium** meristematic, or growth, tissue of a plant; for example, vascular cambium between the xylem and phloem in a stem or the cork cambium beneath the epidermis of the stem *484*

**CAM plant** plant that uses PEP carboxylase to fix carbon dioxide at night; CAM stands for crassulacean acid metabolism *133*

**cancer** condition by which cells grow and divide uncontrollably, detaching from tumors and spreading throughout the body *271*

**capillary** microscopic blood vessel for gas and nutrient exchange with body cells *555*

**carbohydrate** a family of organic compounds consisting of carbon, hydrogen, and oxygen atoms; includes subfamilies monosaccharides, disaccharides, and polysaccharides *44*

**carbon cycle** circulation of carbon between the biotic and abiotic component of both terrestrial and aquatic ecosystems *787*

**carbon dioxide (CO₂) fixation** photosynthetic reaction in which carbon dioxide is attached to an organic compound *130*

**carbonic anhydrase** an enzyme in red blood cells that speeds up the formation of carbonic acid from water and carbon dioxide *614*

**carcinogen** agent that contributes to the development of cancer *273*

**cardiovascular system** animal organ system consisting of the blood, heart, and a series of blood vessels that distribute blood under the pumping action of the heart *555*

**carnivore** a secondary consumer in a food chain that eats other animals *781*

**carpel** simple plant reproductive unit of a simple pistil; consisting of 3 parts, the stigma, style, and ovary *509*

**carrier** an individual who is capable of transmitting an infectious or genetic disease; a protein that combines with and transports a molecule across the plasma membrane *94, 219*

**carrying capacity** the maximum size of a population that can be supported by the environment in a particular locale *747*

**cartilage** a type of connective tissue having a matrix of protein and proteinaceous fibers that is found in animal skeletal systems *541*

**Casparian strip** a waxy layer bordering 4 sides of root endodermal cells; prevents water and solute transport between adjacent cells *481*

**catastrophism** (kah-tas′tro-fizm) belief espoused by Cuvier that periods of catastrophic extinctions occurred, after which repopulation of surviving species took place and gave the appearance of change through time *301*

**cell** the smallest unit that displays properties of life; composed of cytoplasmic regions, possibly organelles, and surrounded by a plasma membrane *3, 61*

**cell plate** (sel plāt) structure across a dividing plant cell that signals the location of new plasma membranes and cell walls *164*

**cell theory** statement that all organisms are made of cells and that cells are produced only from preexisting cells *61*

**cellular respiration** metabolic reactions that provide energy to cells by the step-by-step oxidation of carbohydrates *114, 138*

**cellulose** polysaccharide consisting of covalently bonded glucose chains, hydrogen-bonded within fibrils; important in plant cell walls *45*

**cell wall** a relatively rigid structure composed mostly of polysaccharides that surrounds the plasma membrane of plants, fungi, and bacteria *64*

**central nervous system** the brain and nerve (spinal) cord in animals *634*

**centriole** cell organelle, existing in pairs, that possibly organizes a mitotic spindle for chromosome movement during cell division *78*

**centromere** a constriction where duplicates (chromatids) of chromosomes are held together *157*

**cephalization** (sef′al-i-za′shun) having a well-recognized anterior head with concentrated nerve masses and receptors *421*

**cerebellum** portion of the brain that is dorsal to the pons and medulla oblongata; functioning in muscle coordination to produce smooth graceful motions *645*

**cerebrum** foremost part of the brain consisting of 2 large masses—cerebral hemispheres; the largest part of the brain in humans *644*

**chemical evolution** an increase in the complexity of chemicals that could have led to first cells *355*

**chemiosmotic phosphorylation** (kem″ĭ-os-mot′ik fos″for-ĭ-la′shun) production of ATP by utilizing the energy released when H⁺ ions flow through an ATP synthase complex in mitochondria and chloroplasts *117*

**chemoreceptor** a receptor that is sensitive to chemical stimulation, for example, receptors for taste and smell *653*

**chemoautotroph** bacterium capable of oxidizing inorganic compounds, such as ammonia, nitrites, and sulfides, to gain energy to synthesize carbohydrates *373*

**chiasma** during synapsis in meiosis, the region of attachment between nonsister chromatids of a bivalent due to crossing-over *173*

**chitin** (ki′tin) a strong but flexible, nitrogenous polysaccharide that is found in the exoskeleton of arthropods *46, 437*

**chlorophyll** green pigment that absorbs sunlight energy and is important in photosynthesis *125*

**chloroplast** a membrane-bounded organelle with membranous grana that contain chlorophyll and where photosynthesis takes place *74, 125*

**cholesterol** one of the major lipids found in animal plasma membranes; makes the membrane impermeable to many molecules *50, 88*

**chordate** phylum of animals that includes lancelets, tunicates, fishes, amphibians, reptiles, birds, and mammals; characterized by a notochord, dorsal nerve cord, and gill pouches *443*

**chorion** (kor′e-on) an extraembryonic membrane functioning for respiratory exchange in the eggs of birds and reptiles; contributes to placenta formation in mammals *718*

**chromatid** one of 2 identical parts of a chromosome, produced through DNA replication; within the same chromosome, one chromatid is the sister chromatid of the other *157*

**chromatin** the mass of DNA and associated proteins observed within a nucleus that is not dividing *68, 157, 269*

**chromosome** association of DNA and proteins arranged linearly in genes and visible only during cell division *68, 157*

**chromosome theory of inheritance** theory stating that the genes are on chromosomes accounting for their similar behavior *202*

**cilium (pl., cilia)** short, hairlike projection from the cell membrane, found as one of a group (cilia) *78*

**circadian rhythm** (ser″kah-de′an) a biological rhythm with a 24-hour cycle *530, 731*

**citric acid cycle** a cycle of reactions in mitochondria, that begins and ends with citric acid; produces CO₂, ATP, NADH, and FADH₂; also called the Krebs cycle *144*

**cladistics** (clah-dis′tiks) means of classification that places a common ancestor, and all organisms evolved from it, into one taxon; based on studying homologous structures and other similarities among organisms *342*

**classical conditioning** one type of learning in which an animal learns to give a response to an irrelevant stimulus; also called associative learning *735*

**cleavage** earliest division of the zygote developmentally, without cytoplasmic addition or enlargement *163, 713*

**cleavage furrowing** indenting of the plasma membrane that leads to division of the cytoplasm during cell division *163*

**climax community** mature stage of a community that can sustain itself indefinitely as long as it is not stressed *771*

**cloned** (klōnd) the production of identical copies; in genetic engineering, the production of many, identical copies of a gene *279*

**cnidarian** (ni-dah're-an) phylum of animals with a gastrovascular cavity; includes sea anemones, corals, hydras, and jellyfishes *419*

**cochlea** (kok'le-ah) spiral-shaped structure of the inner ear containing the receptor hair cells for hearing *660*

**codominance** a condition in heredity in which both alleles in the gene pair of an organism are expressed *197*

**codon** 3 nucleotides of DNA or mRNA, codes for a particular amino acid *253*

**coelomate** (sēl'o-māt) characterizes animals with a true coelom which is completely lined with mesoderm *424*

**coenzyme** a nonprotein organic part of an enzyme structure, often with a vitamin as a subpart *112*

**cohesion-tension model** explanation for the transport of water to great heights in a plant by water molecules clinging together as transpiration occurs *497*

**collecting duct** the final portion of a nephron where reabsorption of water can be controlled *625*

**commensalism** a symbiotic relationship in which one species is benefited and the other is neither harmed nor benefited *760*

**communication** an action by a sender that influences the behavior of the recipient *738*

**community** many different populations of organisms that interact with each other *755*

**competitive exclusion principle** theory that no 2 species can occupy the same niche *756*

**complementary base pairing** bonding between particular purines and pyrimidines in DNA *55, 240*

**complement system** series of proteins, produced by the liver and present in the plasma, that produces a cascade of reactions as part of the immune system *576*

**compound** (kom'pownd) a chemical substance having 2 or more different elements in fixed ratio *29*

**compound eye** a type of eye found in arthropods composed of many independent visual units *437, 654*

**condensation** chemical change producing the covalent bonding of 2 monomers with the accompanying loss of a water molecule *44*

**cone** photoreceptors in vertebrate eyes that respond to bright light and allow color vision *655*

**conifer** conifers are one of the 4 groups of gymnosperm plants; cone-bearing trees that include pine, cedar, and spruce *405*

**conjugation** the transfer of genetic material from one cell to another through a cytoplasmic bridge *385*

**connective tissue** animal tissue type that binds structures together, provides support and protection, fills spaces, stores fat, and forms blood cells *540*

**consumer** an organism that feeds on another organism in a food chain; primary consumers eat plants, and secondary consumers eat animals *781*

**control group** a sample that goes through all the steps of an experiment except the one being tested; a standard against which results of an experiment are checked *15*

**corepressor** molecule that binds to a repressor, allowing the repressor to bind to an operator in a repressible operon *268*

**cork** outer covering of bark of trees; made up of dead cells that may be sloughed off *477*

**corpus luteum** (kor'pus lu'te-um) a follicle that has released an egg and increases its secretion of progesterone *701*

**cortex** layer of the young plant root or stem beneath the epidermis; consisting of large, thin-walled parenchyma cells *481*

**cotyledon** (kot-el-ēd'en) seed leaf for embryonic plant, providing nutrient molecules for the developing plant before its mature leaves begin photosynthesis *478*

**covalent bond** (ko-va'lent) a chemical bond in which the atoms share one pair of electrons *31*

**cristae** shelflike folds of the inner membrane of the mitochondrion, projecting into the inner space of the organelle *75*

**Cro-Magnon** hominid who lived 40,000 years ago; accomplished hunters, made compound stone tools, and possibly had language *467*

**crossing-over** an exchange of segments between nonsister chromatids of a bivalent during meiosis *172*

**cultural eutrophication** human activities result in an acceleration of the process by which nutrient poor lakes and ponds become nutrient rich; speeds up aquatic succession, in which a pond or lake eventually fills in and disappears *796*

**cyanobacterium** (si"ah-no-bak-te're-um) formerly called blue-green alga; photosynthetic prokaryote that contains chlorophyll and releases oxygen *66, 375*

**cyclic AMP** an ATP-related compound that promotes chemical reactions in the body; the "second messenger" in peptide hormone activity, it initiates activity of the metabolic machinery *679*

**cyclic photophosphorylation** (fo"to-fos"for-i-la'shun) photosynthetic pathway in which electrons from photosystem P700 pass through the electron transport system, produce ATP, and eventually return to P700 *128*

**cystic fibrosis** a lethal genetic disease affecting the mucus in the lungs and digestive tract *221*

**cytochrome system** a series of electron carrier molecules, part of the electron transport systems of mitochondria and chloroplasts *128, 144*

**cytokinesis** (si"to-ki-ne'sis) the division of the cytoplasm following mitosis and meiosis *157*

**cytokinin** (si"to-ki'nin) plant hormone that promotes cell division; often works in combination with auxin during organ development in plant embryos *526*

**cytoplasm** contents of a cell between the nucleus (nucleoid) and the plasma membrane *64*

**cytosine (C)** (si'to-sin) one of 4 organic bases in the nucleotides composing structure of DNA and RNA *238*

**cytoskeleton** internal framework of the cell, consisting of microtubules, actin filaments, and intermediate filaments *76*

**cytosol** fluid medium that bathes the contents of the cytoplasm *64*

# D

**datum (pl., data)** fact or information collected through observation and/or experimentation *14*

**decomposer** organism, usually a bacterial or fungal species, that breaks down large organic molecules into elements that can be recycled in the environment *373*

**deductive reasoning** a process of logic and reasoning, using "if . . . then" statements *14*

**demographic transition** a decline in death rate, followed shortly by a decline in birthrate, resulting in slower population growth *750*

**denatured** condition of an enzyme when its shape is changed so that its active site cannot bind substrate molecules *111*

**dendrite** the part of a neuron that sends impulses toward the cell body *635*

**denitrification** (de-ni"tri-fi-ka'shun) the process of converting nitrogen compounds to atmospheric nitrogen; is a part of the nitrogen cycle *781*

**deoxyribonucleic acid (DNA)** an organic molecule produced from covalent bonding of nucleotide subunits; the chemical composition of genes on chromosomes *54, 235*

**dependent variable** result or change that occurs when the experimental variable is manipulated *15*

**dermis** deeper, thicker layer of the skin that consists of fibrous connective tissue and contains various structures such as sense organs *544*

**desert** treeless biome where the annual rainfall is less than 25 cm; the rain that does fall is subject to rapid runoff and evaporation *773*

**desertification** (dez"ert-ĭ-fi-ka'shun) transformation of marginal lands to desert *793*

**detritus** (de-tri'tus) falling, settled remains of plants and animals on the land-floor or water bed *781*

**deuterostome** (du'ter-o-stōm") a group of coelomate animals in which the second embryonic opening becomes the mouth; the first embryonic opening, the blastopore, becomes the anus *430*

**diabetes mellitus** (di"ah-be'tēz mĕ-lī'tus) a disease caused by a lack of insulin production by the pancreas *686*

**diaphragm** a dome-shaped muscle separating the thoracic cavity from the abdominal cavity in the animal body and involved in respiration *609*

**diastole** (di-as'to-le) relaxation period of a heart during the cardiac cycle *558*

**dicotyledon** a flowering plant group; members show 2 embryonic leaves, net-veined leaves, cylindrical arrangement of vascular bundles, and other characteristics *407*

**dictyosome** organelle in a plant cell that is called a Golgi apparatus in animal cells *72*

**differentiation** specialization of early embryonic cells with regard to structure and function *712*

**diffusion** the movement of molecules from a region of high to low concentration, requiring no energy and tending toward an equal distribution *89*

**dihybrid** a genetic cross involving 2 traits *190*

**diploid (2N) number** (dip'loid) the condition in which cells have 2 of each type of chromosome *157*

**diplontic cycle** (dip-lon'tik) life cycle in which the adult is diploid *384*

**directional selection** outcome of natural selection in which an extreme phenotype is favored, usually in a changing environment *326*

**disruptive selection** outcome of natural selection in which extreme phenotypes are favored over the average phenotype and can lead to polymorphic forms *325*

**distal convoluted tubule** the portion of a nephron between the loop of the nephron and the collecting duct, where tubular secretion occurs *625*

**DNA** *see* deoxyribonucleic acid

**DNA ligase** (li'gas) an enzyme that links DNA fragments; used in genetic engineering to join foreign DNA to the vector DNA *280*

**DNA marker** genetic variation found within highly repetitive DNA and inherited in a Mendelian fashion; the distinctive pattern of repetition in a person serves as a DNA fingerprint or means of identification *290*

**DNA polymerase** (pol-im'er-ās) enzyme that joins complementary nucleotides to form double-stranded DNA *241*

**DNA probe** known sequences of DNA that are used to find complementary DNA strands; can be used diagnostically to determine the presence of particular genes *283*

**dominance hierarchy** organization of animals in a group that determines the order in which the animals have access to resources *736*

**dominant allele** (ah-lēl') allele of a given gene that hides the effect of the recessive allele in a heterozygous condition *186*

**Down syndrome** a genetic disease caused by trisomy 21 that is characterized by mental retardation and a specific set of physical features *216*

**dryopithecine** (dri"o-pith'e-sin) a possible hominoid ancestor; a forest-dwelling primate with characteristics somewhat like those of living apes *461*

**duodenum** (dū-ah-dē'-num) first part of the small intestine where chyme enters from the stomach *593*

**Duchenne muscular dystrophy** (du-shen') an X-linked, recessive disorder that causes muscular weakness and eventually death at a young age *230*

# E

**echinoderm** (e-kin'o-derm) a phylum of marine animals that includes sea stars, sea urchins, and sand dollars; characterized by radial symmetry and a water vascular system *440*

**ecological pyramid** pictorial graph representing the biomass, organism number, or energy content of each trophic level in a food web from the producer to the final consumer populations *782*

**ecology** the study of the interactions of organisms with their living and physical environment *745*

**ecosystem** a biological community together with the associated abiotic environment *8, 781*

**ectoderm** outermost of an animal's primary germ layers *713*

**effector** organ that makes a response when signaled by a motor neuron *641*

**electromagnetic spectrum** solar radiation divided on the basis of wavelength, with gamma rays having the shortest wavelength and radio waves having the longest wavelength *122*

**electron** negative subatomic particle, moving about in energy levels around the nucleus of the atom *26*

**electron transport system** a mechanism whereby electrons are passed along a series of carrier molecules to produce energy for the synthesis of ATP *114, 144*

**element** the simplest of substances consisting of only one type of atom; for example, carbon, hydrogen, oxygen *26*

**embryo** early developmental stage of a plant or animal, produced from a zygote *510*

**emigrate ( n., emigration)** to move from a geographical area *748*

**endocrine system** one of the major systems involved in the coordination of body activities; uses messengers called hormones which are secreted into the bloodstream *677*

**endocytosis** (en"do-si-to'sis) moving particles or debris into the cell from the environment by phagocytosis (cellular eating) or pinocytosis (cellular drinking) *97*

**endoderm** innermost of an animal's primary germ layers *713*

**endodermis** internal plant root tissue forming a boundary between the cortex and vascular cylinder *481*

**endometrium** (en-do-me'tre-um) a mucous membrane lining the inside free surface of the uterus *700*

**endoplasmic reticulum** (en"do-plas'mik rĕ-tik'u-lum) a system of membranous saccules and channels in the cytoplasm *71*

**endosperm** nutritive tissue in a seed; often triploid from a fusion of sperm cell with 2 polar nuclei *510*

**endospore** a bacterium that has shrunk its cell, rounded up within the former plasma membrane, and secreted a new and thicker cell wall in the face of unfavorable environmental conditions *371*

**endosymbiotic theory** statement that chloroplasts and mitochondria (and perhaps cilia) were originally prokaryotes that evolved a mutualistic relationship with the eukaryotic cell *75*

**energy** capacity to do work and bring about change; occurs in a variety of forms *28, 105*

**environment** surroundings with which a cell or organism interacts *8*

**environmental resistance** the opposing force of the environment on the biotic potential of a population *747*

**enzyme** a protein that acts as an organic catalyst to speed up reaction rates in living systems *50, 108*

**epicotyl** (ep"ĭ-kot'il) plant embryo portion above the cotyledon that contributes to shoot development *515*

**epidermis** covering tissue of roots and leaves of plants, plus stems of nonwoody organisms; outer, protective layer of the skin *476, 544*

**epiglottis** a flaplike covering, hinged to the back of the larynx and capable of covering the glottis or air-tract opening *593, 612*

**epiphyte** (ep'ĭ-fīt) a plant that takes its nourishment from the air because its attachment to other plants gives it an aerial position *503, 778*

**epistatic gene (epistasis)** (ĕp"ĭ-stat'ik jēn) a gene that interferes with the expression of alleles that are at a different locus *198*

**epithelial tissue** (ep-ah-thē'lē-al) animal tissue type forming a continuous layer over most body surfaces (i.e., skin) and inner cavities *539*

**erythrocyte** (e-rith'ro-sit) red blood cell; important for oxygen transport in blood *563*

**esophagus** a muscular tube for moving swallowed food from pharynx to stomach *593*

**essential amino acid** an amino acid that is required by the body but must be obtained in the diet because the body is unable to produce it *149, 598*

**estrogen** female ovarian sex hormone that has numerous effects on the en-dometrium of the uterus throughout the ovarian cycle *700*

**estuary** the end of a river where fresh water and salt water mix as they meet *766*

**ethylene** plant hormone that causes ripening of fruit and is also involved in abscission *527*

**eubacterium** most common type of bacteria, including the photosynthetic bacteria *374*

**euchromatin** diffuse chromatin, which is being transcribed *269*

**eukaryotic cell** typical of organisms, except bacteria and cyanobacteria, having organelles and a well-defined nucleus *64*

**evolution** genetic and phenotypic changes that occur in populations of organisms with the passage of time, often resulting in increased adaptation of organisms to the prevailing environment *300*

**exhalation** stage during respiration when air is pushed out of the lungs *609*

**exocytosis** a process by which cells expel particles or debris in vesicles passing through a plasma membrane to the extracellular environment *97*

**exon** in a gene, the portion of the DNA code that is expressed as the result of polypeptide formation *254*

**exoskeleton** a protective, external skeleton, as in arthropods *437*

**experimental variable** condition that is tested in an experiment by varying it and observing the results *15*

**exponential** referring to a geometrically multiplying, rapid population growth rate *745*

# F

**facilitated transport** process whereby molecules pass through a plasma mem-brane, utilizing a carrier molecule but without the expenditure of energy *94*

**FAD (flavin adenine dinucleotide)** (fla'vin ad'ĕ-nīn di-nu'kle-o-tīd) coenzyme that functions as an electron acceptor in cellular oxidation-reduction reactions *114*

**fat** organic compound consisting of 3 fatty acids covalently bonded to one glycerol molecule *47*

**fatty acid** subunit of a fat molecule, characterized by a carboxyl acid group at one end of a long hydrocarbon chain *47*

**feedback inhibition** process by which a substance, often an end product of a reaction or a metabolic pathway, controls its own continued production by binding with the enzyme which produced it *112*

**fermentation** anaerobic breakdown of carbohydrates that results in products such as alcohol and lactate *139*

**fibroblast** cell type of loose and fibrous connective tissue with cells at some distance from one another and separated by a jellylike substance *540*

**filament** the elongated stalk of a stamen bearing the anther at the tip *510*

**fitness** ability of an organism to survive and reproduce in its local environment *305*

**fixed-action pattern (FAP)** a stereotyped behavior pattern that occurs automatically *733*

**flower** reproductive organ of a flowering plant, consisting of several kinds of modified leaves arranged in concentric rings and attached to a modified stem called the receptacle *407*

**fluid-mosaic model** model for the plasma membrane based on the changing location and pattern of protein molecules in a fluid phospholipid bilayer *86*

**follicle** structures in the ovary of animals that contain oocytes; site of egg production *700*

**follicle-stimulating-hormone (FSH)** gonadotrophic hormone secreted by anterior pituitary that controls the events of the ovarian cycle by being secreted at different rates throughout the cycle *701*

**food chain** a succession of organisms in an ecosystem that are linked by an energy flow and the order of who eats whom *782*

**food web** a complex pattern of interlocking and crisscrossing food chains *782*

**founder effect** the tendency for a new, small population to experience genetic drift after it has separated from an original, larger population *323*

**fruit** flowering plant structure consisting of one or more ripened ovaries that usually contain seeds *407, 511*

**fruiting body** a spore-bearing structure found in certain types of fungi, such as mushrooms *391*

**FSH** see follicle-stimulating hormone

**fungus** saprophytic decomposer; body is made up of filaments called hyphae that form a mass called a mycelium *391*

# G

**gallbladder** attached to the liver; serves as a storage organ for bile *594*

**gamete** (gam'et) haploid sex cell *157*

**gametophyte** haploid generation of the alternation of generations life cycle of a plant; it produces gametes that unite to form a diploid zygote *398*

**ganglion (sing., ganglia)** (gang'gle-on) a knot or bundle of neuron cell bodies outside the central nervous system *640*

**gastrovascular cavity** a blind, branched digestive cavity that also serves a circula-tory (transport) function in animals that lack a circulatory system *419*

**gastrula** (gas'troo-lah) cup-shaped early embryo with 2 primary germ layers, ectoderm and endoderm, enclosing a primitive digestive tract *713*

**gel electrophoresis** (ĭ-lek-tra-fa-rē'sus) technique that allows DNA fragments or proteins to be separated according to differences in the charge and size *290*

**gene** the unit of heredity occupying a particular locus on the chromosome and passed on to offspring 5

**gene flow** sharing of genes between 2 populations through interbreeding 320

**gene locus** the specific location of a particular gene on a chromosome 186

**gene mutation** (jēn mu-ta'shun) a change in the base sequence of DNA such that the sequence of amino acids is changed in a protein 316

**gene pool** the total of all the genes of all the individuals in a population 318

**gene therapy** use of transplanted genes to overcome an inborn error in metabolism 288

**genetic drift** change in the genetic make-up of a population due to chance (random) events; important in small populations or when only a few individuals mate 322

**genetic engineering** alteration of the genome of an organism by technological processes 279

**genome** (jē'nōm) all of the genes of an organism 244

**genotype** (jen'o-tīp) the genes of an organism for a particular trait or traits; for example, BB or Aa 187

**germination** resumption of growth by a seed or any other reproductive structure of a plant or protist 515

**germ layer** developmental layer of body; that is, ectoderm, mesoderm, and endoderm 714

**gibberellin** (gib-ber-el'in) plant hormone producing increased stem growth by cell division and enlargement; also involved in flowering and seed germination 526

**gill** respiratory organ in most aquatic animals; most common in fish as an outward extension of the pharynx 607

**girdling** removing a strip of bark from around a tree 503

**glomerular capsule** a cuplike structure that is the initial portion of a nephron; where pressure filtration occurs 625

**glomerulus** (glo-mer'u-lus) a capillary network within a glomerular capsule of a nephron 625

**glottis** opening for airflow in the larynx 612

**glucagon** (glu'kah-gon) hormone secreted by the pancreas that stimulates the breakdown of stored nutrients and increases the level of glucose in the blood 686

**glucose** monosaccharide with the molecular formula of $C_6H_{12}O_6$ found in the blood of animals, serving as an energy source in most organisms and transported in the blood of animals 45

**glycogen** polysaccharide consisting of covalently bonded glucose molecules, characterized by many side branches, energy storage product in animals 45

**glycolysis** (gli-kol'i-sis) pathway of metabolism converting a sugar, usually glucose, to pyruvate; resulting in a net gain of 2 ATP and 2 NADH molecules 139

**goiter** an enlargement of the thyroid gland caused by a lack of iodine in the diet 683

**Golgi apparatus** an organelle consisting of a central region of saccules and vesicles that modifies and/or activates proteins and produces lysosomes 72

**gonadotrophic hormone (GNRH)** (gō-nad-ah-trō'fik) either FSH or LH, a substance secreted by the anterior pituitary that stimulates the gonads 482

**granum (pl., grana)** (gra'num) in a chloroplast, a stack of flattened membrane sacs, thylakoids, that contain chlorophyll 74, 123

**gravitropism** (grav"i-tro'pizm) directional growth in response to the earth's gravity; roots demonstrate positive gravitropism, stems demonstrate negative gravitropism 529

**greenhouse effect** the reradiation of heat toward the earth caused by gases in the atmosphere 800

**growth hormone (GH)** substance stored and secreted by the anterior pituitary; promotes cell division, protein synthesis, and bone growth 683

**guanine (G)** (gwan'in) one of 4 organic bases in the structure of nucleotides composing the structure of DNA and RNA 54, 238

**guard cell** plant cell type, found in pairs, with one on each side of a leaf stoma; changes in the turgor pressure of these cells regulate the size and passage of gases through the stoma 498

**guttation** (gut-ta'shun) liberation of water droplets from the edges and tips of leaves 497

**gymnosperm** a vascular plant producing naked seeds, as in conifers 405

# H

**habitat** region of an ecosystem occupied by an organism 756

**haploid (N) number** a cell condition in which only one of each type of chromosome is present 157

**haplontic cycle** life cycle in which the adult is haploid 384

**Hardy-Weinberg law** a law stating that the frequency of an allele in a population remains stable under certain assumptions, such as random mating; therefore, no change or evolution occurs 318

**helper T cell** cell that secretes lymphokines, which stimulate all kinds of immune cells 579

**hemoglobin** red respiratory pigment of erythrocytes for transport of oxygen 554, 563, 614

**hemophilia** a disease in which the blood fails to clot; caused by insufficient clotting factor VIII 229

**herbaceous plant** a plant that lacks persistent woody tissue 407

**herbivore** a primary consumer in a food chain; a plant eater 781

**hermaphroditic animal** (her'maf"ro-dit'ik) characterizes an animal having both male and female sex organs 422

**heterochromatin** highly compacted chromatin during interphase 269

**heterogamete** (het"er-o-gam'ēt) a nonidentical gamete, for example, large, nonmotile egg and small, flagellated sperm 385

**heterotroph** (het'er-o-trōf") an organism that cannot synthesize organic compounds from inorganic substances and therefore must take in preformed food 356

**heterozygous** possessing 2 different alleles for a particular trait 187

**homeostasis** the maintenance of internal conditions in cell or organisms, for example, relatively constant temperature, pH, blood sugar 546

**hominid** referring to humans; ancestors of humans who had diverged from the ape lineage 461

**Homo erectus** hominid who lived during the Pleistocene; with a posture and locomotion similar to modern humans 466

**Homo habilis** (ho'mo hah'bǐ-lis) fossil hominid of 2 million years ago; possible first direct ancestor of modern humans 465

**homologous chromosome** a pair of chromosomes that carry genes for the same traits and synapse during prophase of the first meiotic division 171

**homologous structure** in evolution, structure derived from a common ancestor 310

**homologue** (hom'o-log) member of homologous pair of chromosomes 171

**homozygous** possessing 2 identical alleles for a particular trait 186

**hormone** a chemical secreted into the bloodstream in one part of the body that controls the activity of other parts 523

**Huntington disease (HD)** a lethal, genetic disease affecting one in 10,000 people in the United States; affects neurological functioning and generally does not manifest itself until middle age *223*

**hydra** a freshwater cnidarian with only a polyp stage that reproduces both sexually and asexually *420*

**hydrogen bond** a weak bond that arises between a partially positive hydrogen and a partially negative oxygen, often on different molecules or separated by some distance *33*

**hydrolysis** splitting of a compound into parts in a reaction that involves addition of water, with the $H^+$ ion being incorporated in one fragment and the $OH^-$ ion in the other *44*

**hydrophilic** a type of molecule that interacts with water, such as polar molecules, which dissolve in water or form hydrogen bonds with water molecules *34, 43*

**hydrophobic** a type of molecule that does not interact with water, such as nonpolar molecules, which do not dissolve in water or form hydrogen bonds with water molecules *43*

**hydroponics** water culture method of growing plants that allows an experimenter to vary the nutrients and minerals provided so as to determine specific growth requirements *500*

**hydrostatic skeleton** characteristic of animals in which muscular contractions are applied to fluid-filled body compartments *434*

**hypertonic solution** lesser water concentration, greater solute concentration than the cytoplasm of a particular cell; situation in which cell tends to lose water by osmosis *92*

**hypha** a filament of the vegetative body of a fungus *391*

**hypocotyl** (hi″po-kot′il) plant embryo portion below the cotyledon that contributes to development of stem *515*

**hypothalamus** portion of the brain that regulates the internal environment of the body; involved in control of heart rate, body temperature, water balance, and glandular secretions of the stomach and pituitary gland *644*

**hypothesis** a supposition that is established by reasoning after consideration of available evidence and that can be tested by obtaining more data, often by experimentation *14*

**hypotonic solution** greater water concentration, lesser solute concentration than the cytoplasm of a particular cell; situation in which a cell tends to gain water by osmosis *92*

## I

**immune system** includes many leukocytes and various organs and provides specific defense against foreign antigens *577*

**immunity** the ability of the body to protect itself from foreign substances and cells, including infectious microbes *575*

**imprinting** a behavior pattern with both learned and innate components that is acquired during a specific and limited period, usually when very young *733*

**incomplete dominance** a pattern of inheritance in which the offspring shows characteristics intermediate between 2 extreme parental characteristics, for example, a red and a white flower producing pink offspring *197*

**inducer** a molecule that brings about activity of an operon by joining with a repressor and preventing it from binding to the operator *268*

**inducible operon** an operon that is normally inactive but is turned on by an inducer *268*

**induction** the ability of a chemical or tissue to influence the development of another tissue *715*

**inductive reasoning** a process of logic and reasoning, using specific observations to arrive at a hypothesis *14*

**inhalation** stage during respiration when air is drawn into the lungs *609*

**inheritance of acquired characteristics** Lamarckian belief that organisms become adapted to their environment during their lifetime and pass on these adaptations to their offspring *301*

**inorganic** referring to molecules that are not compounds of carbon and hydrogen *42*

**insight learning** ability to respond correctly to a new, different situation the first time it is encountered by using reason *735*

**insulin** hormone secreted by the pancreas that stimulates fat, liver, and muscle cells to take up and metabolize glucose, stimulates the conversion of glucose into glycogen in muscle and liver cells, and promotes the buildup of fats and proteins *686*

**interferon** an antiviral agent produced by an infected cell that blocks the infection of another cell *577*

**interneuron** neuron, within the central nervous system, conveying messages within parts of the central nervous system *636*

**interphase** stage of cell cycle during which DNA synthesis occurs and the nucleus is not actively dividing *159*

**intron** noncoding segments of DNA that are transcribed but removed before mRNA leaves nucleus *254*

**invertebrate** referring to an animal without a serial arrangement of vertebrae or a backbone *416*

**ion** charged derivative of an atom, positive if the atom loses electrons and negative if the atom gains electrons *30*

**ionic bond** an attraction between charged particles (ions) through their opposite charge that involves a transfer of electrons from one atom to another *31*

**isomer** molecules with the same molecular formula but different structure and, therefore, shape *43*

**isotonic solution** a solution that is equal in solute and water concentration to that of the cytoplasm of a particular cell; situation in which cell neither loses nor gains water by osmosis *91*

**isotope** forms of an element having the same atomic number but a different atomic weight due to a different number of neutrons *27*

## J

**jointed appendage** freely moving appendages of arthropods *437*

## K

**karyotype** (kar′e-o-tīp) chromosomes are cut from a photograph of the nucleus just prior to cell division and arranged in groups for display *216*

**kidney** bean-shaped, reddish brown organ in humans that regulates the chemical composition of the blood and produces a waste product called urine *624*

**killer T cell** (kil′er te sel) cell that attacks and destroys cells that bear a foreign antigen *579*

**kingdom** a taxonomic category grouping related phyla (animals) or divisions (plants) *3*

**kin selection** an explanation for the evolution of altruistic behavior that benefits one's relatives *740*

**Klinefelter syndrome** a condition caused by the inheritance of a chromosome abnormality in number; an XXY (trisomy) condition where normally a chromosome pair exists 218

**Krebs cycle** *see* citric acid cycle

***K*-strategist** a species that has evolved characteristics that keep its population size near carrying capacity, for example, few offspring, longer generation time 749

# L

**lacteal** (lak'te-al) a lymphatic vessel in an intestinal villus; aids in the absorption of fats 596

**larynx** voicebox, cartilage-made box for airflow between the glottis and trachea 612

**leaf** usually broad, flat structure of a plant shoot system, containing cells that carry out photosynthesis 482

**learning** change in behavior as a result of experience; there are 3 kinds of learning—imprinting, operant conditioning, and insight learning 733

**leukocyte** (lu'ko-sit) white blood cell; important for immune response 563

**LH** *see* luteinizing hormone

**lichen** (li'ken) a symbiotic relationship between certain fungi and algae which has long been thought to be mutualistic, the fungi providing inorganic food and the algae providing organic food 375, 392

**life cycle** entire sequence of developmental phases from zygote formation to gamete formation 384

**ligament** structure consisting of fibrous connective tissue that connects bones at joints 541, 668

**light-dependent reaction** a reaction of photosynthesis that requires light energy to proceed; involved in the production of ATP and NADH 125

**light-independent reaction** photosynthetic reaction that does not directly require light energy; involved primarily in the conversion of carbon dioxide to carbohydrate utilizing the products of the light-dependent reaction 125

**limbic system** part of the brain beneath the cortex that contains pathways that connect various portions of the brain; allows the experience of many emotions 646

**linkage group** genes that are located on the same chromosome and tend to be inherited together 206

**lipid** (lip'id) a family of organic compounds that tend to be soluble in nonpolar solvents such as alcohol; includes fats and oils 46

**liposome** (lip'o-som) droplet of phospholipid molecules formed in a liquid environment 356

**loop of the nephron** the portion of a nephron between the proximal and distal convoluted tubules where some water reabsorption occurs 625

**lung** respiratory organ mainly found in terrestrial vertebrates, evolving as a vascularized outgrowth of the lower pharyngeal region 608

**luteinizing hormone (LH)** (lūt'ē-ah-nīz-ing) gonadotrophic hormone that controls events of the ovarian cycle by being secreted at different rates throughout the cycle 701

**lymph** tissue fluid collected by the lymphatic system and returned to the general systemic circuit 574

**lymphatic system** mammalian organ system consisting of lymphatic vessels and lymphoid organs 753

**lymphocyte** specialized white blood cell, produced by lymphoid tissue, that fights infection; occurs in 2 forms—T lymphocyte and B lymphocyte 563

**lymphokine** (lim'fo-kin) chemicals secreted by T cells that stimulate immune cells 579

**lysogenic cycle** one of the bacteriophage life cycles in which the virus incorporates its DNA into that of the bacterium; only later does it begin a lytic cycle which ends with the destruction of the bacterium 367

**lysosome** membrane-bounded vesicles that contain hydrolytic enzymes to digest macromolecules 73

**lytic cycle** (lit'ik) one of the bacteriophage life cycles in which the virus takes over the operation of the bacterium immediately upon entering it and subsequently destroys the bacterium 367

# M

**macroevolution** large-scale evolutionary change, for example, the formation of new species 340

**macrophage** (mak'ro-faj) large, phagocytotic white blood cell 563

**Malpighian tubule** (mal-pig'i-an) blind, threadlike excretory tubule near anterior end of insect hindgut 438, 623

**mammal** a class of vertebrates characterized especially by the presence of hair and mammary glands 455

**marsupial** a mammal bearing immature young nursed in a marsupium, or pouch; for example, kangaroo, opossum 457

**mass extinction** recorded in the fossil record as a very large number of extinctions over a relatively short geologic time 343

**matrix** inner space of the mitochondrion, filled with a gellike fluid 75

**mechanoreceptor** a cell that is sensitive to mechanical stimulation, such as that from pressure, sound waves, and gravity 658

**medulla oblongata** (mě-dul'ah ob"long-ga'tah) a brain region at the base of the brainstem controlling heartbeat, blood pressure, breathing, and other vital functions 644

**megaspore** reproductive structure produced by the megasporangium of seed plants; of the 4 produced, one develops into a female gametophyte 405

**meiosis** (mi-o'sis) type of nuclear division that occurs as part of sexual reproduction in which the daughter cells receive the haploid number of chromosomes 157, 171

**meiosis I** (mi-o'sis) first division of meiosis, which ends when duplicated homologous chromosomes separate and move into different cells 173

**meiosis II** (mi-o'sis) second division of meiosis in which sister chromatids separate and move into different cells, resulting in the formation of haploid cells 175

**melanocyte-stimulating hormone** (mah-lan'ah-sīt) substance stored and secreted by the anterior pituitary; causes color changes in many fish, amphibians, and reptiles 683

**memory B cell** a B lymphocyte that remains in the bloodstream after an immune response ceases and is capable of producing antibodies to an antigen 578

**memory T cell** a T lymphocyte that remains in the bloodstream after an immune response ceases 579

**menstruation** periodic shedding of tissue and blood from the inner lining of the uterus 703

**meristem tissue** undifferentiated, embryonic tissue in the active growth regions of plants 475

**mesoderm** middle layer of an animal's primary germ layers 714

**mesoglea** (mes"-o-gle'ah) a jellylike layer between the epidermis and endodermis of cnidarians 419

**mesophyll** inner, thickest layer of a leaf consisting of a palisade layer of elongated cells and a spongy layer of irregularly spaced cells 491

**messenger RNA (mRNA)** a nucleic acid (ribonucleic acid) complementary to genetic DNA and bearing a message to direct cell protein synthesis at the ribosome 251

**metabolic pool** metabolites that are the products of and/or the substrates for key reactions in cells allowing one type of molecule to be changed into another type, such as the conversion of carbohydrates to fats 148

**metabolism** all of the chemical reactions that occur in a cell during growth and repair involve and produce metabolites 4, 107

**metafemale** a female with 3 X chromosomes; most show no obvious physical abnormalities 218

**metamorphosis** change in shape and form that some animals, such as insects, undergo during development 438

**MHC protein (major histocompatibility complex)** protein on the plasma membrane of a macrophage that aids in the stimulation of T cells 580

**microbody** membrane-bounded vesicle containing specific enzymes involved in lipid and alcohol metabolism, photosynthesis, and germination 73

**microsphere** formed from proteinoids exposed to water; has properties similar to today's cells 356

**microspore** reproductive structure produced by a microsporangium of a seed plant; it develops into a pollen grain 510

**microtubule** small cylindrical organelle that is believed to be involved in maintaining the shape of the cell and directing various cytoplasmic structures 76

**mimicry** superficial resemblance of an organism to one of another species; often used to avoid predation 758

**mitochondrion** membrane-bounded organelle of the cell known as the power-house because it transforms the energy of carbohydrates and fats to ATP energy 74

**mitosis** a process in which a parent nucleus reproduces 2 daughter nuclei, each identical to the parent nucleus; this division is necessary for growth and development 157

**model** a suggested explanation for experimental results that can help direct future research 17

**molecule** a unit of a chemical substance formed by the union of 2 or more atoms by covalent or ionic bonding; smallest part of a compound that retains the properties of the compound 29

**mollusk** a vertebrate phylum that includes squids, clams, snails, and chitons and is characterized by a visceral mass, a mantle, and a foot 431

**molt** periodic shedding of the exoskeleton in arthropods 437

**Monera** kingdom that includes bacteria, which are microscopic single-celled prokaryotes 370

**monocotyledon** a flowering plant group; members show one embryonic leaf, parallel leaf veins, scattered vascular bundles, and other characteristics 407

**monohybrid** a genetic cross involving one trait 185

**monotreme** an egg-laying mammal; for example, duckbill platypus or spiny anteater 457

**morphogenesis** (mor"fo-jen'ĕ-sis) movement of early embryonic cells to establish body outline and form 712

**multiple allele** more than 2 alleles for a particular trait 200

**muscle fiber** a cell with myofibrils containing actin and myosin filaments arranged within sarcomeres; a group of muscle fibers is a muscle 671

**muscular (contractile) tissue** animal tissue type, composed of fibers with actin and myosin filaments, that can shorten its length and produce movements 542

**mutagen** an agent, such as radiation or a chemical, that brings about a mutation in DNA 260

**mutation** a change in the composition of DNA, due to either a chromosomal or a genetic alteration 7, 208, 235

**mutualism** a symbiotic relationship in which both species benefit 761

**mycelium** (mi-se'le-um) a tangled mass of hyphal filaments composing the vegetative body of a fungus 391

**mycorrhiza** (mi"ko-ri'zah) a "fungus root" composed of a fungus growing around a plant root, which assists the growth and development of the plant 502

**myelin sheath** (mi'ĕ-lin) covering on neurons that is formed from Schwann cell membrane and aids in transmission of nervous impulses 636

**myofibril** a specific muscle cell organelle containing a linear arrangement of sarcomeres which shorten to produce muscle contraction 671

**myosin** (mi'o-sin) a muscle protein making up the thick filaments in a sarcomere; it pulls actin to shorten the sarcomere, yielding muscle contraction 671

# N

**NAD⁺ (nicotinamide-adenine dinucleotide)** (nik"o-tin'ah-mīd ad'ĕ-nīn di-nu'kle-o-tīd) a coenzyme that functions as an electron and hydrogen ion carrier in cellular oxidation-reduction reactions of glycolysis and cellular respiration 113

**NADP⁺ (nicotinamide-adenine dinucleotide phosphate)** (nik"o-tin'ah-mīd ad'ĕ-nīn di-nu'kle-o-tīd fos'fāt) a coenzyme that functions as an electron and hydrogen ion carrier in cellular oxidation-reduction reactions of photosynthesis 114

**natural selection** the guiding force of evolution caused by environmental selection of organisms most fit to reproduce, resulting in adaptation 305

**Neanderthal** hominid who lived during the last ice age in Europe and the Middle East; made stone tools, hunted large game, and lived together in a kind of society 466

**negative feedback loop** pattern of homeostatic response in which the output is counter to and cancels the input 548

**nematocyst** (nem'ah-to-sist) in the cnidaria, a threadlike fiber enclosed within the capsule of a stinging cell; when released, aids in the capture of prey 419

**nephridium** (nĕ-frid'e-um) for invertebrates, a tubular structure for excretion; its contents are released outside through a nephridiopore 435

**nephron** microscopic kidney unit that regulates blood composition by filtration and reabsorption; over one million per human kidney 625

**nerve** bundle of axons and/or dendrites 640

**neurofibromatosis** (nu"ro-fi-bro'mah-to-sus) disease caused by an autosomal dominant allele, characterized by the proliferation of nerve cells into benign tumors under the skin 223

**neuromuscular junction** region where a nerve end comes into contact with a muscle fiber; contains a presynaptic membrane, synaptic cleft, and postsynaptic membrane 673

**neuron** a nerve cell; composed of dendrite(s), a cell body, and axon 543, 635

**neurotransmitter** a chemical made at the ends of axons that is responsible for transmission across a synapse or neuromuscular junction 639

**neutron** neutral subatomic particle, located in the nucleus of the atom 26

**neutrophil** (nu'tro-fil) white blood cell with granules 563

**niche** total description of an organism's functional role in an ecosystem, from activities to reproduction 756

**nitrogen fixation** process whereby free atmospheric nitrogen is converted into compounds, such as ammonia and nitrates, usually by soil bacteria 373, 788

**nitrogen-fixing bacterium** a bacterium that can convert atmospheric nitrogen into compounds such as ammonia and nitrates; free living in the soil or symbiotic with plants such as legumes 373

**nodule** (nod′ul) structure on plant roots that contains nitrogen-fixing bacteria *502*

**noncyclic photophosphorylation** (fo″to-fos″for-i-la′shun) photosynthetic pathway in which electrons from photosystem P680 pass through the electron transport system and move on to another photosystem (P700) *128*

**nonrenewable resource** resource found in limited supplies that can be used up *803*

**norepinephrine** (nor″ep-ah-nef′ren) a hormone released by the adrenal medulla as a reaction to stress; also a neurotransmitter released by neurons of the central and peripheral nervous systems *639*

**notochord** (no′to-kord) a dorsal, elongated, supporting structure in chordates beneath the neural tube or spinal cord; present in the embryo of all vertebrates *714*

**nuclear envelope** double membrane that separates the nucleus from the cytoplasm *69*

**nuclear power** energy source generation by atomic reactions *803*

**nucleic acid** polymer of nucleotides; both DNA and RNA are nucleic acids *54, 235*

**nucleoid** area in prokaryotic cell where DNA is found *64, 156*

**nucleolus** (pl., nucleoli) dark-staining, spherical body in the cell nucleus that contains ribosomal RNA *68*

**nucleotide** the building block subunit of DNA and RNA consisting of a 5-carbon sugar bonded to a nitrogenous base and phosphorus group *54*

**nucleus** (nu′kle-us) region of a eukaryotic cell, containing chromosomes, that controls the metabolic function of the cell *66*

# O

**oil** lipid containing triglycerides having unsaturated hydrocarbon chains; liquid at room temperature *47*

**olfactory cell** modified neuron that is a receptor for the sense of smell *653*

**omnivore** a food chain organism feeding on both plants and animals *781*

**oncogene** (ong′ko-jen) a cancer-causing gene *273*

**oogenesis** (o″o-jen′ĕ-sis) production of eggs in females by the process of meiosis and maturation *175*

**operant conditioning** trial-and-error learning in which behavior is reinforced with a reward or a punishment *734*

**operator** the sequence of DNA in an operon to which the repressor protein binds *266*

**operon** a group of structural and regulating genes that functions as a single unit *267*

**optic nerve** nerve that carries impulses from the retina of the eye to the brain *655*

**orbital** volume of space around a nucleus where electrons can be found most of the time *28*

**organ** combination of 2 or more different tissues performing a common function *544*

**organ of Corti** specialized region of the cochlea containing the hair cells for sound detection and discrimination *661*

**organelle** small, membranous structure in the cytoplasm having a specific function *64*

**organic** containing carbon and hydrogen; such molecules usually also have oxygen attached to the carbon(s) *42*

**organic soup** accumulation of simple organic compounds in the early oceans *355*

**osmosis** the diffusion of water through a selectively permeable membrane *89*

**osmotic pressure** measure of the tendency of water to move across a selectively permeable membrane into a solution; visible as an increase in liquid on the side of the membrane with higher solute concentration *91*

**osteocyte** bone cell found within the Haversian system of bone tissue *665*

**ovary** in flowering plants, enlarged, base portion of the pistil that eventually develops into the fruit; in animals, an egg-producing organ *407, 509, 700*

**oviduct** tubular portion of the female reproductive tract from ovary to the uterus *700*

**ovulation** bursting of a follicle at the release of an egg from the ovary *700*

**ovule** (o′vul) in seed plants, a structure that contains the megasporangium, where meiosis occurs and the female gametophyte is produced; develops into the seed *407, 510*

**oxidation** a chemical reaction that results in removal of one or more electrons from an atom, ion, or compound; oxidation of one substance occurs simultaneously with reduction of another *33*

**oxidative phosphorylation** (fos″for-i-la′shun) the process of building ATP by an electron transport system that uses oxygen as the final receptor; occurs in mitochondria *145*

**oxidizing atmosphere** an atmosphere with free oxygen; for example, the atmosphere of today *355, 358*

**oxygen debt** use of oxygen to convert lactate, which builds up due to anaerobic conditions, to pyruvate *142*

**oxytocin** hormone made by the hypothalamus and stored in the posterior pituitary; causes the uterus to contract and stimulates the release of milk from mammary glands *681*

**ozone shield** formed from oxygen in the upper atmosphere; protects the earth from ultraviolet radiation *358, 801*

# P

**paleontology** study of fossils that results in knowledge about the history of life *301*

**palisade layer** layer in a plant leaf containing elongated cells with many chloroplasts; along with the spongy layer, it is the site of most of photosynthesis *491*

**pancreas** an abdominal organ that produces digestive enzymes and the hormones insulin and glucagon *686*

**parasitism** (par′ah-si″tizm) a symbiotic relationship in which one species (parasite) benefits for growth and reproduction to the harm of the other species (host) *759*

**parasympathetic nervous system** a division of the autonomic nervous system that is involved in normal activities; uses acetylcholine as a neurotransmitter *641*

**parathyroid gland** embedded in the posterior surface of the thyroid gland; produces parathyroid hormone, which is involved in the regulation of calcium and phosphorus balance *685*

**parathyroid hormone (PTH)** substance secreted from the 4 parathyroid glands that increases the calcium level and decreases the phosphate level in the blood *685*

**parenchyma** (pah-ren′kah-mah) least specialized of all plant cell or tissue types; found in all organs of a plant with many of the cells containing chloroplasts *477*

**parthenogenesis** (par″thĕ-no-jen′-ĕ-sis) development of an egg cell into a whole organism without fertilization *695*

**pelagic division** (pe-laj′ik) the open portion of the sea *768*

**penis** male copulatory organ *697*

**peptide** 2 or more amino acids joined together by covalent bonding *50*

**peptide bond** a covalent bond between two amino acids resulting from a condensation reaction between the carboxyl group of one and the amino group of another *50*

**pericycle** external layer of cells in the vascular cylinder of a plant root, producing branch and secondary roots *481*

**peripheral nervous system (PNS)** made up of the nerves that branch off of the central nervous system *634*

**peristalsis** the rhythmic, wavelike contraction that moves food through the digestive tract *593*

**petal** plant structure belonging to an inner whorl of the flower; colored and internal to the outer whorl of green sepals *509*

**phagocytosis** (fag-ah-si-to'sis) transport process by which amoeboid-type cells engulf large material, forming an intracellular vacuole 97, 381

**pharynx** a common passageway (throat) for both food intake and air movement 612

**phenotype** the visible expression of a genotype; for example, brown eyes, height 187

**phenylketonuria (PKU)** (fen"il-ke"to-nu're-ah) a genetic disorder characterized by the absence of an enzyme that metabolizes phenylalanine; results in severe mental retardation 222

**pheromone** (fer'o-mōn) substance produced and discharged into the environment by an organism that influences the behavior of another organism 738

**phloem** (flo'em) vascular tissue conducting organic solutes in plants; contains sieve-tube cells and companion cells 399, 477

**phospholipid** a molecule having the same structure as a neutral fat except one of the bonded fatty acids is replaced by a phosphorous-containing group; an important component of plasma membranes 48, 87

**photochemical smog** air pollution that contains nitrogen oxides and hydrocarbons which react to produce ozone and peroxylacetyl nitrate 798

**photon** discrete packet of light energy; the amount of energy in a photon is inversely related to the wavelength of light in the photon 122

**photoperiodism** relative lengths of daylight and darkness that affect the physiology and behavior of an organism 530

**photoreceptor** light sensitive receptor cell 654

**photosynthesis** a process whereby chlorophyll-containing organisms trap sunlight energy to build a sugar from carbon dioxide and water 114

**photosystem** a photosynthetic unit where light is absorbed; contains an antenna (photosynthetic pigments) and an electron acceptor 125

**phototropism** (fo-tot'ro-pizm) movement in response to light; in plants, growth toward or away from light 529

**pH scale** measurement scale for the relative concentrations of hydrogen ($H^+$) and hydroxyl ($OH^-$) ions in a solution 36

**phylogenetic tree** (fī-lō-jah-net'ik) diagram with branches illustrating common evolutionary ancestors and the descendants produced from them through evolutionary divergence 340

**phytochrome** a plant photoreversible pigment; this reversion rate of 2 molecular forms is believed to measure the photoperiod 531

**pinocytosis** (pi"no-si-to'sis) process by which vesicles form around and bring macromolecules into the cell 97

**pistil** a female flower structure consisting of an ovary, style, and stigma 407, 509

**pituitary gland** a small gland that lies just below the hypothalamus and is important for its hormone storage and production activities 681

**placenta** a structure formed from an extraembryonic membrane, the chorion, and uterine tissue through which nutrient and waste exchange occur for the embryo and fetus 720

**placental mammal** a mammalian subclass that is characterized by a placenta, which is an organ of exchange between maternal and fetal blood that supplies nutrients to the growing offspring 458

**plankton** fresh- and saltwater organisms that float on or near the surface of the water 766

**plasma** liquid portion of blood; contains nutrients, wastes, salts, and proteins 562

**plasma cell** an activated B cell that is currently secreting antibodies 577

**plasma membrane** membrane that separates the contents of a cell from the environment and regulates the passage of molecules into and out of the cell 64

**plasmid** a self-duplicating ring of accessory DNA in the cytoplasm of bacteria 279

**plasmodesmata** (plaz-mah-dez'mah-tah) in plants, cytoplasmic channels bounded by the plasma membrane and connecting the cytoplasm of adjacent cells 101

**platelet** cell-like disks formed from fragmentation of megakaryocytes that are involved in blood clotting in vertebrate blood 563

**plumule** (ploō'mūl) embryonic plant shoot that bears young leaves 515

**polar body** a nonfunctional product of oogenesis; 3 of the 4 meiotic products are of this type 176

**polar covalent bond** (ko-va'lent) a bond in which the sharing of electrons between atoms is unequal 33

**pollen grain** a male gametophyte in seed plants 404

**pollination** transfer of pollen from an anther to a stigma 404, 510

**pollutant** substance that is added to the environment and leads to undesirable effects for living organisms 793

**polygenic inheritance** a pattern of inheritance in which a trait is controlled by several gene pairs that segregate independently 200

**polymer** macromolecule consisting of covalently bonded monomers; for example, a protein is a polymer of monomers called amino acids 43

**polypeptide** chain of many amino acids, covalently bonded together by peptide bonds 50

**polyploid (polyploidy)** a condition in which an organism has more than 2 sets of chromosomes 208

**polysaccharide** carbohydrate macromolecule consisting of many monosaccharides covalently bonded together 45

**polysome** a string of ribosomes, simultaneously translating different regions of the same mRNA strand during protein synthesis 257

**population** a group of organisms of the same species occupying a certain area and sharing a common gene pool 8, 745

**portal system** circulatory pathway that begins and ends in capillaries, such as the one found between the small intestine and liver 560

**positive feedback loop** pattern of homeostatic response in which the output intensifies and increases the likelihood of response instead of countering it and canceling it 548

**prairie** terrestrial biome that is a temperate grassland, changing from tall-grass prairies to short-grass prairies when traveling east to west across the Midwest of the United States 775

**pressure-flow model** model explaining transport through sieve-tube cells of phloem, with leaves serving as a source and roots as a sink in the summer, and vice versa in the spring 503

**primate** an animal order of mammals having hands and feet with 5 distinct digits and some having an opposable thumb 458

**primitive atmosphere** earliest atmosphere of the earth after the planet's formation 355

**producer** organism at the start of a food chain that makes its own food (e.g., green plants on land and algae in water) 781

**progesterone** female ovarian sex hormone that causes the endometrium of the uterus to become secretory during the ovarian cycle; along with estrogen maintains secondary sex characteristics in females 700

**prokaryote** first type of cell to evolve; found in the kingdom Monera and lacking a well-defined nucleus and organelles 64, 370

**prokaryotic cell** the first, primitive cells on earth, exemplified today by bacteria, which lack a defined nucleus and most organelles 64

**promoter** a sequence of DNA in an operon where RNA polymerase begins transcription *266*

**prostaglandins** (pros-tah-glan'dins) class of chemical messengers, acting locally in the body, with widespread effects; derived from fatty acids and stored in the plasma membrane *690*

**protein** macromolecule having, as its primary structure, a sequence of amino acids united through covalent bonding; many kinds, important in structure and metabolism *50*

**proteinoid** (pro'te-in-oid") abiotically polymerized amino acids that are joined in a preferred manner; possible early step in cell evolution *356*

**protist** organism belonging to the kingdom Protista; for example, protozoans, algae, or slime molds *380*

**protocell** a cell forerunner developed from cell-like microspheres *356*

**proton** positive subatomic particle, located in the nucleus of the atom *26*

**protostome** a group of coelomate animals in which the blastopore (the first embryonic opening) becomes the mouth *430*

**protozoan** animal-like, heterotrophic, unicellular organisms *381*

**proximal convoluted tubule** the portion of a nephron following the Bowman's capsule where selective reabsorption of filtrate occurs *625*

**pseudocoelom** (su"do-se'lom) a coelom that is not completely lined by mesoderm *423*

**pseudopodia** (sud-ah-pod-e-ah) cytoplasmic extensions of amoeboid protists; used for locomotion by these organisms *381*

**pulmonary circuit** pathway of bloodflow between the lungs and heart *560*

**Punnett square** a gridlike graph that enables one to calculate the results of simple genetic crosses by lining gametic genotypes of 2 parents on the outside margin and their recombination in boxes inside the grid *187*

**purine** (pu'rin) a type of nitrogenous base, such as adenine and guanine, having a double-ring structure *54*

**pyrimidine** (pi-rim'i-din) a type of nitrogenous base, such as cytosine, thymine, and uracil, having a single-ring structure *54*

**pyruvate** the end product of glycolysis; its further fate involving fermentation or Krebs cycle incorporation depending on oxygen availability *139*

# R

**radially symmetrical** body plan in which similar parts are arranged around a central axis like spokes of a wheel *419*

**radicle** part of the plant embryo that contains the root apical meristem and becomes the first root of the seedling *515*

**receptor** a cell, often in groups, that detects change or stimulus and initiates a nerve impulse *641, 653*

**recessive allele** (ah-lēl') form of a gene whose effect is hidden by the dominant allele in a heterozygote *186*

**recombinant DNA (rDNA)** DNA that contains genes from more than one source *279*

**reducing atmosphere** an atmosphere with little if any free oxygen, for example, the primitive atmosphere *355*

**reduction** a chemical reaction that results in addition of one or more electrons to an atom, ion, or compound. Reduction of one substance occurs simultaneously with oxidation of another *33*

**reflex** an automatic, involuntary response of an organism to a stimulus *641*

**regeneration** reforming a lost body part from the remaining body mass by active cell division and growth *695*

**regulator gene** in an operon, a gene that codes for a repressor *266*

**renewable resource** resource that can be replenished by physical or biological means *803*

**repetitive DNA** DNA with the same set of base pairs repeated many times, separating genes that direct the synthesis of cytoplasmic proteins *244*

**replication fork** V-shaped region of a eukaryotic chromosome wherever DNA is being replicated *243*

**repressible operon** an operon that is normally active because the repressor must combine with a corepressor before the complex can bind to the operator *268*

**repressor** protein molecule that binds to an operator site, preventing RNA polymerase from binding to the promotor site of an operon *266*

**reproduce** make a copy similar to oneself; for example, bacteria dividing to produce more bacteria, or egg and sperm joining to produce offspring in more advanced organisms *5*

**reptile** class of terrestrial vertebrates with internal fertilization, scaly skin, and an egg with a leathery shell; includes snakes, lizards, turtles, and crocodiles *453*

**resting potential** state of an axon when the membrane potential is about -65 mV and no impulse is being conducted *637*

**restriction enzyme** enzyme that stops viral reproduction by cutting viral DNA; used in genetic engineering to cut DNA at specific points *279*

**retina** innermost layer of the eyeball containing the light receptors—rods and cones *655*

**retrovirus** RNA virus containing the enzyme reverse transcriptase that carries out RNA/DNA transcription; for example, viruses that carry cancer-causing oncogenes and the AIDS virus *368*

**rhizoid** rootlike hairs that anchor a plant and absorb minerals and water from the soil *398*

**rhizome** a rootlike, underground stem *485*

**rhodopsin** (ro-dop'sin) a light-absorbing molecule in rods and cones that contains a pigment and its attached protein *658*

**ribonucleic acid (RNA)** a nucleic acid with a sequence of subunits dictated by DNA; involved in protein synthesis *54, 235*

**ribosomal RNA (rRNA)** type of RNA found in ribosomes; sometimes called structural RNA *251*

**ribosome** cytoplasmic organelle; site of protein synthesis *70*

**RNA** *see* ribonucleic acid

**RNA polymerase** (pol-im'er-ās) enzyme that links ribonucleotides together during transcription *254*

**rod** photoreceptor in vertebrate eyes that responds to dim light *655*

**root** structure of a plant root system; serving to anchor the plant, absorb water and minerals from the soil, and store products of photosynthesis *479*

**root pressure** a force generated by an osmotic gradient that serves to elevate sap through xylem a short distance *497*

**rough ER** complex organelle that is continuous with the nuclear envelope, consisting of membranous channels and studded with ribosomes *71*

***r*-strategist** a species that has evolved characteristics that maximize its rate of natural increase, for example, high birthrate *749*

**RuBP (ribulose biphosphate)** 5-carbon compound that combines with carbon dioxide during the Calvin cycle and is later regenerated by the same cycle *130*

# S

**SA node** (es a nod) mass of specialized tissue in right atrium wall of heart initiates a heartbeat; the "pacemaker" 559

**sac body plan** a body with a digestive cavity that has only one opening, as in cnidarians and flatworms 421

**saltatory conduction** (sal'tah-to"re) movement of a nervous impulse from one node of Ranvier to another along a myelinated axon 639

**saprophyte** organism that carries out external digestion and absorbs the resulting nutrients across the plasma membrane of its cells; indicative of fungi and slime molds 373, 380

**sarcolemma** (sar"ko-lem'ah) plasma membrane of a muscle fiber that forms the tubules of the T system involved in muscular contraction 671

**sarcomere** (sar"ko-mer) a unit of a myofibril, many of which are arranged linearly along its length, that shortens to give muscle contraction 671

**sarcoplasmic reticulum** (sar"ko-plaz'mik rĕ-tik'u-lum) the modified endoplasmic reticulum of a muscle fiber that stores calcium ions, whose release initiates muscle contraction 671

**savanna** terrestrial biome that is a tropical grassland in Africa, characterized by a few trees and a severe dry season 775

**scientific method** a step-by-step process for discovery and generation of knowledge, ranging from observation and hypothesis to theory and principle 14

**sclera** outer, white, fibrous layer of the eye that surrounds the transparent cornea 655

**secondary oocyte** (sek'on-dary o'o-sīt) the largest functional product of meiosis; becomes the egg 176

**seed** a mature ovule that contains an embryo with stored food enclosed in a protective coat 404, 511

**seed coat** covering around the embryonic sporophyte and stored foods in a mature seed; derived from the integument of the ovule 513

**segmented** having repeating body units as is seen in the earthworm 434

**selectively permeable** ability of plasma membranes to regulate the passage of substances into and out of the cell, allowing some to pass through the membrane and preventing the passage of others 89

**semicircular canal** one of 3 half-circle-shaped canals of the inner ear that are fluid filled for registering changes in motion 660

**semiconservative replication** duplication of DNA resulting in a double helix having one parental and one new strand 241

**seminal fluid** (sem'in-al) thick, whitish fluid consisting of sperm and secretions from several glands of the male reproductive tract 697

**sepal** protective leaflike structure enclosing the flower when in bud 509

**sessile filter feeder** an organism that stays in one place and filters its food from the water 417

**sex chromosome** the chromosome that determines the biological sex of an organism 204

**sex-influenced trait** a genetic trait that is expressed differently in the 2 sexes but is not controlled by alleles on the sex chromosomes, for example, pattern baldness 231

**sex-linked gene** unit of heredity, a gene, located on a sex chromosome 204

**sexual reproduction** reproduction involving meiosis and gamete formation; produces offspring with genes inherited from each parent 166, 170

**sexual selection** type of natural selection by which one organism specifically prefers another organism for mating 327

**sickle-cell disease** a genetic disorder in which sickling of red blood cells clogs blood vessels; individuals who are heterozygous for the allele have a greater resistance to malaria; more common among Blacks, but even then quite rare 226

**sister chromatid** one of 2 genetically identical chromosomal units that are the result of DNA replication and are attached to each other at the centromere 157

**slash-and-burn agriculture** practice of cutting and burning forest to use the land for agriculture 795

**smooth ER** complex organelle that is continuous with the nuclear envelope; consists of membranous channels and saccules but not studded with ribosomes 71

**society** a group of individuals of the same species that shows cooperative behavior 740

**sociobiology** the biological study of social behavior 740

**sodium-potassium pump** a transport protein in the plasma membrane that moves sodium out of and potassium into animal cells; important in nerve and muscle cells 94

**solar energy** energy from the sun; light energy 803

**speciation** (spe"se-a'shun) the process whereby a new species is produced or originates 334

**species** a taxonomic category that is the subdivision of a genus; its members can breed successfully with each other but not with organisms of another species 8

**sperm** male sex cell with 3 distinct parts at maturity: head, middle piece, and tail 698

**spermatogenesis** (sper"mah-to-jen'ĕ-sis) production of sperm in males by the process of meiosis and maturation 175

**sphincter** circular muscle in the wall of a tubular structure, such as an artery or the digestive tract, that can open and close the vessel, therefore regulating the amount of material moving through it 555

**spicule** (spik'ūl) skeletal structures of sponges composed of calcium carbonate or silicate 418

**spinal cord** part of the central nervous system, continuous with the base of the brain and housed within the vertebral column 640

**spindle** microtubule structure that brings about chromosome movement during cell division 160

**spongy layer** layer of loosely packed cells in a plant leaf that increases the amount of surface area for gas exchange; along with the palisade layer, it is the site of most of photosynthesis 491

**spore** usually a haploid reproductive structure that develops into a haploid generation; characteristic product of meiosis in plants 384

**sporophyte** diploid generation of the alternation of generations life cycle of a plant; it produces haploid spores, by meiosis, that develop into the haploid generation 398

**stabilizing selection** outcome of natural selection in which extreme phenotypes are eliminated and the average phenotype is conserved 325

**stamen** a pollen-producing flower structure consisting of an anther on a filament tip 407, 510

**starch** polysaccharide consisting of covalently bonded glucose molecules, characterized by few side branches; typical storage product in plants 45

**steady-state society** a society with no yearly increase in population or resource consumption 804

**stem** upright, vertical portion of a plant shoot system, transporting substances to and from the leaves 482

**steroid** biologically active lipid molecule having 4 interlocking rings; examples are cholesterol, progesterone, testosterone 48

**stigma** enlarged, sticky knob at one end of the pistil where pollen grains are received during pollination 407, 509

**stolon** modified, horizontal stem of a plant that is aboveground and produces new plants where its nodes touch the ground 485

**stoma** (stō-ma) small opening with 2 guard cells on the underside of leaf epidermis; their opening controls the rate of gas exchange 476

**stroma** (stro'mah) large, central space in a chloroplast that is fluid filled and contains enzymes used in photosynthesis 74, 123

**structural gene** gene that codes for proteins in metabolic pathways 267

**style** the tubular part of the pistil of a flower where a pollen tube develops from a transferred pollen grain 407, 509

**substrate** the reactant in an enzymatic reaction; each enzyme has a specific substrate 108

**substrate-level phosphorylation** (fos"fōr-i-la'shun) process in which ATP is formed by transferring a phosphate from a metabolic substrate to ADP 117

**succession** an orderly sequence of community replacement, one following the other, to an eventual climax community 769

**suppressor T cell** T lymphocyte that increases in number more slowly than other T cells and eventually brings about an end to the immune response 579

**survivorship** usually shown graphically to depict death rates or percentage of remaining survivors of a population over time 748

**symbiosis** (sim"bi-o'sis) a relationship that occurs when 2 different species live together in a unique way; may be beneficial, neutral, or detrimental to one and/or the other species 759

**symbiotic** (sim"bi-ot'ik) one species having a close relationship with another species; includes parasitism, mutualism, and commensalism 373

**sympathetic nervous system** a division of the autonomic nervous system that is involved in fight or flight responses; uses norepinephrine as a neurotransmitter 641

**sympatric speciation** (spe"se-a'shun) origin of new species in populations that overlap geographically 334

**synapse** a junction between neurons consisting of presynaptic (axon) membrane, the synaptic cleft, and the postsynaptic (usually dendrite) membrane 639

**synapsis** pairing of homologous chromosomes during meiosis I 173

**systematics** means to classify organisms based on their phylogeny or evolutionary history 340

**systemic circuit** pathway of bloodflow ranging from the heart to the tissues throughout the body and back to the heart 560

**systole** (sis'to-le) contraction period of a heart during the cardiac cycle 558

# T

**taiga** (tī'gah) terrestrial biome that is a coniferous forest extending in a broad belt across northern Eurasia and North America 775

**taste bud** an elongated cell that functions as a taste receptor 654

**taxis** movement of an organism toward or away from a stimulus 736

**taxonomy** a system to classify organisms meaningfully into groups based on similarities and differences and using morphology and evolution 340

**Tay-Sachs disease** (ta saks') a lethal, genetic lysosomal storage disease, best known among the U.S. Jewish population, which results from the lack of a particular enzyme; nervous system damage leads to early death 222

**tendon** structure consisting of fibrous connective tissue that connects skeletal muscles to bones 541, 669

**territoriality** behavior used to guarantee exclusive use of a given space for reproduction, feeding, etc. 736

**testcross** a genetic mating in which a possible heterozygote is crossed with an organism homozygous recessive for the characteristic(s) in question in order to determine its genotype 189

**testosterone** male sex hormone produced from interstitial cells in the testis 698

**tetrad** the set of 4 chromatids of a synapsed homologous chromosome pair, visible during prophase of meiosis I; also called bivalent 172

**thalamus** (thal'ah-mus) a lower forebrain region in vertebrates involved in crude sensory perception and screening messages intended for the higher forebrain or cerebrum 645

**theory** a conceptual scheme arrived at by the scientific method and supported by innumerable observations and experimentations 17

**thermal inversion** temperature inversion that traps cold air and its pollutants near the earth with the warm air above it 800

**thigmotropism** (thig-mot'ro-pizm) movement in response to touch; in plants, the coiling of tendrils 529

**thoracic cavity** (tho-ras'ik) internal body space of some animals that contains the lungs, protecting them from desiccation; the chest 610

**thrombocyte** cell fragment in the blood that initiates the process of blood clotting 563

**thylakoid** (thi'lah-koid) flattened disk of a granum; its membrane contains the photosynthetic pigments 74, 125

**thymine (T)** (thi'min) one of 4 organic bases in the nucleotides composing the structure of DNA 54, 238

**thyroid gland** large gland in the neck that produces several important hormones, including thyroxin and calcitonin 683

**thyroxin** a substance, T4, secreted from the thyroid gland, that promotes growth and development in vertebrates; in general, it increases the metabolic rate in cells 683

**tissue** group of similar cells combined to perform a common function 539

**tissue fluid** derivative of the blood plasma from capillaries that bathes the cells throughout the body 566

**T lymphocyte** white blood cell that directly attacks antigen-bearing cells 577

**tonicity** (tō-nis'ĭ-tē) referring to the solute concentration of a solution 91

**trachea** 1. an air tube (windpipe) of the respiratory tract in vertebrates; 2. an air tube in insects 438, 593, 612

**transcription** the process whereby the DNA code determines (is transcribed into) the sequence of codons in mRNA 252

**transfer RNA (tRNA)** active in protein synthesis, and transfers a particular amino acid to a ribosome; at one end it binds to the amino acid and at the other end it has an anticodon that binds to an mRNA codon 251

**transgenic organism** a multicellular organism that contains a transplanted gene 284

**transition reaction** a reaction involving the removal of $CO_2$ from pyruvate; connects glycolysis to the Krebs cycle 143

**translation** the process whereby the sequence of codons in mRNA determines (is translated into) the sequence of amino acids in a polypeptide 252

**transpiration** the plant's loss of water to the atmosphere, mainly through evaporation at leaf stomata 497

**transposon** (trans-po'zun) movable segment of DNA 245

**triplet code** genetic code (mRNA, tRNA) in which sets of 3 bases call for specific amino acids in the formation of polypeptides 253

**trophic level** (trof'ik) feeding level of one or more populations in a food web *782*

**tropical rain forest** forest found in warm climates with plentiful rainfall *777*

**tropism** (tro'pizm) in plants, a growth response toward or away from an external stimulus *529*

**TSH** thyroid stimulating hormone, secreted by the anterior pituitary; stimulates activity in the thyroid gland *683*

**tube-within-a-tube body plan** a body with a digestive tract that has both a mouth and an anus *421*

**tumor suppressor gene** unit of heredity (gene) that codes for proteins that ordinarily suppress cell division, thereby promoting organized cell growth *273*

**tundra** treeless terrestrial biome of cold climates; occurs on high mountains and in polar regions *773*

**turgor pressure** the pressure of the plasma membrane against the cell wall, determined by the water content of the plant cell vacuole; gives internal support to the cell *92*

**Turner syndrome** a condition that results from the inheritance of a single X chromosome; a second sex chromosome is absent: XO *218*

**tympanic membrane** eardrum; membranous region that receives air vibrations in an auditory organ *660*

# U

**uracil** one of 4 nucleotides composing the structure of RNA *54*

**urea** main nitrogenous waste of terrestrial amphibians and mammals *621*

**ureter** (u-re'ter) a tubular structure conducting urine from kidney to urinary bladder *624*

**urethra** (u-re'thrah) tubular structure that receives urine from the bladder and carries it to the outside of the body *624*

**uric acid** main nitrogenous waste of insects, reptiles, birds, and some dogs *621*

**urinary bladder** organ where urine is stored *624*

**urine** liquid waste product made by the nephrons of the kidney through the processes of pressure filtration and selective reabsorption *624*

**uterine cycle** a cycle that runs concurrently with the ovarian cycle; prepares the uterus to receive a developing zygote *703*

**uterus** pear-shaped portion of female reproductive tract that lies between the oviducts and the vagina; site of embryo development *700*

# V

**vaccine** a substance that wakes the immune response without causing illness *580*

**vacuole** organelle that is a membranous sac, particulary prominent in plant cells *74*

**vagina** muscular tube leading from the uterus; the female copulatory organ and the birth canal *700*

**vas deferentia** part of the male reproductive tract that transports sperm cells from the epididymis to the penis *697*

**vascular plant** any organism of the plant kingdom that contains the vascular tissues xylem and phloem as part of its structure *399*

**vector** in genetic engineering, a means to transfer foreign genetic material into a cell, for example, a plasmid *279*

**vein** a blood vessel that arises from venules and transports blood toward the heart *555*

**vena cava (pl., venae cavae)** (ve'nah ka'vah) one of 2 largest veins in the body; returns blood to the right atrium in a 4-chambered heart *560*

**vertebra (pl., vertebrae)** one of many bones of the vertebral column; held to other vertebrae by bony facets, muscles, and strong ligaments *666*

**vertebrate** referring to a chordate animal with a serial arrangement of vertebrae, or backbone *416*

**villus (pl., villi)** (vil'us) small, fingerlike projection of the inner small intestine wall *594*

**viroid** unusual infectious particle consisting of a short chain of naked RNA *368*

**visible light** a portion of the electromagnetic spectrum of light that is visible to the human eye *122*

**vitamin** an organic molecule that is required in small quantities for various biological processes and must be in an organism's diet because it cannot be synthesized by the organism; becomes part of enzyme structure *112, 600*

# W

**white matter** regions of the brain and spinal cord, consisting of myelinated nerve fibers *642*

**woody angiosperm** characterizes angiosperm trees with hardwood *407*

# X

**X-linked gene** gene located on the X chromosome that does not control a sexual feature of the organism *204*

**xylem** a vascular tissue that transports water and mineral solutes upward through the plant body *399, 477*

**XYY male** a genetic condition in which affected males have 2 Y chromosomes, are taller than average, and may exhibit learning difficulties *218*

# Y

**yolk sac** one of 4 extraembryonic membranes in vertebrates; important in the development of fishes, amphibians, reptiles, and birds, it is largely vestigial in mammals *719*

# Z

**zero population growth** no growth in population size *751*

**zygote** the diploid (2N) cell formed by the union of 2 gametes, the product of fertilization *510*

# CREDITS

## Photographs

### History of Biology

**Leeuwenhoek**: Bettmann Newsphotos; **Darwin**: © 78 John Moss/ Black Star; **Pasteur**: The Bettmann Archive; **Koch**: Bettmann Newsphotos; **Pavlov**: UPI/Bettmann; **Lorenz**: UPI/Bettmann; **McClintock**: AP/Wide World Photos, Inc.; **Jarvik**: UPI/Bettmann Newsphotos; **Gallo**: Rueters/Bettmann Newsphotos

### Part Openers

**Part 1**: © Carolina Biological Supply Co.; **Part 2**: © Gunter Ziesler/Peter Arnold, Inc.; **Part 3**: © Douglas Faulkner/Photo Researchers, Inc.; **Part 4**: © Astrid and Hahns-Frieder Michler/Science Photo Library/Photo Researchers, Inc.; **Part 5**: © Runk/Schoenberger/Grant Heilman; **Part 6**: © Carl R. Sams II/Peter Arnold, Inc.

### Chapter 1

**Opener**: © Gerard Lacz/Peter Arnold, Inc.; **1.1a**: © Rod Allin/Tom Stack & Associates; **1.1b,c**: © Kjell B. Sanved; **1.Aa**: © Eric Grave/Photo Researchers, Inc.; **1.Ab**: © John Cunningham/ Visuals Unlimited; **1.Ac**: © Rod Planck/Tom Stack & Associates; **1.Ad**: © Farrell Grehan/ Photo Researchers, Inc.; **1.Ae**: © Leonard Lee Rue Enterprises; **1.2**: © John Gerlach/Animals Animals; **1.3a**: © Stephen Kraseman/Photo Researchers, Inc.; **1.3b**: © Mitch Reardon/Photo Researchers, Inc.; **1.3c**: © David Fritts/Animals Animals; **1.4a**: © Biophoto Associates/Photo Researchers, Inc.; **1.4b**: © Philip Zito; **1.4c**: © Breck Kent/Animals Animals; **1.4d**: © Simion/ IKAN/Peter Arnold, Inc.; **1.6a**: © Ed Robinson/ Tom Stack & Associates; **1.6b**: © F. Stuart Westmorland/Tom Stack & Associates; **1.6c**: © Dave Fleetham/Tom Stack & Associates; **1.6d**: © C. Ressler/Animals Animals; **1.6e**: © Tom Stack/Tom Stack & Associates; **1.7a**: © Richard Thom/Tom Stack & Associates; **1.7b**: © Zig Lesczynski/Animals Animals; **1.7c**: © C.W. Schwartz/Animals Animals; **1.7d**: © Oxford Scientific Films/Animals Animals; **1.7e**: © Wendy Neefus/Animals Animals; **1.7f**: © Tom McHugh/Photo Researchers, Inc.

### Chapter 2

**Opener**: © Jim Olive/Peter Arnold, Inc.; **2.5**: © Kenneth E. Glander, Ph.D.

### Chapter 3

**Opener**: © Werner Muller/Peter Arnold, Inc.; **3.4**: © SIU/Journalism Services; **3.8c**: © Charles Kinger/Micro/Macro Stock Photography; **3.9**: © Douglas Faulkner/Sally Faulkner Collection; **3.12a**: © Clyde Smith/Peter Arnold, Inc.; **3.12b**: © Holt Confer/Grant Heilman; **3.12c**: © JoAnn Kalish/Peter Arnold, Inc.

### Chapter 4

**Opener**: © CNRI/Science Photo Library/Photo Researchers, Inc.; **4.1a**: © John Gerlach/Tom Stack & Associates; **4.1b**: © Leonard Lee Rue III/Photo Researchers, Inc.; **4.1c**: © H. Pol/CNRI/ Science Photo Library/Photo Researchers, Inc.; **4.7b**: © Stanley Flegler/Visuals Unlimited; **4.7c**: © Don Fawcett/Photo Researchers, Inc.; **4.8b**: © Biophoto Associates/Photo Researchers, Inc.; **4.11**: © S. Maslowski/Visuals Unlimited; **4.Ba**: © James Karales/Peter Arnold, Inc.; **4.Cb**: Courtesy Dr. Wolfgang Beerman.

### Chapter 5

**Opener**: © Dr. Jeremy Burgess/Science Photo Library/Photo Researchers, Inc.; **5.1a**: © Bob Coyle; **5.1b**: © Carolina Biological Supply; **5.1c**: © B. Miller/Biological Photo Service; **5.1d**: © Ed Reschke; **5.Aa**: © M. Abbey/Visuals Unlimited; **5.Ab**: © M. Schliwa/Visuals Unlimited; **5.Ac**: © Drs. Kessel/Shih/Peter Arnold, Inc.; **5.Ba–c**: © David Phillips/Visuals Unlimited; **5.3a**: © Henry Aldrich/Visuals Unlimited; **5.4b**: © Richard Rodewald/Biological Photo Service; **5.5b**: © W. P. Wergin, University of Wisconsin, courtesy E. A. Newcomb/Biological Photo Service; **5.6b**: © L. L. Sims/Visuals Unlimited; **5.7b**: © D. W. Fawcett/Photo Researchers, Inc.; **5.8a**: © W. Rosenberg/Biological Photo Service; **5.9a**: © David Phillips/Visuals Unlimited; **5.11**: © K. G. Murti/Visuals Unlimited; **5.12**: © S. E. Frederick, courtesy of E. H. Newcomb/Biological Photo Service; **5.13a**: © Dr. Herbert Israel/Cornell University; **5.14a**: © Dr. Keith Porter; **5.16b,c**: © Don Fawcett/Photo Researchers, Inc.; **5.16d**: © Don Fawcett/John Heuser/Photo Researchers, Inc.; **5.17b**: © Photo Researchers, Inc.; **5.18a**: © Biophoto Associates/Science Source/Photo Researchers, Inc.

### Chapter 6

**Opener**: © CNRI/Science Photo Library/Photo Researchers, Inc.; **6.2a**: © W. Rosenberg/ Biological Photo Service; **6.2d**: © Don Fawcett/ Photo Researchers, Inc.; **6.10a**: © Eric Grave/ Photo Researchers, Inc.; **6.10b,c**: © Thomas Eisner; **6.13**: © Francis Leroy; **6.14b**: © Mark S. Bretscher; **6.17a**: © Ray F. Evert.

### Chapter 7

**Opener**: © William Weber/Visuals Unlimited; **7.1**: © Joe DiStefano/Photo Researchers, Inc.; **7.5**: Courtesy Dr. Alfonso Tramontano; **7.8**: © Doolittle, R. F., University of California, from Scientific American, "Proteins," Oct. 1985, p. 95.

### Chapter 8

**Opener**: © John Kaprielian/Photo Researchers, Inc.; **8.1**: © Jeff Lepore/Photo Researchers, Inc.; **8.3b,c**: © Gordon Leedale/Biophoto Associates; **8.8a**: © Dr. Melvin Calvin; **8.11a**: © Ray F. Evert; **8.13**: © Pat Armstrong/Visuals Unlimited.

### Chapter 9

**Opener**: © Gunter Ziesler/Peter Arnold, Inc.; **9.1a**: © Dieter Blum/Peter Arnold, Inc.

### Chapter 10

**Opener**: © Carolina Biological Supply; **10.1a**: © Brian Parker/Tom Stack & Associates; **10.1b**: © John Cunningham/Visuals Unlimited; **10.2b**: © John J. Cardamone Jr./University of Pittsburg, Medical Microbiology/Biological Photo Service; **10.B**: © Dr. Stephen Wolfe; **10.3a**: BioPhoto Associates/Science Source/Photo Researchers, Inc.; **10.5b, 10.6b, 10.7b**: © Ed Reschke; **10.8a**: © Francis Leroy; **10.9b**: © Ed Reschke; **10.10**: R. G. Kessel and C. Y. Shih: *Scanning Electron Microscopy* © 1976, Springer Verlag, Berlin; **10.11a-d**: © Dr. Andrew Bajer; **10.12a**: © B. A. Palevity and E. H. Newcomb, University of Wisconsin/Biological Photo Service

**25.10b:** © John Cunningham/Visuals Unlimited; **25.10c:** © Field Museum of Natural History, Chicago; **25.11a:** © Karlene Schwartz; **25.11b:** © Milton Rand/Tom Stack & Associates; **25.12b:** © Cabisco/Visuals Unlimited; **25.12c:** © John Cunningham/Visuals Unlimited; **25.A:** Courtesy of the Center for Plant Conservation/Dr. Charles McDonald

## Chapter 26
**Opener:** © E. Robinson/Tom Stack & Associates; **26.1a:** © BioMedia Associates; **26.1b:** © Stephen Krasemann/Peter Arnold, Inc.; **26.6a:** © Bill Cartsinger/Photo Researchers, Inc.; **26.6b:** © Ron Taylor/Bruce Coleman, Inc.; **26.6c:** © Carolina Biological Supply Company; **26.8a:** © Dr. Fred Whittaker; **26.10b:** © Fred Marsik/Visuals Unlimited; **26.11:** © Jim Solliday/Biological Photo Service; **26.12:** © Markell, E.K. and Voge, M.: *Medical Paristology,* 4th ed., W.B. Saunders Co. 1981

## Chapter 27
**Opener:** © J. Alcock/Visuals Unlimited; **27.3b:** © William Jorgenson/Visuals Unlimited; **27.4a:** © William Ferguson; **27.4c:** © William Jorgenson/Visuals Unlimited; **27.6a:** © Michael DiSpezio/Images; **27.6c:** © Geral Corsi/Tom Stack & Associates; **27.7a:** © Michael DiSpezio/Images; **27.7c:** © Gary Milburn/Tom Stack & Associates; **27.9:** © Michael DiSpezio/Images; **27.10a:** © Robert Evans/Peter Arnold, Inc.; **27.10b:** © Tom McHugh/Photo Researchers, Inc.; **27.10c:** © Dwight Kuhn; **27.10d:** © John McGregor/Peter Arnold, Inc.; **27.10e, f:** © John Fowler/Valan Photos; **27.11a:** © James H. Carmichael/Peter Arnold, Inc.; **27.11c:** © Fred Bavendam/Peter Arnold, Inc.; **27.Ab:** © Edward S. Ross; **27.14a:** © Michael DiSpezio/Images; **27.15b:** © Oxford Scientific Films, Ltd.

## Chapter 28
**Opener:** © John Cancalosi/Peter Arnold, Inc.; **28.3:** © Field Museum of Natural History, Chicago; **28.4a:** © Tom Stack/Tom Stack & Associates; **28.4b:** © Douglas Faulkner/Sally Faulkner Collection; **28.5a:** © Paul L. Janosi/Valan Photos; **28.5b:** © Thomas Kitchin/Valan Photos; **28.6a:** © Tom McHugh/Steinhart Aquarium; **28.7a-d:** © Jane Burton/Bruce Coleman, Inc.; **28.8a:** © American Museum of Natural History; **28.8b:** © John Cunningham/Visuals Unlimited; **28.9a:** © Andrew Odum/Peter Arnold, Inc.; **28.9b:** © E.R. Degginger/Bruce Coleman, Inc.; **28.10a:** © Wolfgang Bayer Productions, Inc.; **28.11:** © American Museum of Natural History; **28.12:** © R. Austing/Photo Researchers, Inc.; **28.13:** © Jim David/Photo Researchers, Inc.; **28.14:** © Renee Stockdale/Animals Animals; **28.15:** © Stouffer Enterprises, Inc./Animals Animals; **28.18a:** © Martha Reeves/Photo Researchers, Inc.; **28.18b:** © Bios/Peter Arnold, Inc.; **28.18c:** © George Holton/Photo Researchers, Inc.; **28.18d:** © Tom McHugh/Photo Researchers, Inc.; **28.23:** © American Museum of Natural History

## Chapter 29
**Opener:** © M.I. Walker/Science Source/Photo Researchers, Inc.; **29.2a:** © Ed Reschke/Peter Arnold, Inc.; **29.2b:** © John Cunningham/Visuals Unlimited; **29.2c:** © Biophoto Associates/Photo Researchers, Inc.; **29.3a:** © Biological Photo Service; **29.3b, c:** © Biophoto Associates/Photo Researchers, Inc.; **29.4b:** © Biological Photo Service; **29.5b:** © George Wilder/Visuals Unlimited; **29.8b:** © Carolina Biological Supply Company; **29.9:** © Dwight Kuhn; **29.10:** © John Cunningham/Visuals Unlimited; **29.11a:** © G.R. Roberts; **29.11b:** © E.R. Degginger/Color-Pic; **29.11c:** © David Newman/Visuals Unlimited; **29.12:** © J.R. Waaland, University of Washington/Biological Photo Service; **29.13a:** © Carolina Biological Supply Company; **29.13b, c:** © Runk/Schoenberger/Grant Heilman; **29.14:** © Carolina Biological Supply Company; **29.16b:** © Carolina Biological Supply Company; **29.B:** © Earl Roberge/Photo Researchers, Inc.; **29.19b:** © J.H. Troughton and F.B. Sampson; **29.21a:** © Martha Cooper/Peter Arnold, Inc.; **29.21b:** © Larry Mellichamp/Visuals Unlimited; **29.21c:** © Carolina Biological Supply Company

## Chapter 30
**Opener:** © Cabisco/Visuals Unlimited; **30.6a:** © David Phillips/Visuals Unlimited; **30.8a-c:** © Dr. Mary E. Doohan; **30.10a:** © Gordon Leedale/BioPhoto Associates; **30.10b:** © Donald Marx/USDA Forest Service

## Chapter 31
**Opener:** © D. Newman/Visuals Unlimited; **31.3:** © J. Robert Waaland/Biological Photo Service; **31.4a:** © BioPhoto Associates/Photo Researchers, Inc.; **31.4b:** © Ed Reschke/Peter Arnold, Inc.; **31.4c:** © David Scharf/Peter Arnold, Inc.; **31.Aa:** © Nicholas Smythe/Photo Researchers, Inc.; **31.Ab:** © H. Eisenbeiss/Photo Researchers, Inc.; **31.Ac:** © Anthony Mercieca/Photo Researchers, Inc.; **31.Ad:** © Donna Howell/Animals Animals; **31.6a (1):** © Ralph A. Reinhold/Animals Animals; **31.6a (2):** © W. Ormerod/Visuals Unlimited; **31.6b (1):** © C.S. Lobban/Biological Photo Service; **31.6b (2):** © Biological Photo Service; **31.6c (both):** © Dwight Kuhn; **31.7:** © Ted Levin/Earth Scenes

## Chapter 32
**Opener:** © D. Newman/Visuals Unlimited; **32.8:** Courtesy R.J. Weaver; **32.4:** © Tom McHugh/Photo Researchers, Inc.; **32.5:** © Robert E. Lyons/Visuals Unlimited; **32.6a-d:** © Kiem Tran Thanh Van and her colleagues; **32.7:** © Runk/Schoenberger; **32.10:** © John Cunningham/Visuals Unlimited; **32.11a, b:** © Tom McHugh/Photo Researchers, Inc.; **32.14:** © Frank B. Salisbury

## Chapter 33
**Opener:** © Michael Gabridge/Visuals Unlimited; **33.2a-e:** © Ed Reschke; **33.2f:** © Ed Reschke/Peter Arnold, Inc.; **33.3a,b, 33.4b,c, 33.6a-c, 33.7:** © Ed Reschke

## Chapter 34
**Opener:** © Manfred Kage/Peter Arnold, Inc.; **34.1a:** © Eric Grave/Photo Researchers, Inc.; **34.1b:** © Carolina Biological Supply Company; **34.1c:** © Michael DiSpezio/Images; **34.A:** The Bettmann Archive; **34.Ba:** © Lewis Lainey; **34.12b:** © Science Photo Library/Photo Researchers, Inc.; **34.15a:** © Stuart I. Fox

## Chapter 35
**Opener:** © Manfred Kage/Peter Arnold, Inc.; **35.2:** © Keetin, Biological Science, Norton Publishing; **35.4:** © Boehringer Ingleheim International GmbH, Courtesy of Lennart Nilsson; **35.7:** © R. Feldman/Dan McCoy/Rainbow; **35.8:** © Boehringer Ingleheim International GmbH, courtesy Lennart Nilsson

## Chapter 36
**Opener:** © Matt Meadows/Peter Arnold, Inc.; **36.1a:** © David Dennis/Tom Stack & Associates; **36.1b:** © Harry Rogers/Photo Researchers, Inc.; **36.1c:** © Grant Heilman Photography; **36.14a,b:** Courtesy of the World Health Organizations; **36.14c,d:** Courtesy of the Atlanta Centers for Disease Control

## Chapter 37
**Opener:** © CNRI/Science Photo Library/Photo Researchers, Inc.; **37.10a:** © Lennart Nilsson: Behold Man, Little Brown and Company, Boston; **37.A, 37.B:** © Martin Rotker

## Chapter 38
**Opener:** © CNRI/Science Photo Library/Photo Researchers, Inc.; **38.2b:** © Robert Myers/Visuals Unlimited

## Chapter 39
**Opener:** © John Allison/Peter Arnold, Inc.; **39.13:** © Dan McCoy/Rainbow

## Chapter 40
**Opener:** © Runk/Schoenberger/Grant Heilman; **40.4:** © T. Norman Tait/BioPhoto Associates; **40.6d:** © Dr. Frank Werblin, University of California at Berkeley; **40.A (both):** Courtesy Professor J.E. Hawkins, Robert S. Preston, Kresage

## Chapter 41
**Opener:** © SIU/Visuals Unlimited; **41.1:** © Joe McDonald/Bruce Coleman, Inc.; **41.9b:** © H.E. Huxley; **41.11a:** © Victor Eichler/Bio Art

## Chapter 42
**Opener:** © Manfred Kage/Peter Arnold, Inc.; **42.6:** © Bettina Cirone/Photo Researchers, Inc.; **42.9:** © Lester Bergman and Associates

## Chapter 43
**Opener:** © John Mitchell/Photo Researchers, Inc.; **43.1b:** © John Shaw/Bruce Coleman, Inc.; **43.2a:** © Manfred Kage/Peter Arnold, Inc.; **43.2b:** © Matt Meadows/Peter Arnold, Inc.; **43.5:** © BioPhoto Associates/Photo Researchers, Inc.; **43.8b:** © Baganvandoss/Photo Researchers, Inc.

## Chapter 44
**Opener:** © Karl Switak/Photo Researchers, Inc.; **44.12a:** © Edelmann/First Days of Life/Black Star

## Chapter 45

**Opener:** © Gregory Scott/Photo Researchers, Inc.; **45.1a-d:** © Dr. Rae Silver; **45.5:** © C.C. Lockwood/Animals Animals; **45.6a:** © Steve Kaufman/Peter Arnold, Inc.; **45.6b:** Nina Leen, Life Magazine © 1964 Time, Inc.; **45.11:** © Susan Kuklin/Photo Researchers, Inc.

## Chapter 46

**Opener:** © Bruce Berg/Visuals Unlimited; **46.4a:** © Paul Janosi/Valan Photos

## Chapter 47

**Opener:** © Stephen Dalton/Photo Researchers, Inc.; **47.4a:** © Dr. Gregory Antipa and H.S. Wesenbergand; **47.5a:** © Alan Carey/Photo Researchers, Inc.; **47.6a:** © Runk/Schoenberger/Grant Heilman; **47.6b:** © National Audubon Society/Photo Researchers, Inc.; **47.6c:** © Z. Leszczynski/Animals Animals; **47.7:** © Hans Pfletschinger/Peter Arnold, Inc.; **47.8:** © Cliff B. Frith/Bruce Coleman, Inc.' **47.9a-c:** © Dr. Daniel Jantzen

## Chapter 48

**Opener:** © S.J. Krasemann/Peter Arnold, Inc.; **48.3:** © Douglas Faulkner/Sally Faulkner Collection; **48.7a:** © Stephen Krasemann/Peter Arnold, Inc.; **48.7b:** © Richard Ferguson/William Ferguson; **48.7c:** © Mary Thatcher/Photo Researchers, Inc.; **48.7d:** © Karlene Schwartz; **48.9:** © W.H. Hodge/Peter Arnold, Inc.; **48.10:** © Stephen Krasemann/Peter Arnold, Inc.; **48.12:** © R.S. Virdee/Grant Heilman; **48.13:** © Norman Owen Tomalin/Bruce Coleman, Inc.; **48.Aa:** © Stephen Dalton/Photo Researchers, Inc.; **48.Ab:** © Erwin and Peggy Bauer/Bruce Coleman, Inc.; **48.Ac:** © Bruce Coleman/Bruce Coleman, Inc.

## Chapter 49

**Opener:** © Wyman Meinzer/Peter Arnold, Inc.; **49.1:** © Gallbridge/Visuals Unlimited; **49.8:** © Bob Coyle; **49.11:** © Jacques Jangoux/Peter Arnold, Inc.

## Chapter 50

**Opener:** © Larry Brock/Tom Stack & Associates; **50.1:** © Link/Visuals Unlimited; **50.2a:** © Sydney Thomson/Animals Animals; **50.2b:** © Nichols/Magnum; **50.2c:** © G. Prance/Visuals Unlimited; **50.3:** © Gary Milburn/Tom Stack & Associates; **50.7a,b:** © Dr. John Skelly; **50.10:** NASA; **50.12:** © Thomas Kitchin/Tom Stack & Associates

## Line Art

### Chapter 2

**2.3:** Copyright © Mark Lefkowitz.

### Chapter 3

**3.1:** Copyright © Mark Lefkowitz.

### Chapter 4

**4.C:** Figure from *Cell Biology* by L. J. Kleinsmith and V. M. Kish. Copyright © 1988 by Harper & Row, Publishers, Inc. Reprinted by permission of HarperCollins Publishers.

## Chapter 7

**7.12:** From Stuart Ira Fox, *Human Physiology*, 3d ed. Copyright © 1990 Wm. C. Brown Communications, Inc., Dubuque, Iowa. All Rights Reserved. Reprinted by permission.

## Chapter 10

**10.A, 10.8b, 10.12b-d:** Copyright © Mark Lefkowitz.

## Chapter 13

**13.6:** Copyright © Mark Lefkowitz.
**13.14:** From Robert F. Weaver and Philip W. Hedrick, *Genetics*. Copyright © 1989 Wm. C. Brown Communications, Inc., Dubuque, Iowa. All Rights Reserved. Reprinted by permission.

## Chapter 14

**14.3:** From Robert F. Weaver and Philip W. Hedrick, *Genetics*. Copyright © 1989 Wm. C. Brown Communications, Inc., Dubuque, Iowa. All Rights Reserved. Reprinted by permission.
**14.9b:** From E. Peter Volpe, *Biology and Human Concerns*, 3d ed. Copyright © 1983 Wm. C. Brown Communications, Inc., Dubuque, Iowa. All Rights Reserved. Reprinted by permission.

## Chapter 15

**15.8:** From A. H. Wang, et al., *Nature,* Vol. 282:684. Copyright © 1979 Macmillan Magazines, Ltd., London, England. Reprinted by permission.

## Chapter 17

**17.4b,c:** Copyright © Mark Lefkowitz.

## Chapter 18

**18.10:** From J. C. Stephens, "Mapping the Human Genome: Current Status" in *Science,* 250:237, 12 October 1990. Copyright 1990 by the AAAS.

## Chapter 19

**19.6, 19.13:** Copyright © Mark Lefkowitz.

## Chapter 20

**20.5:** From D. Hartl, *A Primer of Population Genetics.* Copyright © 1981 Sinauer Associates, Inc., Sunderland, MA. Reprinted by permission. Data from P. Buri, "Gene Frequency in Small Populations of Mutant *Drosophila*" in *Evolution* 10:367-402, 1956.
**20.7a:** From E. Peter Volpe, *Biology and Human Concerns,* 3d ed. Copyright © 1983 Wm. C. Brown Communications, Inc., Dubuque, Iowa. All Rights Reserved. Reprinted by permission.

## Chapter 21

**21.17, 21.19:** Copyright © Mark Lefkowitz.

## Chapter 23

**23.5 (left):** From Helena Curtis and N. Sue Barnes, *Biology,* 5th ed. Copyright © 1989 Worth Publishers, New York, NY. Reprinted by permission.

## Chapter 24

**24.10b:** Copyright © 1974 Kendall/Hunt Publishing Company.

## Chapter 26

**26.2, 26.3, 26.5:** Copyright © Mark Lefkowitz.

## Chapter 27

**27.1, 27.5:** Copyright © Mark Lefkowitz.

## Chapter 28

**28.10:** Adapted from figure in *Introduction to Embryology*, Fifth Edition, by B. I. Balinsky and B. C. Fabian, copyright © 1981 by Saunders College Publishing, reprinted by permission of the publisher.
**28.19, 28.20, 28.21, 28.22:** Copyright © Mark Lefkowitz.

## Chapter 29

**29.7:** From Kingsley R. Stern, *Introductory Plant Biology,* 4th ed. Copyright © 1988 Wm. C. Brown Communications, Inc., Dubuque, Iowa. All Rights Reserved. Reprinted by permission.

## Chapter 30

**30.7:** From Kingsley R. Stern, *Introductory Plant Biology,* 4th ed. Copyright © 1988 Wm. C. Brown Communications, Inc., Dubuque, Iowa. All Rights Reserved. Reprinted by permission.

## Chapter 31

**31.9, 31.10:** From Kingsley R. Stern, *Introductory Plant Biology,* 4th ed. Copyright © 1988 Wm. C. Brown Communications, Inc., Dubuque, Iowa. All Rights Reserved. Reprinted by permission.
**A.1a:** Reproduced by permission of the National Research Council of Canada from the *Canadian Journal of Botany,* Vol. 39, pages 891-900, 1961.
**A.1b:** From H. W. Woolhouse, *Symposia Society for Experimental Biology,* 21:179. Copyright © 1967 Cambridge University Press, New York, NY. Reprinted by permission.

## Chapter 33

**33.4a:** From John W. Hole, Jr., *Human Anatomy and Physiology,* 5th ed. Copyright © 1990 Wm. C. Brown Communications, Inc., Dubuque, Iowa. All Rights Reserved. Reprinted by permission.
**33.9:** From Kent M. Van De Graaff and Stuart Ira Fox, *Concepts of Human Anatomy and Physiology,* 2d ed. Copyright © 1989 Wm. C. Brown Communications, Inc., Dubuque, Iowa. All Rights Reserved. Reprinted by permission.

## Chapter 34

**34.5, 34.6:** Copyright © Mark Lefkowitz.
**34.B (b):** From Kent M. Van De Graaff and Stuart Ira Fox, *Concepts of Human Anatomy and Physiology,* 2d ed. Copyright © 1989 Wm. C. Brown Communications, Inc., Dubuque, Iowa. All Rights Reserved. Reprinted by permission.

## Chapter 36

**36.6, page 605:** Copyright © Mark Lefkowitz.

## Chapter 37

**37.4, 37.7, 37.8:** Copyright © Mark Lefkowitz. **37.9:** From John W. Hole, Jr., *Human Anatomy and Physiology*, 5th ed. Copyright © 1990 Wm. C. Brown Communications, Inc., Dubuque, Iowa. All Rights Reserved. Reprinted by permission. **37.11:** From A. J. Vander, et al., *Human Physiology*. Copyright © McGraw-Hill, Inc., New York, NY. Reprinted by permission. **page 605:** Copyright © Mark Lefkowitz.

## Chapter 38

**38.7, 38.8, 38.9:** Copyright © Mark Lefkowitz.

## Chapter 39

**39.6, 39.7, 39.8, 39.10, 39.12:** Copyright © Mark Lefkowitz.

## Chapter 40

**40.1:** Copyright © Mark Lefkowitz.

## Chapter 41

**41.11b,c:** From John W. Hole, *Human Anatomy and Physiology*, 5th ed. Copyright © 1990 Wm. C. Brown Communications, Inc., Dubuque, Iowa. All Rights Reserved. Reprinted by permission.

## Chapter 42

**42.2:** Copyright © Mark Lefkowitz.

## Chapter 43

**43.3:** From John W. Hole, Jr., *Human Anatomy and Physiology*, 5th ed. Copyright © 1990 Wm. C. Brown Communications, Inc., Dubuque, Iowa. All Rights Reserved. Reprinted by permission. **43.4, 43.5c:** From Kent M. Van De Graaff and Stuart Ira Fox, *Concepts of Human Anatomy and Physiology*, 2d ed. Copyright © 1989 Wm. C. Brown Communications, Inc., Dubuque, Iowa. All Rights Reserved. Reprinted by permission. **43.7, 43.8a, 43.10:** From John W. Hole, *Human Anatomy and Physiology*, 5th ed. Copyright © 1990 Wm. C. Brown Communications, Inc., Dubuque, Iowa. All Rights Reserved. Reprinted by permission. **43.11:** From Kent M. Van De Graaff and Stuart Ira Fox, *Concepts of Human Anatomy and Physiology*, 2d ed. Copyright © 1989 Wm. C. Brown Communications, Inc., Dubuque, Iowa. All Rights Reserved. Reprinted by permission. **page 713:** From John W. Hole, Jr., *Human Anatomy and Physiology*, 5th ed. Copyright © 1990 Wm. C. Brown Communications, Inc., Dubuque, Iowa. All Rights Reserved. Reprinted by permission.

## Chapter 44

**44.13:** From John W. Hole, Jr., *Human Anatomy and Physiology*, 5th ed. Copyright © 1990 Wm. C. Brown Communications, Inc., Dubuque, Iowa. All Rights Reserved. Reprinted by permission. **44.14:** From *The Human Body: Growth and Development*. Copyright © Marshall Editions Limited, London, England. Reprinted by permission.

## Chapter 45

**45.5b:** From J. P. Hailman, *Behavior Supplement #15*. Copyright © E. J. Brill, Publisher, Leiden, Holland. Reprinted by permission.

## Chapter 46

**46.2:** From *Laboratory Studies in Biology: Observations and Their Implications* by Chester A. Lawson, et al. Copyright © 1955 W. H. Freeman and Company, New York, NY. Reprinted by permission. **46.4b:** From Victor B. Scheffer, "The Rise and Fall of a Reindeer Herd" in *Scientific Monthly*, Vol. 73:356-362. Copyright © 1951 by the AAAS.

## Chapter 47

**47.5b:** Figure from *Fundamentals of Ecology*, Third Edition, by Eugene P. Odum, copyright © 1971 by Saunders College Publishing, reprinted by permission of the publisher.

## Illustrators

### Part 1

**Kathleen Hagelston:**
5.16a.

**Illustrious, Inc.:**
text art, pages 20, 39, 47; 4.9, 4.B, 4.C.a, text art, page 50; 4.16a-d, text art, pages 58, 72; 5.18b-e, 5.19a-b, 6.11, 6.12, 7.9, 7.13, 7.14, text art, page 119; 8.9, text art, pages 130, 133, 136, 140, 142; 9.7, text art, page 144; 9.10, text art, page 151.

**Carlyn Iverson:**
2.7a, text art, page 82; 8.2, 8.3a, 8.7, 9.1b, 9.2.

**Laurie O'Keefe:**
5.10.

**Mark Lefkowitz:**
2.3, 3.1.

**Rolin Graphics:**
text art, page 15; 2.2, 2.4, 2.6, text art, pages 33, 34, 36; 3.2, 3.5, 3.6, 3.7, 3.8a-b, 3.A, 3.10, 3.11, 3.13, 3.14, 3.15, text art, page 43; 4.2, 4.3, 4.4, text art, page 48; 4.5, 4.6, text art, pages 54, 58; 4.10, 4.13, 4.14, 4.15, 4.17, 4.18, 4.20, text art, page 65; 5.2a-b, 5.3b, text art, page 74; 5.4b, 5.5a, 5.7a, 5.7c, 5.9b, 5.15, 5.17a, text art, page 85; 6.1a-b, text art, page 86; 6.2b-c, 6.3, text art, page 87; 6.4, text art, page 88; 6.5, 6.6, 6.8, 6.9, text art, page 97; 6.A.a-f, 6.17b, text art, pages 102, 103, 106; 7.3a-b, text art, page 109; 7.6a-b, 7.7, text art, pages 110, 113; 7.10, text art, page 114; 7.11, text art, pages 117, 122; 8.4, text art, pages 127, 128; 8.8b, text art, page 131; 8.10, 8.12, text art, pages 138, 140; 9.3, 9.4, text art, page 142; 9.5, 9.6, 9.8, text art, page 145; 9.9, text art, pages 146, 150; 9.11, 9.12.

**Rolin/Iverson:**
4.19.

### Part 2

**Molly Babich:**
text art, page 184

**Kathleen Hagelston:**
18.3

**Hagelston/Margorie Leggitt:**
14.2B

**Illustrious, Inc.:**
11.1, text art, page 171; 13.3, 13.5, 13.8, 13.11, text art, page 209; 14.12A, 15.5B, 15.6, 15.7, 15.8, 16.2, 16.3B&C, 16.4, text art, page 253; 16.7, 16.11A, 16.15, text art, pages 269, 274; 18.2, text art pages 280, 289, 18.10, 18.11, text art, page 294.

**Illustrious, Inc./Leonard Morgan:**
10.5A, 10.6A, 10.7A, 10.9A, 11.3, 11.7, text art, page 167.

**Carlyn Iverson:**
14.1, 16.10A, 18.B.

**Laurie O'Keefe:**
14.8A-C, 17.7, 17.8

**Mark Lefkowitz:**
10.8B, 10.A, 10.12B, 12.2, 13.6, 15.2A, 17.4B&C

**Precision Graphics:**
14.9B

**Rolin Graphics:**
text art, pages 155, 157; 10.2A, text art, page 160; 10.4, 11.2, text art, page 174; 11.4B-D, 11.6, 12.3, text art, page 187; 12.4, 12.5, text art, page 189; 12.6, 12.7, 12.8, 12.9, text art, page 192; 13.1, text art, page 201; 13.9, 13.12, 13.16, 14.3, 14.6, 14.7, 14.B.A-C, 14.14, text art, page 236; 15.3, text art, page 237; 15.4A, text art, page 240; 15.A, text art, pages 243, 249; 16.6, 16.8B, 16.9A-E, text art, page 266; 16.14, 17.1, text art, pages 258, 266; 17.A, 17.B, 17.2, 17.5, 18A, 18.8, 18.12B.

**Rolin/Iverson:**
16.1

### Part 3

**Molly Babich:**
22.6, 24.14, text art, page 401; 27.A, 28.17

**Illustrious, Inc.:**
text art, page 309; 20.2A&B, text art, page 318; 20.5, 20.8, 20.12, 21.3, 21.5, text art, page 340; Tbl. 21.3, text art, pages 364, 446, 470; fig. A.

**Carlyn Iverson:**
21.6A&B, 21.7, 21.16B

**Margorie Leggitt:**
28.16

**Mark Lefkowitz:**
19.6A&B, 19.13, 21.17, 21.19, 23.1C1, 24.2, 24.4, 26.2, 26.3, 26.5, 26.7, 27.1, 27.5, 27.12, 28.19, 28.20, 28.21, 28.22.

**Laurie O'Keefe:**
21.8, 21.9, 21.10, 21.12, text art, page 450

**Rolin Graphics:**
19.1A, 20.3A&B, 20.9A, text art, page 344; 21.11, 21.14, 21.15, 21.16A, 22.1, 22.5, 23.3, 23.5A, text art, page 377; 24.8, 24.9, 24.12, 24.16, 24.18A, 24.19A, text art, page 395; 25.4A, 25.5, 25.7, text art, page 405; 25.12A, text art, page 414; 27.2, 27.4B, 27.6B.

**Rolin/Iverson:**
25.3

**Part 4**

**Kathleen Hagelston:**
29.16A

**Illustrious, Inc.:**
29.A, text arts, page 493; Fig. A

**Mark Lefkowitz:**
31.2, 31.8A-F

**Rolin Graphics:**
29.1, 29.3A2, 29.3B2, 29.3C2, 29.6, 29.14B, 29.15, 29.17A-B, 29.19A&C, 29.20, 30.1, 30.2, 30.3A-B, 30.5B, text art, page 498; 30.6B&C, text art, page 500; 30.9, 30.11, 30.12, text art, pages 504-505; 31.1, text art, pages 509, 521; 32.1, 32.2A-C, 32.3, 32.8, text art, page 529; 32.12A&B, 32.13.

**Rolin/Iverson:**
31.5

**Part 5**

**Molly Babich:**
34.2, 38.6B

**Chris Creek:**
35.3, 38.1A

**Anne Greene:**
405, text art, page 663; 44.10A-E

**Kathleen Hagelston:**
43.12, 43.13

**Hagelston/Leggitt:**
36.13

**Hagelston/O'Keefe:**
35.5

**Illustrious, Inc.:**
text art, page 539; 33.1, 33.2A, text art, page 545, 558; 34.11B, 35.A, text art, page 596; 44.2, text arts, page 717; Figs. A&B

**Carlyn Iverson:**
33.8, 33.12, 34.2, 34.3, 34.11A, 35.6, 36.7, 38.11, 39.14A-E, 40.10, 40.11A-D, 41.9A, 43.7B, 44.11, 44.12B

**Laurie O'Keefe:**
33.5, 35.1

**Mark Lefkowitz:**
34.5, 34.6, 36.2, 36.6, text art, page 604; 37.4, 37.5, 37.7, 37.8, text art, page 618; 38.5, 38.6A, 38.7, 38.8, 38.9, 39.1, 39.6, 39.7, 39.8, 39.10, 39.12, 40.1, 42.2

**Ron McClean:**
43.8A

**Steve Moon:**
33.9, 43.11

**Diane Nelson:**
34.B.B, 44.13

**Rolin Graphics:**
33.6A.2, 33.6B.2, 33.6C.2, 33.11, 34.7, 34.8, 34.12A, 34.13, 34.14, 35.9A-B, 35.10, text art, pages 591, 593, 594, 595; 36.12, text art, page 605; 37.10B&C, text art, pages 614, 616; 38.1B, 38.2A, 38.3, 38.10, 38.12, text art, page 631; 39.3, 39.4, 39.5, 40.2A-D, 40.6C, 40.7A-D, 40.8, 41.2, 41.3, 41.4, 41.5, 41.6, text art, pages 671, 672, 675; 42.3A&B; 42.4, 42.5, 42.7A-C, 42.8A-C, 42.12, 42.13, text art, page 693; 43.1A, 43.6, 43.9, 44.5, 44.9

**Mike Schenk:**
42.11, 43.5C

**Tom Waldrop:**
text art, page 550; 39.2A, 39.9, 41.11B&C, 42.A, 43.3, 43.4A&B, 43.7A, 43.10, text art, page 710

**Part 6**

**Illustrious, Inc.:**
45.A, 45.12, 46.3A, 46.8, 50.6

**Carlyn Iverson:**
49.5

**Mark Lefkowitz:**
49.10, 49.12

**Rolin Graphics:**
45.2A-C, 45.3, 45.4, 45.5B, 45.7A&B, 45.8, 45.9, 46.1, 46.2, 46.3B, 46.4B, 46.5, 46.6A-C, text art, page 753; 47.1, 47.2, 47.3, 47.4B, 47.5B, 47.A.A-D, 48.1, 48.2, 48.4, 48.5, 48.6B, 48.8, 48.11A-C, 49.1A, 49.2A&B, 49.3, 49.4, 50.8, 50.9B, 50.11

## How Do You Do That? Boxes

p. 292, Holly Ahern; p. 345, Steve Miller; p. 487, Dr. Kingsley Stern; p. 581, Dr. Andy Anderson; p. 737, Lee Drickamer

# INDEX

Amino, 43
Amino acids, 43, 50–51, 112, 148, 237, 261, 598, 621
Amino groups, 620
Aminoguanidine, 724
Amish, 323, 324
Ammonia, 373, 620–21, 627, 629
Ammonium, 788
Amniocentesis, 218, 226, 227
Amnion, 455, 719, 720
*Amoeba proteus,* 7, 62, 381
Amoeboid cells, 418
Amoeboids, 381
Amphibians, 5, 309, 417, 443, 449, 450, 452–53, 557, 609, 610, 727
Amphioxus, 444
Amplification, PCR, 285
Amplitude, 662
Ampulla, 441
Amylase, 591, 594, 614
Amyloplasts, 74
*Anabaena,* 255, 371
*Anabaena cylindrica,* 371
Anabolic steroids, 688–89
Anaerobes, 372
Anaerobic proces, 142
Anagenesis, 340, 348
Analogous structures, 340
Analogues, 292
Analyses
    PCR, 285
    RFLP, 290–92
Anaphase, 161, 162, 163, 165, 172, 173, 174–75
Anatomy
    of cell wall, 67
    of chambered nautilus, 95
    and comparative anatomy, 310–12
    of muscles, 669–72
    of nucleus, 70
    of plant cell, 68
    of prokaryotes, 66
Andes Mountains, 299
Androgens, 680
Anemia, 284
Anemones, 417, 420, 421, 553
ANF. *See* Atrial natriuretic factor (ANF)
Angelfish, 451
Angiosperms, 401, 404, 407–8, 410–11, 513
Angiotensin, 629, 686
Angiotensinogen, 628, 629
Anhydrase, 614
Animal behavior, 730–43
Animal cells
    anatomy of, 67
    cytokinesis in, 163, 164
    extracellular matrix of, 99
    junctions between cells, 99–101
    mitosis in, 160–64
    osmosis in, 93
Animalia, 4, 5
Animals, 3, 5
    behavior of, 730–43
    characteristics of, 416
    and chromosome number, 157
    development in, 711–26
    feeding modes, 588–89
    and five kingdoms system, 360–62
    life cycle of, 170
    mollusks to chordates, 429–46

organization and homeostasis, 538–51
organs and organ systems, 544–46
phylogenetic tree, 417
reproduction in, 694–710
sponges to roundworms, 415–28
structure and form, 537–728
structure and function, 726–28
tissue types, 539–44
tracking of, 737
and transgenic animals, 287–88
Animal starch, 45, 597
Animal viruses, 368
Annelids, 417, 431, 434–37, 437, 443, 665
Annulus, 402
*Anopheles,* 326, 383, 384
Answers to questions, A-7_-A-18
Ant, 439
Antagonistic pairs, 669
Antarctica, 316
Antennae, 440
Anterior pituitary, 680, 681, 682
Antheridia, 398, 399, 403
Anthers, 407, 409, 509, 510
Anthophyta, 4, 410–11
Anthropoids, 459–61
Antibiotics, 370, 706, 707
Antibiotic therapy, 706, 707
Antibodies, 563, 578–79, 581, 582–84
Antibody-mediated immunity, 578–79
Antibody titer, 580
Anticodon, 257
Antidiuretic hormone (ADH), 628, 629, 680, 681
Antigen-presenting cell (APC), 580
Antigens, 220, 563, 577, 580, 581
Antipredator defenses, 758, 759
Antisense RNA, 268–69
Antiviral drugs, 370
Ants, 740, 762
Anus, 591, 592
Anvil, 660, 661
Aorta, 560
Aortic bodies, 612
APC. *See* Antigen-presenting cell (APC)
Apes, 461, 462, 463, 735
Aphids, 502
Appendages, 437, 666
Appendicitis, 597
Appendicular skeleton, 666, 668
Apples, 209
Aquatic communities, 766–69
Aquatic environments, and respiratory systems, 607–8
Aquatic food chain, 782
Aqueous humor, 655
Arachnids, 438
ARC. *See* AIDS related complex (ARC)
Archaebacteria, 374
*Archaeopteryx,* 309, 454, 456
Archaic humans, 465–67
Archegonia, 398, 399, 403
Arginine, 253
Argon, 345
Aristotle, 300
Arm, 667, 668
Armadillo, 303
Arnold Arboretum, 412
Arteries, 555, 567
Arterioles, 555
Arthropods, 417, 431, 437–40, 443, 654, 665

Artificial selection, 306
Artiodactyla, 459
*Ascaris,* 170, 423–24, 425
Ascomycetes, 392
Ascomycota, 391
Asexual reproduction, 166, 384, 386, 387, 515–19, 695
Asparagine, 253
Aspartame, 286
Aspartate, 112
Ass, 341
Assortative mating, 321
Assortment, independent, 190, 192, 203
Aster, 160
Asteroidea, 441
Asymptomatic carrier, 707
Atmosphere, primitive, 355
Atomic force microscope (AFM), 64
Atomic number, 26, 27
Atomic symbol, 26
Atomic weight, 26–28
Atoms, 25, 26–29
ATP. *See* Adenosine triphosphate (ATP)
Atria, 555
Atrial natriuretic factor (ANF), 284, 629
Atrioventricular (AV) node, 559
Atrioventricular (AV) valves, 557
Auditory canal, 659
Auditory communication, 739–40
*Aurelia,* 421
Australia, 310, 353, 358
*Australopithecus afarensis,* 464, 465
*Australopithecus africanus,* 465
*Australopithecus boisei,* 465
*Australopithecus robustus,* 465
Autoimmune diseases, 584–85. *See also* Acquired immune deficiency syndrome (AIDS)
Autonomic nervous system, 641–42, 643
Autopolyploids, 472
Autosomal chromosome abnormalities, 216–18, 222–24
Autosomal recessive inheritance, 221–23
Autosomes, 204, 222
Autotrophic bacteria, 372–73
Autotrophism, 358
Autotrophs, 356
Auxins, 523–26
Availability, and bioavailability, 603
Axial skeleton, 666, 668
Axillary buds, 483
Axons, 543, 544, 635–40
Azidothymidine (AZT), 370, 708

## B

Baboons, 738
Bacillus, 238, 371, 372
*Bacillus subtilis,* 238
Bacteria, 5, 42, 66, 245, 365, 782
    and autotrophic bacteria, 372–73
    and bacteriophages, 236, 237
    and bioremediation, 286
    classification of, 374–76
    denitrifying bacteria, 789
    and heterotrophic bacteria, 373
    nitrifying bacteria, 788
    nitrogen-fixing bacteria, 788
    and photosynthesis, 358
    and selection, 326

shapes of, 371
and STD, 706–7
transformation of, 235–37, 282, 283
and transgenic bacteria, 284–86
Bacteriophages, 236, 237, 367–68
Balance, 659, 660
*Balanus balanoides,* 755–56
Baldness, 700
Ball-and-socket joints, 668
Baltimore oriole, 334, 335
Bananas, 209
Banting, Frederick, 687
Barash, David P., 18–19, 731
Barnacles, 755–56
Barn owl, 10
Barr, M. L., 220
Barracuda, 9
Barr body, 220, 269
Basal body, 67
Basal nuclei, 646
Base pairing, 55, 56, 288, 292
Base-pair map, 288, 292
Bases, 36, 37–38, 54, 238, 240
Basidia, 392, 393
Basidiomycetes, 392–93
Basidiospores, 393
Basilar membrane, 660
Basophils, 565
Bass, 645
Bates, Henry Walter, 308, 758
Batesian mimicry, 758
Bateson, William, 198
Bats, and pollination, 519
B cells, 563, 577–79, 583
B-DNA, 243–44
Beach, 767
Beadle, George, 249, 250
Beagle, 306
*Beagle,* 299, 302, 307
Bean, 517
Beaumont, William, 593, 594
Bees, and pollination, 518
Beets, 481
Behavior, 5
    animal behavior, 730–43
    and communication, 738–40
    and ecology, 729–808, 807–8
    genetic bases and learned behavior, 732–35
    innate types of, 732
    mechanisms of, 731–32
    modification of, 734
    and natural selection, 735–38
    physiological control of, 731
    and reproduction, 736–38
Behavioral inheritance, 224–25
Behavioral isolation, 333
Bell-shaped curve, 748
Beltian bodies, 762
Beneden, Pierre-Joseph van, 170
Benthic division, 768–69
Bernard, Claude, 18, 546
Bernstein, Julius, 637
Best, Charles, 687
Bicarbonate ion, 614
Bilaterally symmetrical, 419, 421, 426, 435, 443, 448
Bilateral symmetry, 422
Bile, 594

Binary fission, 154–68, 371
Binnig, Gerd, 63
Binomial names, 4
Bioavailability, 603
Biochemistry, and comparative biochemistry, 311–12
Biogenesis theory, 18
Biogeochemical cycle, 787
Biogeography, 303–5, 309–10
Biological clock, 530, 732
Biological diversity, loss of, 795
Biological evolution, 357–60
Biological magnification, 784
Biology
    and cell biology, 62–64
    and measurements, 65–66
    and molecular biology, 250–52
    sociobiology, 740–41
    unifying theories of, 18
Biomass, pyramid of, 782–84
Biomes, 773–75, 769–78
Bioremediation, 286
Biosphere, 8, 765–79, 781
Biosynthesis, and metabolic pool, 148–49
Biotechnology, and rDNA, 278–95
Biotic components, 781
Biotic environment, 745
Biotic potential, 745–46
Biotin, 602
Bipolar cells, 656
Birds, 2, 5, 299, 309, 325, 327, 332, 334, 335, 343, 443, 449, 454–55, 456, 519, 694, 733, 734, 736, 756, 759
Birds of prey, 454
Birth, 722
Birth control, 705, 706
Birth defects, detecting, 226
Birthrate, 745
Birth weight, 325
Bisphosphoglycerate, 141, 143
*Biston betularia,* 326
Bivalent, 171, 172, 174, 186, 208
Bivalves, 431, 433
Bivalve shells, 431
Black bread molds, 391
Blackman, F. F., 125, 126
Blade, 489
Blastocoel, 713
Blastocyst, 718, 719
Blastopore, 430, 431
Blastula, 712, 713
Blending theory of inheritance, 183
Blind spot, 655
Blood, 31, 36, 541–42, 543, 555
    of humans, 562–69
    pH, 629
    and pregnancy, 704–5
    and Rh blood factor, 226
    and systemic blood flow, 561–62
    and trypanosome infection, 383
    vessels, 560–61
Blood cells, 541, 563–65
Blood clotting, 563–65
Blood groups, 569
Bloodhound, 306
Blood pressure, 284, 561, 562, 566
Blood Rh factor, 468
Blood Rh system, 567, 568–69
Blood types, 225–26, 567–69

Bluebirds, 18
B lymphocytes, 574, 577
Bobcat, 10, 416
Bobolink, 736
Body, temperature control, 548–49
Body cavities, 426, 546, 548
Body cells, 220
Body plans, 421, 422, 423, 426, 434, 435, 443, 448
Bog moss, 399
Boiling, 33, 35
Bonds
    chemical, 29
    covalent, 31–33, 53
    hydrogen, 33, 35, 47, 53
    ionic, 30–31, 53
    peptide, 50
    polar covalent, 32–33
Bone, and bones, 541, 542, 666, 668
Bone marrow, 564, 574
Bonner, J., 531
Bony fishes, 448–52
Book lungs, 438
Booster, 582
Boston terrier, 306
Bottleneck effect, 324
Botulism, 373
Boveri, Theodor, 202
Bovine growth hormone, (bGH), 282–83, 287
Bowman's capsule, 625, 626
Brachydactyly, 223
Bracken fern, 397, 410–11
Bracket fungi, 392, 393
Brain
    conscious brain, 646
    and drugs, 647
    human, 644–49
    and neurotransmitters, 647–49
    sizes among vertebrates, 645
    unconscious brain, 644–46
Brain hormone, 677–78
*Branchiostoma,* 444
*Brassica oleracea,* 307
Bread molds, 391
Breasts, 705
Breathing, 610, 612, 726–28
Breeders, 741
Bright-field light microscope, 62
Bristlecone pines, 407
Brittle stars, 442
Broad-bean plant, 499
Broad billed birds, 454
Bromine, 501
Bronchi, 611
Bronchioles, 611
Brown algae, 388, 389
Bryophytes, 398–99, 410–11
Bubble baby, 288
Budding, 368, 695
Buds, 482–84, 591, 654, 720
Buffers, 36
Buffon, Comte de, 300
Bulbourethral glands, 697
Bulbs, 485, 491
Bulk, 45, 46
Bullhorn acacia, 762
Bullock's oriole, 335
Bundle sheath, 491
Bundle sheath cells, 131

Burgess Shale, 430
Bursae, 668
Buttercup, 202
Butterflies, 105, 429, 439, 518–19

# C

Cabbage, 307, 471
Cacti, 349, 412, 490
Cairns-Smith, Graham, 356, 357
Calcareous skeleton, 421
Calcitonin, 680, 685
Calcium, 26, 30, 112, 500, 563, 601–3, 627
Calcium carbonate, 359
California condor, 412
California cypress, 323
Callus, 266
Calorie, 34
Calvin, Melvin, 129, 130
Calvin cycle, 129–31, 132, 807
*Camarhynchus pallidus,* 304
*Camarhynchus psittucula,* 304
Cambium, 479, 484
Cambrian, 360, 401, 416, 430, 450
Camera-type eyes, 433, 434, 655
cAMP, 679
CAM plants, 133–34
Cancer, 38, 284
    causes of, 273–74
    and diet, 600
    as genetic control failure, 271–75
    and PCR, 285
    prevention of, 274–75
    and smoking, 615
*Candida albicans,* 393
Canines, 588
Cann, Rebecca, 467
Cannabis delirium, 648
Cannabis psychosis, 648
*Cannabis sativa,* 409
Cannon, Walter, 546
Capillaries, 91, 555, 566–67, 568
Capillary exchange, 566–67, 568
Capsid, 367, 368
Capsule, 66, 399
Capuchin monkey, 312, 342
Carajas Mountains, 795
Carapace, 439
Carbaminohemoglobin, 614
Carbohydrates, 43, 44–46, 57, 122, 125, 148, 598–99
Carbon, 26, 27, 28, 29, 31, 42–43, 345
Carbon cycle, 787–88
Carbon dioxide, 122, 125, 126, 140
    and Calvin cycle, 129–31
    fixation, 130, 131, 132, 133, 134
    transport of, 614–16
Carbonic acid, 614
Carbonic amylase, 614
Carbonic anhydrase, 614
Carboniferous, 401, 403, 405, 442, 449, 450, 452
Carbon monoxide, 799
Carboxyl, 43
Carcinogens, 273
Cardiac cycle, 558, 559
Cardiac muscles, 543
Cardiovascular disease, 560, 566–67
Cardiovascular system, 555
Carnivores, 107, 459, 588, 781

Carotid bodies, 612
Carpals, 667
Carpel, 509
Carrier protein, 90
Carriers, 88, 92–99, 219, 229, 707
Carrots, 287, 481
Carrying capacity, 746
Cartilage, 448, 451, 541, 665
Cartilaginous disk, 665
Cartilaginous fishes, 448, 451
Casparian strip, 480, 481, 495
Castle, W. E., 177
Catastrophism, 301
Cats, 220, 310, 416, 645
Cat's cry syndrome, 210, 218
Cavendish Laboratories, 240
CCK-PZ. *See* Cholecystokinin-pancreozymin
        (CCK-PZ)
Cech, Thomas, 356
Cecum, 438
Cell biology, 62–64
Cell body, 543, 544, 636
Cell cycle, 159–60
Cell division, 156–59, 166
Cell doctrine, 61
Cell fractionation, 64, 126
Cell-mediated immunity, 579–80
Cell plate, 164
Cell recognition protein, 90
Cell reproduction
    binary fission and mitosis, 154–68
    and meiosis, 169–81
Cells, 3, 23–152
    and aging, 724
    and amoeboid cells, 418
    animal, 67, 160–64
    blood. *See* Blood cells
    bundle sheath, 131
    and cell biology, 62–64
    and cellular energy, 104–20
    and cellular metabolism, 107–15
    and collar cells, 417
    and cytoskeleton, 76, 77
    daughter, 160, 179
    embryonic cells, 226, 227
    and epidermal cells, 417, 474
    eukaryotic, 64, 65, 66–76, 81, 360
    extracellular matrix of, 99
    fetal cells, 226, 227
    flame cells, 623
    and flame-cell system, 422
    and fungal cells, 375, 376
    germ, 177
    guard cells, 474
    internal environment of, 620
    junctions between, 99–101
    liver, 41
    and lysogenic cells, 368
    mesophyll, 131
    and movement, 76, 77, 89–99
    and multicellularity, 360
    olfactory cells, 653–54
    organic molecules in, 57
    origin of first cell, 355
    osmosis in, 91, 92, 93
    photosynthetic cells, 375
    plant, 164, 165
    plasma cells, 577
    prokaryotic, 64, 66, 81

    and prokaryotic cells, 357–58
    shape, 76, 77
    sizes of, 65–66
    structure and function, 60–83
    surface modifications, 99–101
    variety of, 61–66, 67
Cell theory, 18, 61
Cellular energy, 104–20
Cellular metabolism, 107–15
Cellular respiration, 114, 137–52
Cellulase, 527
Cellulose, 45–46, 47, 57, 488
Cell wall, 64, 66, 67, 68, 69, 81, 99
Cenozoic, 344, 401, 450, 460, 461
Center for Plant Conservation, 412
Centigrade scale, A-6
Centimeter, 65
Centipedes, 417, 438–40
Central nervous system (CNS), 634, 635, 640,
        642–49, 719
Central vacuole, 68
Centrifugation, 64
Centrioles, 67, 69, 78, 81
Centromere, 157, 159, 160
*Cepaea nemoralis,* 325
Cephalization, 421, 422, 434
Cephalochordata, 444
Cephalopods, 433, 652
Cephalothorax, 440
Cerebellum, 645
Cerebrum, 645
Ceredase, 284
Cervix, 701
*Cervus elaphus,* 341
Cestoda, 422
Cetacea, 459
CFTR. *See* Cystic fibrosis transmembrane
        conductance regulator (CFTR)
Chakrabarty, 286
Chambered nautilus, 95
Chance, and probability, 188, 190
Channel protein, 90
Channels, 88
Chaparral, 775
Characteristics, 184, 301
Chargaff, Erwin, 237–38
Chargaff's rules, 237–38, 240
Chase, Martha, 237
Cheetah, 137
Cheilosis, 601
Chemical bonds, 29
Chemical communication, 738
Chemical cycle, 787–89
Chemical elements, table of, A-3
Chemical evolution, 355–57
Chemical reactions, 33
Chemiosmotic ATP synthesis, 128–29, 146
Chemiosmotic phosphorylation, 117
Chemiosmotic theory, 117, 146
Chemistry
    basics, 24–40
    and comparative biochemistry, 311–12
    of life, 41–59
    and warm chemistry, 115
Chemoautotrophs, 373
Chemoreceptors, 612, 653–54
Chemotaxis, 736
Chernobl, 803
Chiasmata, 173, 175

Chick, 311, 712
Chiggers, 438
Chihuahua, 306
Childhood, 722
Chilopoda, 438, 439
Chimpanzee, 312, 342, 461, 463, 645, 735, 740
Chinese cabbage, 307
Chipmunk, 10
*Chironomus,* 270, 295–96
*Chironomus pallidivittatus,* 295–96
*Chironomus tentans,* 295–96
Chiroptera, 459
Chitin, 45–46, 47, 437, 665
Chlamydia, 706, 707
*Chlamydomonas,* 384, 385, 386
*Chlorella,* 130
Chlorides, 627
Chlorine, 26, 30
Chlorofluorocarbons (CFCs), 801
Chlorophyll, 124, 125, 126, 371, 384
Chlorophyta, 384
Chloroplasts, 68, 69, 74–75, 81, 123–25, 476, 489
CHNOPS, 26, 42
Cholecystokinin-pancreozymin (CCK-PZ), 596, 680
Cholesterol, 50, 57, 87, 223, 566–67, 599–600
Cholesterol blood level, 566–67
Chondrichthyes, 448
Chordae tendineae, 557
Chordates, 4, 417, 429, 431, 443–44, 448, 665
Chorion, 455, 718
Chorionic villi sampling, 226, 227
Choroid, 655, 657
Chromatids, 157, 159, 160, 208
Chromatin, 67, 68, 69, 70, 157, 269–70
Chromosome mapping, 206–8
Chromosome mutations, 208–11, 316–17
Chromosome number
    changes in, 208–9
    and organism types, 157
Chromosome puffs, 269, 295
Chromosomes, 66, 68, 69, 157
    and autosomal chromosome abnormalities, 216–18, 222–24
    and chromosome mapping, 206–8
    and chromosome number, 208–9
    and chromosome puffs, 269, 295
    comparison of meiosis and mitosis, 178, 179
    contents of, 158, 159, 160
    duplicated, 159, 160
    eukaryotic, 157–59
    and genes, 196–214
    and genetic mapping, 288–92
    and human chromosomes, 216–19
    independent assortment of, 176
    and life cycles of algae, 385
    and meiosis, 171–76
    and mutations, 208–11
    and pangenesis, 177
    and polytene chromosomes, 269
    prokaryotic, 156–57
    and sex chromosomes, 204, 218–19, 220, 222, 228–31
    structure changes, 209–11
Chromosome sex determination, 204
Chromosome theory of inheritance, 202
*Chrysanthemum,* 471
Chrysophyta, 388–89
*Chthamalus stellatus,* 755–56

Chyme, 593
Cicada, 665
Cigarettes, 615
Cilia, 69, 78, 79, 81
Ciliary body, 657
Ciliates, 381–82
Ciliophora, 381
Cilium, 67
Circadian rhythms, 530, 731
Circulatory system, 546, 552–71
    closed circulatory system, 448, 552, 553–54
    invertebrates, 553–54
    and open circulatory system, 431, 553–54
    vertebrates, 555–57
Cirrhosis, 597
Citric acid cycle, 144
Cladistics, 342–43
Cladogenesis, 340, 347–48
Cladogram, 342, 343
Clams, 417, 431, 433, 590
Class, 4, 340
Classical conditioning, 735
Classifications
    of bacteria, 374–76
    and five kingdoms system of classification, 360–62
    hierarchy of, 340
    of leaves, 490
    of organisms, 4–5, A-1–A-2
    of primates, 459
    of skeletons, 665
Clathrin, 98
Clavicle, 666
Cleavage, 713, 716, 718
Climax community, 769
Clitellum, 435
Clitoris, 700, 701
Cloaco, 621
Clonal selection theory, 577, 578
Cloning, 279–81, 286–87, 288, 577, 578
Closed circulatory system, 448, 552, 553–54
*Clostridium botulinum,* 373
Clotting, 284, 563–65
Clotting factor, 284
Clover, 409
Clownfishes, 761
Club fungi, 392
Club mosses, 405, 410–11
Cnidarians, 417, 419–21, 426, 553, 665
Coastal communities, 766–68
Cobamide coenzymes, 112
Cocaine, and abuse, 648–49
Coccolithophore, 359
Coccus, 371, 372
Coccyx, 667
Cochlea, 659, 660
Code, and genes, 252–54
Codominance, 197
Codon, 257, 259
Coelocanth, 451, 452
Coelom, 424, 434, 443, 665, 714
Coelomates, 431
Coenzyme A, 143, 145
Coenzymes, 57, 112–14, 143, 145
Coevolution, of plants and pollinators, 518–19
Cofactor, 112
Cohesion, 34, 35, 497–98
Cohesion-tension model of xylem transport, 497–98

Coitus interruptus, 706
Cole, K. S., 637
Coleoptile, 517, 523, 525
Collagen, 57, 540, 724
Collagen fibers, 540
Collar cells, 417
Collecting duct, 625, 626
Collenchyma cells, 476, 477
Colon, 597
Colonial green algae, 385
Color, of skin, 224, 225
Color blindness, 228–29
Color vision, 658
Colostrum, 705
Columbia University, 204
Columnar epithelium, 539
Commelinales, 4
Commensalism, 373, 374, 759, 760–61
Commensalistic bacteria, 373, 374
Common descent, 308–10
Communal groups, 741
Communication, and behavior, 738–40
Community, 3, 744, 754–64, 766–69, 773, 781
Companion cells, 478
Comparative anatomy, 310–12
Comparative biochemistry, 311–12
Comparative embryology, 311
Compartmentalization, 66
Competition, 755–56
Competitive exclusion principle, 756
Competitive inhibition, 111–12
Complementary base pairing, 55, 56, 239, 240
Complementary DNA (cDNA), 280, 289, 368, 369
Complement system, 576–77
Complete gut, 590–91
Complete linkage, 207
Complete proteins, 598
Complete ventilation, 610
Complex carbohydrates, 598
Compound eye, 437, 440, 654, 655
Compound fruits, 514
Compounds, and molecules, 29–33
Concentration, and substrates, 111
Concentration gradient, 89
Concepts and critical thinking, answers to, A-7–A-18
Conclusions, 14
Condensation, and hydrolysis, 44
Conditioning, classical and operant, 734–35
Condom, 706
Condor, 412
Conduction, 639
Cones, 397, 406, 655, 656, 657, 658
Configurations, 29
Conidia, 391
Coniferophyta, 410–11
Conifers, 397, 405–7, 410–11
Conjugation, 282, 283, 371, 382, 385–86
Connective tissue, 539, 540–42
Connell, Joseph, 755
Conscious brain, 646
Conservation, 412, 776–77
Constant regions, 578
Consumers, 8, 781, 783
Contact inhibition, 271
Continental drift, 344
Continuous feeders, 589
Contour farming, 786
Contour feathers, 454

Contractile tissue, 542–43
Contractile vacuoles, 381
Contraction, 669, 671
Contrast microscopy, 63
*Contributions to the Theory of Natural Selection,* 308
Control
    of body temperature, 548–49
    and gene action, 265
    physiological control of behavior, 731
    of pregnancy, 705–6
    of water excretion, 626–30
Control group, 15
Controlled experiment, 15, 16
Convergent evolution, 348
Convoluted tubules, 625, 626
Cooling, and origin of first cell, 355
Copepods, 437
Copper, 26
*Coprinus comatus,* 5
Coral, 9, 415, 420, 421, 451, 767
Coral reefs, 9, 767
Cordgrass, 335
Corepressor, 268
Corey, Robert, 51
Cork, 476, 479, 484
Cork cambium, 479, 484
Corms, 485, 486
Corn, 60, 61, 238, 245, 481, 483, 517
Cornea, 656, 657
Cornell University, 117
Coronary arteries, 560
Corpus luteum, 703, 704
Correns, Karl, 202
Cortex
    adrenal, 624, 680, 685–86
    of root, 481
Cortisol, 680, 685
Cotyledons, 516
Countercurrent flow, 607
Countercurrent mechanism, 626–30
Country. *See* Less developed country (LDC) *and* More developed country (MDC)
Courtship, 738
Covalent bonding, 31–33, 53
Covered seeds, 407
C plants, 131–32, 133
Crabs, 42, 417, 439, 440
Cranial nerves, 640
Cranium, 666
Crassulaceae, 133
Craving, 649
Crayfishes, 440
Creatine phosphate, 672, 673
Creatinine, 627
Creation, 300
Cretaceous, 401, 450
Cretinism, 684
CRH, 686
Crick, Francis H. C., 18, 238, 239, 240, 241, 242, 243
Cri du chat syndrome, 210, 218
Crill, 437
Crinoidea, 442
Cristae, 75
Critical period, 733
Crocodiles, 453, 455
Cro-Magnon, 467
Crop, 590

Cross-bridges, 671
Crossing-over, 172, 174, 176, 178, 207, 208, 317
Crossopterygii, 451
*Crotalus atrox,* 305
Crustaceans, 437, 439, 440, 661
Cuboidal epithelium, 539
Cucumbers, 490
Cultural eutrophication, 796
Culture, of plant, 516
Cup fungi, 392–93
Curtis, J. J., 637
Cushing syndrome, 686
Cuticle, 57
Cuticular plate, 661
Cuttings, 516
Cuvier, George, 301
Cyanobacteria, 5, 66, 358, 371, 372, 375–76, 788
Cycadophyta, 410–11
Cycads, 405, 406, 410–11
Cycles
    biogeochemical cycle, 787
    Calvin cycle, 807
    carbon cycle, 787–88
    chemical cycle, 787–89
    nitrogen cycle, 788–89
    of materials and ecosystems, 780–91
Cyclic AMP, 679
Cyclic behavior, 731–32
Cyclic electron pathway, 126–29
Cyclic GMP, 658
Cyclic photophosphorylation, 128
Cypress, 323
Cysteine, 51, 253
Cysticercus, 422
Cystic fibrosis, 221–22, 226, 284
Cystic fibrosis transmembrane conductance regulator (CFTR), 221–22
Cytochrome, 125, 128, 144
Cytochrome system, 128, 144
Cytokinesis, 157, 160
    in animal cells, 163, 164
    and mitosis, 160–64
    in plant cells, 164, 165
Cytokinins, 523, 524, 526–27
Cytoplasm, 64, 66, 543, 544, 553, 591
Cytoplasmic cleavage, 163
Cytosine, 238, 239, 240, 251, 253
Cytoskeleton, 69, 76, 77, 81
Cytosol, 64, 67, 139
Cytotoxic T cells, 579, 580

## D

*Dacrydium,* 338
Dalmation, 306
Dalton, John, 26
Dandelion, 515
Danielli, J., 85
Danner, David, 723
Daphne Major, 340
Dark-field illumination, 63
Darter, 412
Darwin, Charles, 18, 176, 177, 183, 186, 298–314, 316, 324, 523
Darwin, Erasmus, 300
*Dasyurus,* 310
Data, 14, 17
Dating of fossils, 344–45
Daughter cells, 160, 179, 716

Daughter molecules, 240
Daughter nuclei, 160, 179
Daughter strands, 242
Davson, H., 85
Day-neutral plants, 531
DDI. *See* Dideoxyinosine (DDI)
DDT, 326, 784, 795
Deamination, 597
Decapods, 440
Decay series, 345
Decibels (db), 662
Deciduous forests, 776, 783
Decomposers, 8, 373
Deductive reasoning, 14
Deep-sea ecosystems, 807–8
Deer, 341, 729
Deer tick, life cycle of, 760–61
Defenses, 758, 759, 575–80
Deforestation, 801
Degradative reaction, 107, 114, 115
Degrees of dominance, 197–98, 222, 226
Dehydration synthesis, 44, 45
Dehydrogenase, 108
Deletion, 210, 211
Delirium, 648
Demographic transition, 750
Denaturation, 53, 111
Dendrites, 543, 544, 635–40
Denitrification, 788–89
Denitrifying bacteria, 789
Density, 35
Density-dependent effects, 746
Density-independent effects, 746
Deoxyribonucleic acid (DNA), 5, 7, 19, 45, 54–56, 57, 64, 66, 67–68, 75, 81, 234–47, 248–63, 264–77, 278–95
    and aging, 724
    and apes, 461
    and autoimmune diseases, 585
    and bacteria classification, 374
    B-DNA and Z-DNA, 243–44
    and binary fission, 156–57
    and biological evolution, 357–60
    and blastula, 713
    cDNA, 280, 289, 368, 369
    and chromosome contents, 158, 163, 196
    and chromosome mutations, 316–17
    and comparative biochemistry, 311–12
    compared to RNA, 252
    and cytokinins, 526
    DNA animal viruses, 368
    DNA fingerprinting, 291
    DNA ligase, 280
    DNA markers, 290–92
    DNA polymerase, 241
    DNA probe, 283, 284, 291
    DNA sequencing, 292
    and fruit fly chromosomes, 211
    gDNA, 281
    gel electrophoresis, 290, 292
    and gene expression, 261
    and gene mutations, 316
    and genetic code, 252–54
    and HIV, 581
    homeobox DNA, 717
    and hormones, 681, 687
    and meiosis and mitosis, 172, 173, 179
    mitochondrial DNA, 467–68
    and modern humans, 467–68

and molecular biology, 250–52
and molecular systematics, 340–41
and pangenesis, 177
and plants, 412
plasmid DNA or viral DNA, 279–80, 282–83, 289
and prokaryotic reproduction, 371
rDNA and biotechnology, 278–95
recombinant DNA, 412
and regulation of gene activity, 264–77
and repetitive DNA, 244–45
replication, 240, 241–43
RNA. *See* Ribonucleic acid (RNA)
and RNA-first hypothesis, 356
and search for genetic material, 235–37
structure of, 237–40, 249
target DNA, 285
viral DNA, 368, 369
and viruses, 366, 367, 368, 369
Deoxyribose, 45, 54
Dependence, 649
Dependency age group, 748, 749
Dependent variables, 15
Depolarization, 637, 639
Depolymerize, 76
Dermal tissue, 475, 476–77
Dermatitis, 601
Dermis, 544, 545
Descent, 300, 308–10
Descent with modification, 300
Descriptive research, 15–17
Desertification, 793
Deserts, 773
Desiccation, 398
Desmosomes, 99, 101, 540
Detritus, 781, 782
Detritus food chains, 782
Deuteromycota, 393
Deuterostomes, 430, 431
Development
    after birth, 722–24
    in animals, 711–26
    of embryo, 513–15
    of fetus, 721–22
    of gametophytes, 510
    of humans, 665, 718–24
    of human skeleton, 665
    in plants, 522–34
    of pollen grain, 511
Devonian, 401, 448, 450, 452
de Vries, Hugo, 202
Diabetes mellitus, 284, 625, 686–88
Dialysis, 629
Diaphragm, 609, 610, 655, 706
Diarrhea, 597
Diastole, 558
Diastolic pressure, 561
Diatoms, 388–89, 390
Dicot, 407, 410–11, 478, 479, 479–81, 516
Dictyosome, 68, 69, 72
Dideoxyinosine (DDI), 370, 708
*Didinium,* 757
Diet
    and cancer, 274–75, 600
    recommended daily, 599
    and sodium, 603
Differential reproduction, 307, 324
Differentiation, 712, 715–18
Diffusion, 89–92, 93, 99

Digestive enzymes, 594, 595
Digestive juice, 686
Digestive system, 546, 587
    animal feeding modes, 588–89
    and digestive tracts, 589–97
    and hormones, 596
    and nutrition, 587–605
Digestive tracts, 589–97
Digits, 667
Dihybrid crosses, 190–93
Dihybrid genetics problems, 190–92
Dihybrid inheritance, 190–93
Dihybrid testcross, 192–93
Dihydroxyacetone, 43
Dimorphism, 327
Dinoflagellates, 388
Dinosaurs, 449, 450, 454
Diploid, 403, 409, 471
Diploid chromosomes, 157
Diploid number, 157
Diploidy, 327–28
Diplontic cycle, 384, 385
Diplopoda, 438, 439
Dipnoi, 451
Directional selection, 324, 326
Disaccharides, 45, 57
Discontinuous feeders, 589
Disease, STD. *See* Sexually transmitted disease (STD)
Disks, 666
Disruptive selection, 324, 325–26
Distal convoluted tubule, 625, 626
Distribution of phenotypes, 224
Divergent evolution, 340, 347–48
Divering birds, 454
Diversity
    and diatoms, 390
    and evolution, 297–472, 471–72
    of fruits, 514–15
    of life, 11
    and loss of biological diversity, 795
    of plants, 396–414
    in rain forests, 777
    and reptiles, 454
    and roots, 481–82
    and vertebrates, 447–71
Division, 4, 340
Dixon, Henry, 497
DNA. *See* Deoxyribonucleic acid (DNA)
Dogs, 306, 457
Dogwood, 7
*Dolichotis patagonium,* 303
Domesticated horse, 341
Dominance, 184, 185, 197, 222, 223–24, 226
    allele, 186
    degrees of, 197–98, 222, 226
    hierarchies, 736, 755
    modification, 198
    simple, 193
Donor, and electrons, 30
Dopamine, 288, 647
Dormancy, 528
Double fertilization, 407, 510, 513
Double helix, 239, 240, 244, 252, 254
Double jointed, 668
Douglass, A. E., 487
Doves, 731
Down, John L., 216
Down feathers, 454

Down syndrome, 216–18, 231
Dragonfly, 3
*Drosophila melanogaster,* 190–92, 204–7, 210, 211, 235, 238, 244, 245, 269, 270, 282, 290, 318, 322, 717, 746
*Drosophila pseudoobscura,* 319–20
Drugs
    and abuse, 648–49
    antibiotics and antivirals, 370
    and brain, 647
Dryopithecines, 461, 462
Duchenne muscular dystrophy, 220, 222, 230–31, 284
Ducks, 454
Dump sites, 796
Duodenum, 593
Duplicated chromosomes, 159, 160
Duplication, 210, 211
Dwarfism, 223, 323, 683
Dynamic equilibrium, 660
Dynein, 79
Dystrophin, 230

# E

Ear, of human, 659–62
Eardrum, 661
Earlobes, 187–88
Earthworms, 417, 435–36, 554, 588, 591, 609, 623, 782
Ear wax, 57
Ecdysone, 677, 678
Echinoderms, 417, 431, 440–43
Echinoidea, 442
Echolocation, 736
Ecological niche, 299
Ecological pyramid, 782–84
Ecology, 781
    and behavior, 729–808, 807–8
    and bioremediation, 286
    and ecological pyramid, 782–84
    of populations, 744–53
EcoRI, 280
Ecosystems, 3, 8–11, 780–91, 807–8
Ectoderm, 421, 713–14
Edward syndrome, 217, 218
Edward VII, King, 229
Eel, 9
Effectors, 547, 641
Efficiency, and energy transformation, 148
Egg, 7, 169, 177, 204, 386, 695
    hard-shelled, 455
    and hormones, 700–704
    and reptiles, 455
    yolk in, 712–13
EI. *See* Environmental impact (EI)
Ejaculation, 697
*Elaphe obsoleta,* 320, 321, 334
Eldredge, Niles, 339
Electrocardiogram (ECG), 559
Electrochemical gradient, 94
Electromagnetic spectrum, 122, 123
Electron model, of water, 32
Electrons, 26, 27, 28
    and acceptor and donor, 30
    configurations, 29
    and oxidation-reduction reaction, 33
    and photosynthesis, 126–29
    and sharing, 31

Isogametes, 385
Isolating mechanisms, 332–34
Isoleucine, 253
Isomers, 43
Isotonic solution, 91
Isotopes, 27–28, 55, 345
Itano, Harvey, 249
IUD, 706
Ivanowsky, Dimitri, 366

## J

Jack fish, 9
Jacob, Francois, 267, 268, 282
Jagendorf, Andre, 117
Jaundice, 597
Jawless fishes, 448, 449, 450
Jeffrey pine tree, 525
Jellyfishes, 417, 420, 421
Jet propulsion, 433
Johns Hopkins University, 279
Joining, 241
Jointed appendages, 437
Joints, of skeleton, 668
J-shaped growth curve, 745
Jumping genes, 245
Junctions
    between cells, 99–101
    tight and gap, 540
Jurassic, 401, 449, 450, 455
Juvenile hormone, 678
Juxtaglomerular apparatus, 628, 629

## K

Kamen, Martin, 126
Kangaroo, 310
Kaposi's sarcoma, 708
Karyotypes, 216, 226, 471
Kcalorie, 109
Kelp, 389
Keratin, 51
Ketone, 43
Kettlewell, H. B. D, 14, 15, 326
Keystone predator, 757
Keystone species, 757
Kidneys, 624–30, 686
Killer whale, 31
Kilocalorie, 34
Kinetin, 526
Kinetochore microtubules, 163, 174
Kinetochores, 162, 163, 174
Kingdoms, 3, 4–5, 340, 360–62, A-1_-A-2
King's College, 238
Kin selection, 740
Kleckner, Nancy E., 268
Klinefelter syndrome, 217, 218, 219, 220
Knee, 668
Kneecap, 668
Knowlton cactus, 412
Knudson, Alfred, 274
Koala, 198
Krause end bulbs, 658
Krebs, Hans, 144
Krebs, John, 736
Krebs cycle, 143, 144, 147, 148
Krill, 448
K-strategists, 749
Kurosawa, 526

## L

Labia majora, 700, 701
Labia minora, 700, 701
Laboratories, 13, 15
Laboratory of Tree-Ring Research, 487
Lac operon, 267–68
Lactase, 108
Lactate, 143, 148, 672, 673
Lactating breast, 705
Lacteal, 596
Lactose, 45, 108
Ladder-type nervous system, 421–22
Lady fern, 410–11
Lagging strand, 242
Lagomorpha, 459
Lakes, 766
Lamarck, Jean Baptiste de, 301–2
Lamellae, 68, 74, 607
*Laminaia,* 389
Lampreys, 448, 450
Lancelet, 444
Lancet, 712, 713
Land, 793–96
Lansteiner, Karl, 567
Lantern fly, 759
Laparoscope, 226, 227
Large intestine, 592, 597
Larvae, 7
Larynx, 611, 612
Lateral line system of fishes, 659
Latitude, and altitude, 778
Laurasia, 348
Lavoisier, Antoine, 607
Law of independent assortment, Mendel's, 190,
    192, 203
Law of segregation, Mendel's, 186, 188, 203
Laws of probability, 188, 190
"Law which Has Regulated the Introduction of
    New Species, On the," 308
LDC. *See* Less developed country (LDC)
LDL. *See* Low-density lipoprotein (LDL)
Lead, 345
Leading strand, 242
Leaf. *See* Leaves
Learned behavior, 732–35
Learning, 646–47, 733, 735
Least flycatcher, 332
Leaves, 474, 482, 489–92
    classification of, 490
    epidermis, 476
    modifications of, 490
    senescence, 527, 534–36
    types and uses of, 491–92
    vegetative organs and major tissues, 492
    veins, 489, 491
Leclerc, Georges Louis, 300
Ledenberg, Joshua, 317
Leder, Philip, 253
Lederberg, Joshua, 282, 283
Leeches, 417, 436
Leeuwenhoek, Anton van, 61
Leg, 667, 668
Legumes, 374, 511
Lemur, 312
Length, units of, A-4
Lens, 655, 656, 657
Lenticels, 484
Lesch-Nyhan syndrome, 228

*Lesquerella pallida,* 412
Less developed country (LDC), 750–51
Leucine, 251, 253
Leukocytes, 541, 563–65, 572, 576
Lewis, Warren, 716
Lewontin, R. C., 319
LH. *See* Luteinizing hormones (LH)
Lichens, 375, 392
Life
    characteristics of, 3–8
    chemistry of, 41–59
    elements in living things, 26
    genetic basis of, 153–296
    and organization, 3, 4–5, 8
    origin of, 353–64
    unity and diversity of, 11
Life cycles
    of algae, 384
    alternation of generations, 398, 399
    of angiosperm, 513
    of animals, 170
    of deer tick, 760–61
    of ferns, 403
    of flowering plants, 407–8, 508–13
    of fungi, 170
    of moss, 398–99, 400
    of mushroom, 393
    of pine, 405, 408
    of plants, 170, 399
    of *Plasmodium vivax,* 384
    of *Rhizopus,* 391
    of *Taenia,* 423
    of *Ulva,* 388
    of viruses, 367–70
Li-Fraumeni syndrome, 274
Ligaments, 541, 668
Ligand, 98
Light
    and sunlight, 122–23
    and vision, 655–58
Light-dependent reactions, 125–29
Light-independent reactions, 125, 129–36
Light microscope, 62
Lignin, 99
Lily, 397, 410–11, 507, 510
Limb buds, 720
Limbic system, 645, 646–49
Limnetic zone, 766
Linkage, genetic, 206–8
Linkage group, 206
Linkage map, 288, 292
Linnaeus, Carolus, 4, 300, 340
Linnean Society, 308
Lions, 6, 153, 589, 696
Lipase, 108, 594
Lipids, 57, 108, 599–600
    and fatty acids, 46–50
    and fluid-mosaic model, 87–89, 90
Lipoproteins, 98, 599–600
Liposomes, 357
Lipscomb, David, 661
Liquids, 34, 35
Littoral zone, 766
Liver, 41, 596–97, 672, 673
Liverworts, 398, 410–11
Lizards, 4, 453
Lobe-finned fishes, 451, 452
Lobes of brain, 645, 646
Lobsters, 437–38, 440

Lobules, 705
Local osmosis, 95
Locus, 186, 220
*Loligo,* 434, 637
*Loligo pedlei,* 434
Long bone, anatomy of, 542
Long-day plants, 531
Loons, 454
Loop of Henle, 625, 626
Loop of the nephron, 625, 626
Loose connective tissue, 540–41, 541
Lorenz, Konrad, 732–35, 734
Love Canal, 795
Low-density lipoproteins (LDL), 98, 599–600
Lucy, 465
Lungfishes, 451
Lungs, 606, 612
    amphibians and reptilians, 609, 610
    and book lungs, 438
    and cancer from smoking, 615
    disorders of, 616
    and gas exchange and transport, 612–16
    saclike lungs, 448
Luteal phase, 702, 703
Luteinizing hormones (LH), 680, 683, 698, 700,
    701, 702, 703, 704
Lycophyta, 410–11
*Lycopodium,* 402
Lycopod trees, 402
Lyell, Charles, 302, 308
Lyme disease, 284, 438, 760–61
Lymph, 574
Lymphatic duct, 574
Lymphatic system, 547, 573–75
Lymphatic vessels, 568, 573–74
Lymph nodes, 574
Lymphocytes, 563, 565, 577
Lymphoid organs, 574–75
Lymphokines, 579
Lynx, 5, 758
Lyon, Mary, 220
Lysine, 51
Lysogenic cells, 368
Lysogenic cycle, 367, 368
Lysosomes, 67, 69, 73, 81, 222, 223
Lytic cycle, 367, 368

# M

MacArthur, Robert, 756
Macaw, 777
McClintock, Barbara, 244, 245
Macleod, J. J., 687
*Macrapus,* 310
*Macrocystis,* 389
Macroevolution, 340–47
Macromolecules, 43, 55, 99, 356
Macrophages, 97, 99, 563, 565, 575, 576
Madagascar periwinkle, 412
Magnesium, 26, 29, 30, 112, 494, 500, 627
Magnification, biological, 784
Major histocompatibility complex (MHC) protein,
    580, 582, 583, 584
Malaria, 226, 284, 326, 328, 383, 384
Malay Archipelago, 308
Male gametophyte, 405, 508
Male reproductive system, 696–700
Malignant tumors, 272
Malleus, 660

Malpighi, Marcello, 503, 624
Malpighian tubules, 438, 441, 623–24
Maltase, 108
Malthus, Thomas, 305, 308
Maltose, 45, 108
Mammals, 4, 5, 7, 297, 310, 343, 349, 417, 443,
    449, 450, 455–58, 459
    body cavities of, 548
    and respiration, 609, 610
    spermatogenesis and oogenesis in, 176
Mammary glands, 455, 456
Manatees, 297, 307
Mangold, Hilde, 716
Manic depressive psychosis, 228
Man-of-war, 420
Mantle, 431, 434
Maples, 333
Mapping
    and chromosome mapping, 206–8
    and genetic mapping, 288–90
*Maranta leuconeura,* 530
*Marchantia,* 398
Marfan syndrome, 223
Marigolds, 10, 655
Marijuana, 409, 648
Marine annelids, 435
Marine Iguana, 299
Markers, and DNA markers, 290–92
Marrow, 564, 574
Marshes, 767
Marsh marigold, 655
Marsupials, 310, 457
Maryland darter, 412
Masked messengers, 271
Mass extinctions, 343–44
Mast cells, 584
Master gland, 683
Mastodon, 301
Materials, and ecosystems, 780–91
Mating, 321–22, 332–34
Matrix, 75, 99
Matter, 26
Matthei, J. Heinrich, 253
Maturation promoting factor (MPF), 160
Mature mRNA, 254
Maturing face, 72
Maximum sustained yield, 748
Mayr, Ernst, 334
M. D. Anderson Hospital and Tumor Institute, 274
MDC. *See* More developed country (MDC)
Measurements, and biology, 65–66
Mechanical isolation, 333
Mechanoreceptors, 658–62
*Medicago,* 484
Medulla, adrenal, 624, 680, 685
Medulla oblongata, 644
Medusae, 419, 421
Megaspores, 404, 405, 407, 509, 510
Meiosis, 157, 169–81, 203, 207, 208, 209, 210,
    219, 317, 335, 385, 386, 393, 400, 405,
    513, 695
Meissner's corpuscles, 658
Melanocyte-stimulating hormone (MSH), 680, 683
Melatonin, 680, 689–90
Membranes, structure and function, 84–103
Memory, 645, 646–47
Memory B cells, 578
Memory T cells, 579
Mendel, Gregor, 18, 182–95, 197, 198, 199, 202,
    282, 316

Mendelian patterns of inheritance, 182–95
Mendel's law of independent assortment, 190,
    192, 203
Mendel's law of segregation, 186, 188, 203
Meninges, 642
Menisci, 668
Menopause, 724
Menses, 703
Menstruation, 702, 703
Mental retardation, 217, 218, 221, 228, 231
Menthol, 409
Meristem tissue, 475
Meristem tissues, 479
Merkel's disks, 658
Mermaid's wineglass, 69
Merrick, Joseph, 223
Meselson, Matthew, 241–43
Mesoderm, 713–14
Mesoglea, 419, 421, 553
Mesophyll, 131, 489, 491
Mesozoic, 344, 401, 404, 450, 454, 457
Messenger RNA (mRNA), 251, 252, 253, 254,
    255, 261, 270, 271, 368, 681, 715, 723
Metabolic pathways, 108, 139
Metabolic pool, and biosynthesis, 148–49
Metabolism, 4, 6, 8
    and cellular metabolism, 107–15
    and glucose metabolism, 139
    inborn error of metabolism, 249
    and prokaryotic metabolism, 372–74
Metabolites, 249
Metacarpals, 667
Metafemale syndrome, 217, 218, 219, 220
Metamorphosis, 438, 452, 453, 677
Metaphase, 161, 163, 164, 172, 173, 174, 175
Metastasis, 272
Metatarsals, 668
Metatheria, 457
Methamphetamine (ice), and abuse, 649
Methyl group, 112
Metochondrion, 139
Metric system, A-4–A-6
MHC protein. *See* Major histocompatibility
    complex (MHC) protein
Mice, 10, 16
Microbodies, 67, 69, 73–74, 81
Microevolution diagram, 320
Microfilaments, 78
Micrometer, 65
Micropyle, 515
Microscopes, 62–64
Microspheres, 356, 357
Microsporangium, 404
Microspores, 404, 405, 509, 510
Microtubule organizing center (MTOC), 76–78,
    160
Microtubules, 67, 68, 76–78, 79, 160, 161, 163,
    174
Microvilli, 67, 654
Midbrain, 645
Middle ear, 659
Middle lamella, 68
Midges, 270, 295–96
Miescher, Friedrich, 235
Milk, 45, 705
Miller, Stanley, 355
Millimeter, 65
Millipedes, 10, 438–40
Mimicry, 758
*Mimosa pudica,* 530

Noise, 661–62
Nonbreeders, 741
Noncompetitive inhibition, 111–12
Noncyclic electron pathway, 126–29
Noncyclic photophosphorylation, 128
Nondisjunction, 209, 210, 218, 219
Nongonoccal urethritis (NGU), 707
Nonrandom mating, 321–22
Nonrenewable energy sources, 802–4
Nonvascular plants, 398–99, 410–11
Norepinephrine (NE), 639, 641, 647, 680, 685
Norplant, 706
*Nostoc,* 5
Notochord, 443, 714
*Notocytes caurinus,* 4
*Notorcytes typhlops,* 4
Nuclear envelope, 67, 68, 69, 70, 81
Nuclear pores, 68, 69, 70
Nuclear power, 803
Nucleic acids, 43, 45, 54–56, 57, 235, 366
Nuclein, 235
Nucleoid, 64, 66, 156
Nucleoli, 67, 68, 69, 70, 81
Nucleoplasm, 68, 70
Nucleotides, 43, 54–56, 57
Nucleus, 66, 67, 69, 70, 160, 179
Nudibranches, 432
Numbers, pyramid of, 782–84
NurtraSweet, 286
Nutrition, 5, 6, 8, 587, 597–600
    and carbohydrates, 598–99
    clam and squid compared, 590
    and digestive system, 587–605
    essential elements for plant nutrition, 499
    in plants, 494–506
    and proteins, 597–98

## O

Objective questions, answers to, A-7–A-18
Obligate anaerobes, 372
Observations, 14
Obstructive jaundice, 597
Occipital lobe of brain, 645, 646
Oceans, 9, 768–69
    and deep-sea ecosystems, 807–8
    and ocean floor, 768–69
    and ocean water pollution, 796–97
Ocelot, 777
*Ocillatoria,* 375
Octet rule, 29
Octopuses, 431, 433, 652
*Oedogonium,* 386
Oils, 47–48, 57, 403, 803
Old World monkeys, 312, 459–61
Oleate, 48
Olfactory cells, 653–54
Oligocene, 461
Oligochaeta, 435
Omnivores, 588, 781
Oncogenes, 273–74
One gene-one enzyme hypothesis, 249
One gene-one polypeptide hypothesis, 249–50
Onion, 491
Onychophran, 437
Oocyte, 176, 719
Oogenesis, 175, 176, 219, 700
Oomycota, 389
Ooplasmic segregation, 715

Oparin, Alexander, 355, 356, 357
Open circulatory system, 431, 553–54
Operant conditioning, 734–35
Operon model, 267–68
Operons, 267–68
Ophiuroidea, 442
*Ophrys,* 3
Opium, 647, 649
Opportunistic species, 749
Optic nerve, 655, 657
Oral herpes, 284
Orangutan, 447, 461, 463
Orbital model, of water, 32
Orbitals, 28, 29
Order, 4, 340
Ordovician, 401, 448, 450
Organelles, 64, 65, 66–80
Organic molecules, 42–44, 57, 355
Organic soup, 355
Organic substances, transport of, 503–4
Organisms, classification of, 4–5, A-1–A-2
Organization
    of animals, 538–51
    levels of, 25
    and life, 3, 4–5, 8
Organ of Corti, 659, 660, 661
Organoid, 289
Organs, 3, 25
    of animals, 544–46
    of excretion, 623–24
    formation of, 719–20
Organ systems, 3, 25
    of animals, 544–46
    of humans, 546–47
Orgasm, 697
Orientation, 736
Origin of life, 353–64
*Origin of Species, On the,* 307–8
Oriole, 334, 335
Oryx, 341
*Oryx gazella,* 341
Oscilloscope, 637
Osculum, 418
Osmosis, 89–91, 92, 93, 95, 495, 621–23
Osmotic pressure, 91
Osmotic regulation, 621–23
Ossicles, 659, 660
Osteichthyes, 448
Osteoblasts, 665
Osteoclasts, 665
Osteocytes, 665
Osteoporosis, 601, 724
Ostriches, 454
Otolaryngologist, 661
Otoliths, 660
Otosclerosis, 659
Outdoor air pollution, 798–800
Outer ear, 659
Oval window, 659, 660
Ovarian cycle, 700–702, 703
Ovaries, 407, 409, 509, 680, 688, 700, 701, 702
Overkill, 758
Oviducts, 700, 701
Ovipositor, 441
Ovulation, 700, 703, 719
Ovules, 404, 407, 509, 510
Owls, 10, 454
Oxford University, 326
Oxidation, 33, 107, 114, 115

Oxidation-reduction reaction, 33
Oxidative phosphorylation, 145, 147–48
Oxidizing atmosphere, 355, 358–60
Oxygen, 26, 31, 32, 122, 140
    and osmosis, 91
    and skin breathing, 726–28
    transport of, 614–16
Oxygen debt, 142, 672, 673
Oxytocin, 680, 681
Oysters, 431, 433
Ozone shield, 358, 799, 800, 801–2

## P

Pacemaker, 559
Pacinian corpuscles, 658
Pain, 647
Paine, Robert, 757
Pain receptors, 658
Paleocene, 461
Paleontology, 301
Paleozoic, 344, 401, 404, 450, 452
Palisade layer, 489, 491
*Palm Trees of the Amazon,* 308
PAN, 799
Pancreas, 676, 680, 686–88
Pancreatic amylase, 594
Pancreatic islets, 676
Panda, 331, 335
Pangaea, 344, 347, 348, 404
Pangenesis, 177, 186
Panther, 412
Pantothenic acid, 112, 602
Paper, history of, 488
Paramecia, 63, 155, 380, 381, 553, 757
Parapodia, 435
Parasitic bacteria, 373, 374
Parasitic lampreys, 448, 450
Parasitism, 373, 374, 759–60
Parasympathetic nervous system, 641, 642, 643
Parathyroid glands, 680, 684, 685
Parathyroid hormone (PTH), 680, 684, 685
Parathyroids, 680, 684, 685
Parenchyma cells, 476, 477
Parental imprinting of genes, 198, 199
Parietal lobe of brain, 645, 646
Parkinson's disease, 288, 647
Parrotfish, 9, 451
Parthenogenesis, 695
Passive immunity, 582
Pasteur, Louis, 18
Patagonian Desert, 299
Patagonian hare, 303
Patau syndrome, 217, 218
Pathways
    and metabolic pathways, 108
    noncyclic electron and cyclic electron, 126–29
Patterns
    of evolution, 347–48, 349
    FAP. *See* Fixed-action pattern (FAP)
    of inheritance, Mendelian, 182–95
    of reproduction, 695–96
Pauling, Linus, 51, 249
Pavlov, I., 735
PCB, 784
PCR. *See* Polymerase chain reaction (PCR)
Peacock, 315
Peas, 182, 183
Peat moss, 399

Rat, 312
Rat snake, 320, 321, 334
Rattlesnakes, 305
Ray-finned fishes, 450, 451
Rays, 448, 451
RB gene. *See* Retinoblastoma (RB) gene
RDA. *See* Required dietary (daily) allowance (RDA)
Reabsorption, 625
Reactants, and products, 33
Reactions
    chemical, 33
    light-dependent and light-independent, 125–36
    and transition reactions, 143, 144, 145, 147
Reactivation, of proteins, 53
Reasoning, 14, 735
Receptacle, 509
Receptor-mediated endocytic cycle, 98–99
Receptor protein, 90
Receptors, 88, 547, 612, 641, 653–62
Receptor sites, 639
Recessive, 184, 185, 197, 222
Recessive allele, 186
Recessive inheritance, autosomal, 221–23
Reciprocal crosses, 185
Recombinant DNA (rDNA), 278–95, 412
Recombinant gametes, 207, 208
Recombinant phenotypes, 207, 208
Recombination, 175, 278–95, 317–18, 412
Recycling, of NAD and FAD, 146
Red algae, 388
Red blood cells, 541, 563–65, 614
Red bone marrow, 564
Red chow, 306
Red deer, 341
Red grouper, 9
Red maple, 333
Red tide, 388
Reducing atmosphere, 355
Reduction, 33, 130
Redwoods, 407, 410–11
Reefs, 9, 451, 767
Reflex arc, 641
Reflexes, 641
Refraction, 657
Refractory period, 697
Regeneration, 131, 695
Regulation
    of enzyme activity, 111
    of gene action, 264–77
    osmotic regulation, 621–23
Regulator gene, 266, 267
Reindeer, 341, 747
Reinforcement, 734
Rejection, of tissue, 584
Relative dating method, 345
Relaxation, 669
Remediation, 286
Renal pelvis, 624
Renewable energy sources, 802–4
Renin, 628, 629, 686
Renin-angiotensin-aldosterone system, 686
Repair enzymes, 261
Repetitive DNA, 244–45
Replacement reproduction, 751
Replica-plating technique, 317
Replication, 203, 240
    and first replication system, 357
    and replication fork, 243

Repolarization, 639
Reporting of experiments, 14
Repressible operon, 268
Repressor, 266, 267
Reproduction, 5, 7, 8
    in animals, 694–710
    and asexual reproduction, 166, 384, 386, 387, 515–19, 695
    and behavior, 736–38
    cell. *See* Cell reproduction
    and comparisons among plants, 410–11
    and differential reproduction, 307, 324
    female reproductive system, 700–704
    of humans, 696–705
    patterns of, 695–96
    in plants, 507–21
    and prokaryotic reproduction, 371–72, 373
    replacement reproduction, 751
    and sexual reproduction, 166, 169, 170, 384, 386, 387, 695–96
Reproductive age group, 748, 749
Reproductive behavior, 736–38
Reproductive isolating mechanism, 332–34
Reproductive system, 547
    female, 700–704
    male, 696–700
Reptiles, 5, 309, 343, 349, 417, 443, 449, 453–54, 455, 609, 610, 718
Required dietary (daily) allowance (RDA), 598, 600, 601
Research, descriptive, 15–17
Resistance, environmental, 746–48
Resolution, 62
Respiration
    and cellular respiration, 114, 137–52
    and glycolysis, 137–52
    and health, 616
    and origin of first cell, 355
    and photorespiration, 132–33
    and tracheae, 441
Respiratory distress of newborn, 284
Respiratory gas-exchange surfaces, 607–10
Respiratory system, 546, 606–18
Responses, 4–5, 8
Responsibility, and science, 19
Resting potential, 637–39
Restricted fragment length polymorphism (RFLP), 290–92
Restriction enzymes, 279–80
Retardation, 217, 218, 221, 228, 231
*Reticulitermes*, 333
Retina, 222, 655, 656, 657
Retinal, 658
Retinoblastoma (RB) gene, 274, 723
Retroviruses
    RNA, 368, 369
    as vectors, 289
Reverse transcriptase, 368
Revision, 15
RFLP. *See* Restricted fragment length polymorphism (RFLP)
R groups, 51, 52, 53
Rhea, 299
Rhesus monkey, 342
Rh factor, and system, 226, 468, 567, 568–69
*Rhizobium*, 374
Rhizobium, 502
Rhizoids, 398, 403
Rhizomes, 401, 485, 486, 515

*Rhizopus*, 391
*Rhodnius*, 677
Rhodopsin, 658
*Rhynia*, 400, 401, 409
Rhyniophytes, 400
Rhythmic behavior, 731–32
Rhythm method, 706
Rhythms, of plants, 530
Riboflavin, 112, 601, 602
Ribonuclease, 53, 108
Ribonucleic acid (RNA), 45, 54–56, 57, 68, 69, 108, 235, 245, 248, 251, 252, 253, 254, 255, 264–77
    and aging, 723
    antisense RNA, 268–69
    and bacteria classification, 374
    and biological evolution, 357–60
    compared to DNA, 252
    and cytokinins, 526
    and genetic code, 252–54
    and leaf senescence, 535
    and molecular biology, 251–52
    mRNA, 251, 252, 253, 254, 255, 261, 270, 271, 368, 681, 715, 723
    and regulation of gene control, 264–77
    RNA animal viruses, 368
    RNA-first hypothesis, 356
    RNA polymerase, 254
    RNA processing or RNA splicing, 254–56
    RNA replicase and RNA transcriptase, 368
    RNA retroviruses, 368, 369
    rRNA, 245, 251, 256–57, 261, 264
    tRNA, 251, 254, 257–58, 261
    and viruses, 366, 367, 368, 369
Ribose, 45, 54
Ribosomal RNA (rRNA), 68, 245, 251, 256–57, 261, 264
Ribosomes, 66, 67, 68, 69, 70, 81
Ribozymes, 108
Ribs, 666
Ribulose bisphosphate (RuBP), 130, 131, 132
Rich, Alexander, 243
Rickets, 601
Ringdove, 731
RNA. *See* Ribonucleic acid (RNA)
Robertson, J. D., 86
Robins, 18
Rockhopper penguins, 2
Rocks, dating methods, 345
Rocky Mountain spotted fever, 438
Rocky shore, 767
Rodentia, 459
Rods, 655, 656, 657, 658
Rohrer, Heinrich, 63
Roid mania, 689
Romania, 316
Root epidermis, 476
Root hairs, 494
Root pressure, 497
Roots, 494
    and diversity, 481–82
    and mineral uptake, 502–3
    types of, 481–82
    vegetative organs and major tissues, 492
    woody roots, 486
Root system, 475, 479–82
Root tip, 60, 480
*Rosa*, 5
Rostrum, 440

Roughage, 45, 46
Rough endoplasmic reticulum (ER), 67, 68, 69,
    71–74
Roundworms, 415, 417, 423–26
Rous, Peyton, 273
Rous sarcoma virus (RSV), 273
Royal Society, 61
rRNA. *See* Ribosomal RNA (rRNA)
r-strategists, 749
RSV. *See* Rous sarcoma virus (RSV)
Ruben, Samuel, 126
Rubidium, 345
Rubisco, 130, 131, 132
RuBP. *See* Ribulose bisphosphate (RuBP)
Ruffini's endings, 658
Rugae, 594
Runner's high, 647
Rusts, 393

# S

Sac body plan, 421, 422, 426
Saccule, 72, 659, 660
Sac fungi, 391–92
Sachs, Julius, 126
Saclike lungs, 448
Sacrum, 667
St. Martin, Alexis, 593, 594
St. Matthew Island, 747
Salivary amylase, 591
Salivary glands, 591
*Salix,* 481
Salk, Jonas, 708
Salmon, 736
*Salmonella typhimurium,* 283, 365
Salt, and salts, 31, 33, 42
    and diet, 599
    and water balance of fishes, 622
Saltatory conduction, 639
Salt marshes, 767
Saltwater communities, 766–69
Sampling error, 322
San Andreas Fault, 344
Sand dollars, 442
Sandpipers, 454
Sandwich model of membrane structure, 85
Sandworms, 417, 435
Sandy beach, 767
Sanger, Frederick, 51, 292
Saprophytes, 373, 380
Sarcodinia, 381
Sarcolemma, 671
Sarcoma, 708
Sarcoma virus, 273
Sarcomeres, 671, 673
Sarcoplastic reticulum, 671
*Sargasum natans,* 389
Saturated fatty acids, 47
Saussure, Nicholas de, 126
Savanna, 775
Scala naturae, 300
Scallops, 431, 433
Scanning electron microscope (SEM), 62, 63
Scanning tunneling microscope (STM), 63–64
Scapula, 666
SCF. *See* Stem cell growth factor (SCF)
Schally, A. V., 682
*Schistosoma,* 422
Schleiden, Matthias, 18, 61, 126

Schwann, Theodor, 18, 61
Schwann cells, 636, 639
SCID. *See* Severe combined immune deficiency
    syndrome (SCID)
Science, and social responsibility, 19
Scientific method, 13–21
Scientific model, 17
Scintillations, 55
Sclera, 655, 657
Sclerenchyma cells, 476, 477
Scolex, 422, 423
Scorpions, 438, 439
Scrotum, 696
Scurvy, 601
Sea anemones, 417, 420, 421, 553
Sea cucumbers, 417, 442
Sea ecosystems, 807–8
Sea fan, 415
Sea lilies, 442
Sea skates, 307
Sea slugs, 432
Sea squirt, 443
Sea stars, 417, 441–42
Sea turtles, 307
Sea urchins, 417, 442
Seawater, 31
Seaweeds, 386–88
Secondary growth, 404, 478–79, 484–85
Secondary oocyte, 176
Secondary structure, 51, 52, 53
Secondary succession, 770
Secondary xylem, 488
Secretin, 680
Secretions, 596, 625–26
Secretory phase, 703
Sedgwick, Adam, 302
Seed. *See* Seeds
Seed coat, 513
Seed cones, 406
Seedless grapes, 524
Seedless vascular plants, 400, 410–11
Seedlings, 517, 522
Seed plants, 397, 404–9, 410–11
Seeds, 7, 397, 404–9, 410–11, 510–13, 515, 516
Segmentation, 423, 434, 436, 443, 444
Segmented worms, 434, 436
Segregation, 186, 188, 203, 715
Selection, 305–8, 324–27, 735
    and behavior, 735–38
    types of, 325
Selectively permeable membranes, 89
Selective reabsorption, 625
Selfish behavior, 735
Self-mutilation, 228
Self-replication, and origin of first cell, 355
SEM. *See* Scanning electron microscope (SEM)
Semen, 697–98
Semicircular canals, 659, 660
Semiconservative replication, 240, 241–43
Semilunar valves, 557
Seminal fluid, 697–98
Seminal vesicles, 697
Seminiferous tubules, 698
Senescence, 527, 534–36
Sense organs, 652–63
Sensors, 547
Sensory neurons, 636
Sepals, 407, 409, 509
Septum, 557

Sequence-tagged site (STS), 292
Sere, 769–72
Serine, 51, 253
Serotonin, 647
Serous membrane layer, 592
Serum, 563
Sessile filter feeders, 417
Setae, 435
Severe combined immune deficiency syndrome
    (SCID), 288
Sex chromosomes, 204, 218–19, 220, 222, 228–31
Sex determination, 204, 220
Sex hormones, 50, 680
Sex-influenced traits, 231
Sex-linked genes, 204, 222, 228–31
Sexual dimorphism, 327
Sexually transmitted disease (STD), 284, 706–7
Sexual reproduction, 166, 169, 170, 384, 386, 387,
    695–96
Sexual selection, 327
*Seymouria,* 309, 453
Sharks, 448, 451
Sheldon, Peter, 340
Shelf fungi, 393
Sherrington, Charles, 639
Shoot system, 475, 482–89
Shore dwellering birds, 454
Shores, 767
Short-day plants, 531
Shrews, 459
Shrimps, 440
Shrubs, 410–11
Sickle-cell disease, 222, 226–28, 249–50, 251,
    284, 328
Sickle-cell hemoglobin, 249, 250, 251
Side effects, immunological, 584–85
Sieve plate, 441, 478
Sieve-tube cells, 478
Signal, 715
Sign stimulus, 733
Silurian, 401, 450
Simple dominance, 193
Simple eye, 437
Simple fruits, 511, 514
Simple Mendelian inheritance, 197
Simple sugars, and polysaccharides, 44–46
Singer, S., 86
Single lens of eye, 655
Sinoatrial (SA) node, 559
Sinuses, 666
Siphons, 433, 443
Sister chromatids, 157, 159, 160
Skates, 448
Skeletal muscles, 543
Skeletal system, 546, 665
Skeletons
    appendicular skeleton, 666, 668
    axial skeleton, 666, 668
    bones of, 666, 668
    calcareous skeleton, 421
    of cartilage, 448
    classification of, 665
    endoskeleton, 440, 665
    exoskeleton, 437, 665
    of humans, 665–68
    hydrostatic skeleton, 434
    internal skeleton, 418
    joints of, 668
    muscles of, 666
    musculoskeletal system, 664–75

Skin
  breathing of, 726–28
  color, 224, 225
  of humans, 544–45, 658
Skin gills, 441
Skinner, B. F., 734
Skull, 666
Slash-and-burn agriculture, 795
SLE. *See* Systemic lupus erythematosus (SLE)
Sliding filament theory, 671
Slime molds, 380, 389
Slugs, 431, 432
Small intestine, 592, 593, 594–96
Small nectar feedering birds, 454
Smell, 653–54
Smith, Hamilton, 279
Smith, Jim, 716
Smog, 798–800
Smoking, 615
Smooth endoplasmic reticulum (ER), 67, 68, 69, 71–74
Smooth muscle fibers, 542
Smooth muscle layer, 592
Smuts, 393
Snails, 302, 325, 417, 431, 537, 646
Snakes, 320, 321, 334, 453, 587
Snowshoe hare, 758
Social responsibility, 19
Society, 740–41, 804
Sociobiology, 740–41
Sodium, 26, 30, 603, 627
Sodium chloride, 30, 33
Sodium gates, 639
Sodium ions, 639
Sodium-potassium pump, 94, 95, 96
Soft palate, 654
Soil erosion, 793
Soil pores, 496
Soil profiles, 774
Soil water, 496
Solar energy, 114, 803
Solar radiation, 122, 123
Solutions, 91–92, 93
Solvent, 33
Somatic nervous system, 641
Somatotropic hormone, 680
Sonar, 736
Songbirds, 454
Sori, 402, 403, 404
Sound, 661–62, 739–40
Southern Illinois University, 356
Soybean plant, 374
Space-filling model, of water, 32
*Spartina,* 335
Spawning, 736
Special creation, 300
Speciation, 332–40
Species, 4, 8, 332
  category in classification, 340
  keystone species, 757
  and strategies for existence, 749
Specific defense, 577–80
Specimen, 63
Spectrum, 122, 123
Spemann, Hans, 715–16
Sperm, 79, 169, 177, 204, 386, 403, 695, 698–700
Spermatogenesis, 175, 176, 219, 698
Sphagnum, 399
S phase, 159

Sphenophyta, 410–11
Sphincters, 555, 593
Spicules, 418
Spiders, 417, 438, 588
Spinal cord, 640, 641
Spinal nerves, 640
Spinal reflex, 641
Spindle, 154, 160, 161, 162, 166
Spines, 490
Spiracles, 441
Spirillum, 371, 372
*Spirogyra,* 385–86, 387
Spleen, 574–75
Sponges, 5, 391, 399, 403, 404, 415, 417–19, 426
Spongy layer, 489, 491
Spontaneous, 106
Sporazoans, 383, 384
Spores, 384, 385, 386, 391, 393, 396, 404, 509
Sporophyte, 398, 399, 400, 403, 409, 410–11, 509, 513
Sporozoans, 381, 383
Spot desmosomes, 99, 101, 540
Spruces, 410–11
Squamous epithelium, 539
Squids, 417, 431, 433–34, 590, 637
Squirrel, 10, 122
Sry gene, 220
S-shaped growth curve, 746
Stabilizing selection, 324, 325, 326
Staghorn fern, 410–11
Stahl, Franklin, 241–43
Staining, 216, 220, 374
Stalk, of plant, 399
Stamens, 407, 409, 510
Stanley, W. M., 366
Stapes, 660
*Staphylococcus aureus,* 370, 374
Starch, 45, 46, 57, 68, 126, 596, 597
Starch grain, 68
Starfishes, 441, 757
Stasis, 348
Static equilibrium organs, 660
Statocysts, 661
Statolith, 661
STD. *See* Sexually transmitted disease (STD)
Steady state, 804
Steady-state society, 804
Steam, 34, 35
Stearate, 48
Stem cell growth factor (SCF), 563
Stem elongation, 526
Stem reptiles, 449
Stems, 482
  herbaceous stems, 483–84
  modifications of, 486
  monocot stems, 483
  primary growth of, 482–84
  secondary growth of, 484–85
  types and uses, 485–89
  vegetative organs and major tissues, 492
  woody stems, 485, 486
Stereocilia, 661
Sterility, 334, 705
Sterilization, 373
Sternum, 666
Steroids, 47, 48–50, 57, 678, 679–81, 688–89
Stickleback, 451, 738
Sticky ends, 280
Stigma, 407, 409

Stimuli, 4–5, 8, 733
Sting ray, 451
Stirrup, 660, 661
STM. *See* Scanning tunneling microscope (STM)
Stohr, Oskar, 225
Stolons, 485, 486
Stomach, 592, 593–94
Stomata, 474, 476, 489, 498–99, 528
Stop codon, 259
Strategy for existence, 748–50
Stratified epithelium, 539
*Streptococcus pneumoniae,* 235
*Streptomyces,* 370
Streptomycin, 370
Stress, 566
Striated muscle fibers, 542
Strobili, 396
Stroma, 74, 123, 124, 125
Stromatolites, 357, 358
Strontium, 345, 500
Structural gene, 266, 267
Structures
  and comparative anatomy, 310
  homologous and analogous, 340
Struggle for existence, 305
STS. *See* Sequence-tagged site (STS)
Style, 407, 409, 509
Stylet, 502
Subatomic particles, 27
Subcutaneous, 544, 545
Substrate concentration, 111
Substrate-level phosphorylation, 117, 140, 141, 142, 147
Substrates, and enzymes, 108–10
Succession, 769–72
Suckers, 516
Sucrose, 45, 57
Sudoriferous glands, 544
Sugar diabetes, 625
Sugar maple, 333
Sugars, 43, 57
  and diet, 599
  simple sugars and polysaccharides, 44–46
Sulfa, 370
Sulfates, 627
Sulfhydryl, 43
Sulfur, 26, 500
Sulfur dioxide, 37–38
Summation, and integration, 639
Sun, 4, 8
Sunflowers, 316, 501
Sunlight, 122–23
Suppressor T cells, 579
Surface water pollution, 796, 797
Surroundings, 106
Survival of the fittest, 308
Survivorship, 748
Sustained yield, 748
Sutherland, Earl W., 679
Sutton, Walter S., 202, 206
Swallowing, 593
Swallowtail butterfly, 439
Swamps, 401, 402, 405
Sweat glands, 544
Sweetener, 15, 16
Swim bladder, 450
Swimmerets, 440
Swordfish, 451
Symbiosis, 360, 373, 502, 759–62

*Trichonympha collaris,* 383
Triglycerides, 48, 57
Triiodothyronine, 680
Trilobite, 438
Trinity College, 340
Triplet code, 253, 311–12
Trisomy, 209, 210, 217, 218
tRNA. *See* Transfer RNA (tRNA)
Trophic levels, 782
Trophoblast, 718
Tropical conservation, 776–77
Tropical Forest Resources Assessment Project, 777
Tropical rain forests, 777–78, 788, 793–95
Tropism, 529
Trout, 451
Trp operon, 268
True-breeding plants, 183
True coelom, 424
Trypanosome infection, 383
Trypsin, 110, 111, 594
Tryptophan, 253, 265, 268
TSH. *See* Thyroid-stimulating hormone (TSH)
T-tubules, 671
Tubal ligation, 706
Tube cell, 510
Tubers, 486
Tube-within-a-tube body plan, 421, 423, 424, 435, 443, 448
Tubeworms, 373, 435
Tubular secretion, 625–26
Tubules, 623–24, 625, 626
Tumor angiogenesis factor, 690
Tumor necrosis factor, 284
Tumors, 272, 273
Tumor supressor genes, 273, 274
Tundra, 765, 773–75
Tune, 451
Tunicate, 443
Turbellaria, 421
Turgor pressure, 92, 93, 498
*Turitella,* 302
Turner syndrome, 217, 218, 219
Turnips, 481
Turtles, 307, 453
T virus, 234, 236
Twins, 225
Tympanic membrane, 659, 660
Tympanum, 441
Tyrosine, 51, 253, 274
Tyrosine phosphatases, 274

# U

Ulcers, 584
Ulna, 667
Ultimate causes, 731
Ultrasound, 226, 227
Ultraviolet (UV) radiation, 260, 267, 273, 544
*Ulva,* 386, 388, 398
Unconscious brain, 644–46
United Nations (UN), 777
United States Bureau of Land Management, 793
United States Department of Agriculture, 531, 598, 786, 793
United States Department of Health and Human Services, 598
United States Environmental Protection Agency (EPA), 802

United States National Museum Natural History, 776
Unit membrane model of membrane structure, 86
Units of length, A-4
Units of volume, A-5
Units of weight, A-5
Unity of life, 11
Universal solvent, 33
University of Arizona, 487
University of California, Berkeley, 467
University of Colorado, 356
University of Halle, 637
University of Tennessee Noise Laboratory, 661
University of Toronto, 687
University of Vienna, 183
Unsaturated fatty acids, 47
Unwinding, 241
Uptake of water, 495–96
Uracil, 251, 253
Uranium, 345
Urban sprawl, 793
Urchins, 417, 442
Urea, 108, 621, 627
Urease, 108
Ureter, 624
Urethra, 624, 697
Urethritis, NGU. *See* Nongonoccal urethritis (NGU)
Uric acid, 621, 627
Uridine, 55
Urinary bladder, 624
Urine, 624, 625–30
Urochordata, 443
Uropods, 440
Uterine cycle, 703
Uterus, 700, 701
Utricle, 659, 660

# V

Vaccines, 283–84, 580, 708
Vacuoles, 41, 67, 68, 69, 74, 81, 381, 553
Vagina, 700, 701
Vaginal infection, 393
Vagus nerves, 640
Valine, 51, 251
Valve, cross section of, 562
Van Helmont, Jean-Baptiste, 126
Variability of bases, 238, 240
Variable regions, 578
Variables, experimental and dependent, 15
Variation, 305, 324, 327–28
Varicose veins, 561
Variety, in cells, 61–66, 67
Vascular bundles, 483–84
Vascular cambium, 479, 484
Vascular cylinder, 481
Vascularization, 272
Vascular pathways, 560–61
Vascular plants, 399–404
Vascular system, 441
Vascular tissue, 475, 477–78, 481
Vas deferentia, 697
Vasectomy, 706
Vasopressin, 680
Vectors, 279, 289
Vegetarian birds, 454
Vegetation, 10
Vegetative organs and major tissues of plants, 492

Vegetative propagation, 515–19
Veins, of body, 555
Veins, of leaf, 489, 491
Venae cavae, 557, 560
Ventilation, complete and incomplete, 610
Ventral solid nerve cord, 437
Ventricles, 642
Venus's-flytrap, 490, 491
Vertebrae, 666, 667
Vertebral column, 448, 664, 666, 667
Vertebrates, 416
    brains among, 645
    circulatory system, 555–57
    and diversity, 447–71
    endocrine system, 678–90
    evolution of, 448–58
    phylogenetic tree of, 449
    sexual reproduction, 696
Vesicles, 67, 69, 71
*Vespula arenaria,* 758
Vessel elements, 477–78
Vestigial structures, 311, 667
*Vicia fava,* 499
Victoria, Queen, 229, 230
Vidarabine, 370
Vietnam War, 524, 793
Villi, 594, 595
Viral DNA, 279–80, 282–83, 289
Viral encephalitis, 370
Viral hepatitis, 597
Virchow, Rudolf, 61
Viroids, 368–70
Viruses, 13, 89–90, 234, 236, 237, 366–70
    and animal viruses, 368
    and cancer, 271–75
    and cloning, 279–80
    HIV. *See* Human immunodeficiency virus (HIV)
    HPV. *See* Human papilloma virus (HPV)
    and life cycles, 367–70
    and Monera, 365–78
    and retroviruses as vectors, 289
    and STD, 706
    and vectors, 279, 280, 282–83
    and in vivo therapy, 288
Visceral mass, 431, 432, 434
Visible light, 122, 123
Vision, 655–58
Visual communication, 738
Vitamins, 112, 113, 600–603, 658
*Vitis vinifera,* 524
Vitreous humor, 655
Voice box, 611, 612
Voltage, 637
Volume, units of, A-5
*Volvox,* 380, 385, 387
Von Baer, Karl E., 714
von Seysenegg, Erich Tschermak, 202
Vulva, 700, 701

# W

Wading birds, 454
Wallace, Alfred Russel, 305, 308
Wallace's Line, 308
Wang, Andrew, 243
Warblers, 756
Warm chemistry, 115
Warming, global, 800